Environmental Footprints and Eco-design of Products and Processes

Series Editor

Subramanian Senthilkannan Muthu, Chief Sustainability Officer, Green Story Inc., Tung Chung, Hong Kong

Indexed by Scopus

This series aims to broadly cover all the aspects related to environmental assessment of products, development of environmental and ecological indicators and eco-design of various products and processes. Below are the areas fall under the aims and scope of this series, but not limited to: Environmental Life Cycle Assessment; Social Life Cycle Assessment; Organizational and Product Carbon Footprints; Ecological, Energy and Water Footprints; Life cycle costing; Environmental and sustainable indicators; Environmental impact assessment methods and tools; Eco-design (sustainable design) aspects and tools; Biodegradation studies; Recycling; Solid waste management; Environmental and social audits; Green Purchasing and tools; Product environmental footprints; Environmental management standards and regulations; Eco-labels; Green Claims and green washing; Assessment of sustainability aspects.

Ketan Kuperkar • Dinesh Kumar
Sapna Raghav • Anil Kumar Jangir
Mohammad Shahid
Editors

Deep Eutectic Solvents

Volume One: Fundamentals and Applications

Springer

Editors
Ketan Kuperkar
Sardar Vallabhbhai National Institute of Technology
Surat, Gujarat, India

Dinesh Kumar
School of Chemical Sciences
Central University of Gujarat
Gandhinagar, Gujarat, India

Sapna Raghav
Department of Chemistry
SJJTU
Jhunjhunu, Rajasthan, India

Anil Kumar Jangir
Department of Chemistry
SJJTU
Jhunjhunu, Rajasthan, India

Mohammad Shahid
Department of Chemistry
Koneru Lakshmaiah Education Foundation
Vaddeswaram, Guntur
Andhra Pradesh, India

ISSN 2345-7651 ISSN 2345-766X (electronic)
Environmental Footprints and Eco-design of Products and Processes
ISBN 978-3-032-03800-5 ISBN 978-3-032-03801-2 (eBook)
https://doi.org/10.1007/978-3-032-03801-2

This Springer imprint is published by the registered company Springer Nature Switzerland AG
The registered company address is: Gewerbestrasse 11, 6330 Cham, Switzerland

Preface

Deep eutectic solvents (DESs) have emerged as a promising class of green solvents due to their simple preparation, low cost, and environmentally friendly nature. Formed through hydrogen bonding between readily available components, DESs offer flexible and tunable properties that make them useful across a wide range of scientific and industrial fields. This book, *Deep Eutectic Solvents—Volume One: Fundamentals and Applications*, brings together foundational knowledge and recent developments in the area. It is designed to serve as a helpful resource for students, researchers, and professionals who are interested in understanding how DESs work and how they are being applied to real-world problems.

Chapter 1 introduces the history, classification, and synthesis of DESs. It outlines how the concept evolved over time, the various types of DESs that have been developed, and methods used to prepare them, including newer approaches like microwave and ultrasonic techniques. Chapter 2 focuses on the physicochemical properties of DESs, such as viscosity, conductivity, density, and surface tension. It discusses how factors like component ratios, temperature, and water content influence these properties and their relevance to practical applications. Chapter 3 explores the use of DESs in biotechnology, particularly in enhancing enzyme performance, supporting drug delivery, and aiding in the extraction of valuable compounds from natural sources. The chapter highlights DESs as an important tool for sustainable bioprocessing and innovation in life sciences. Chapter 4 discusses the altered micellization behavior observed when amphiphilic molecules interact with DESs. It looks at the thermodynamics and interfacial properties of these systems, and how DESs can influence the formation and stability of micelles. Chapter 5 covers the toxicity and environmental impact of DESs, providing an overview of biological effects, biodegradability, and green assessment tools. It emphasizes the need for more toxicological data and safe design of DESs for broader use. Chapter 6 reviews the use of DESs in extraction and separation techniques, such as solid–liquid and liquid–liquid extraction. It explains the role of DESs in improving solubility, selectivity, and efficiency in separating valuable compounds from complex mixtures. Chapter 7 examines the use of DESs as green alternatives in catalysis, including homogeneous, heterogeneous, and electrocatalytic systems. The chapter presents case studies where DESs have improved reaction outcomes and contributed to more sustainable chemical transformations. Chapter 8 outlines how DESs are applied in the synthesis of inorganic materials, especially metal and metal oxide nanoparticles. It highlights how DESs support controlled particle growth and stability in processes relevant to catalysis, electronics, and energy storage. Chapter 9 examines the extraction of phenolic compounds from plant sources using eutectic solvents. It analyzes how solvent composition and emerging technologies like microwave and ultrasound-assisted extraction affect efficiency and bioactivity retention. Chapter 10 discusses the use of DESs and natural deep eutectic solvents (NADES) in modern sample preparation and microextraction methods. The chapter also reviews recent studies demonstrating the efficiency and versatility of DESs in analytical workflows. Chapter 11 presents the current status and recent developments in the use of hydrophobic DESs (HDES) for drug extraction and separation. It highlights advances in selectivity, efficiency, and green performance of HDES systems in pharmaceutical analysis. Chapter 12 highlights the role of DESs in advanced materials engineering, including their application in electrodeposition, metal recycling, nanomaterial synthesis, and polymer chemistry. The chapter shows how DESs support the development of functional materials for sustainable technologies.

Together, these chapters provide a well-rounded overview of DESs, from their basic principles to advanced uses in research and industry. We hope this volume encourages further exploration into DESs and helps advance greener and more efficient scientific practices.

Surat, Gujarat, India — Ketan Kuperkar
Gandhinagar, Gujarat, India — Dinesh Kumar
Jhunjhunu, Rajasthan, India — Sapna Raghav
Jhunjhunu, Rajasthan, India — Anil Kumar Jangir
Vaddeswaram, Guntur, Andhra Pradesh, India — Mohammad Shahid

Contents

Editors and Contributors

About the Editors

Dinesh Kumar is a distinguished faculty member at the Central University of Gujarat, School of Chemical Sciences, Gujarat, India. An alumnus of the University of Rajasthan, Jaipur, he earned his master's and doctorate degrees in Chemistry from its esteemed Department of Chemistry. Professor Kumar's scholarly excellence has been recognized through numerous national and international awards and fellowships, placing him among the top 2% of scientists worldwide for four consecutive years: 2021, 2022, 2023, and 2024. His pioneering research mainly focuses on synthesizing innovative nanomaterials, such as capped metal nanoparticles (MNPs), core-shell nanoparticles (NPs), and biopolymer-based nanoadsorbents and nanosensors. These are created to detect and remove hazardous inorganic pollutants, including heavy metal ions, from water, significantly advancing the field of water purification. Additionally, Professor Kumar has developed advanced hybrid nanomaterials using various biopolymers—pectin, chitin, cellulose, and chitosan—and created photocatalytic materials for environmental cleanup. His work also involves creating mixed metal oxide-based hierarchical nanostructures with potential energy applications. As a prolific author, Professor Kumar has published over 143 papers in reputable international journals, written five books, and contributed more than 125 book chapters. His impact is reflected in his h-index of 41, i10 index of 120, and nearly 6500 citations.

Ketan Kuperkar currently holds the position of Associate Professor in the Department of Chemistry at Sardar Vallabhbhai National Institute of Technology (SVNIT), Surat, Gujarat, India. His research interests include surfactant science, polymers, colloids, materials chemistry, and deep eutectic solvents (DESs). He has an h-index of 20, has published over 100 articles, and has authored 4 book chapters. His work focuses on preparing stable DESs using a selective hydrogen bond donor (HBD) and a hydrogen bond acceptor (HBA) in a fixed molar ratio. He aims to control water mixing with neat DESs as a quick and efficient method to tune the chemical and thermodynamic properties of DESs. He is also investigating the structural solvation and micellization dynamics of the added surfactants in the prepared DESs, including water at varying concentrations, from low to high.

Sapna Raghav has a very good academic record. She completed her PhD from the Department of Chemistry at Banasthali University, Rajasthan, in 2019. Her research interest lies in environmental chemistry, especially in water decontamination, sensing, and material science. She has more than 25 research articles, 40 book chapters, and 3 books published in international journals, and an h-index of 16 and i10-index of 19. She has attended many national and international conferences. She has one year of experience as a Women Scientist in Intellectual Property Rights in TIFAC, DST, New Delhi. She also has four and half years of teaching experience and has guided 35 students on their M.Sc. dissertation projects. She is currently working as an Assistant Professor and Head in the Department of Chemistry at Shri Jagdishprasad Jhabarmal Tibrewala University, Jhunjhunu, Rajasthan.

Anil Kumar Jangir was born in Rajasthan, India, and currently holds the position of Assistant Professor at a private university located in Rajasthan. He pursued his M.Sc. (Chemistry) at the University of Rajasthan. He qualified for the CISR-NET (AIR-43) and GATE examination in chemistry. Later, he joined the Department of Chemistry, Sardar Vallabhbhai National Institute of Technology (SVNIT), Surat, Gujarat, India, to pursue his doctoral studies under the supervision of Dr. Ketan Kuperkar. His research revolves around investigating thermophysical and spectroscopic approaches for designing novel deep eutectic solvents (DESs) and cosolvent-modified DESs for various significant research problems and their uses in substantial applications in green chemistry. He has also published several research articles and a book in high-impact international journals, such as the *New Journal of Chemistry, Journal of Molecular Liquids,* and *RSC Advances*.

Mohammad Shahid is an Assistant Professor in the Department of Chemistry, Koneru Lakshmaiah Education Foundation, India. Before joining KLEF, he worked at various esteemed institutions, including Jamia Millia Islamia, New Delhi; Soochow University, China; the University of Glasgow, UK; the Institute of Chemical Technology, Mumbai; Dr. K N Modi University, Tonk; and Mewar University, Chittorgarh. He earned his Master's degree in Chemistry from Shibli National College, Azamgarh, in 2006 and his Ph.D. in Chemistry from Jamia Millia Islamia, New Delhi, in 2014. Dr. Shahid has received several prestigious fellowships to support his research, including grants from UGC, India, a China Postdoctoral Fellowship, an EU Marie Curie Fellowship, and a Dr. D. S. Kothari Fellowship. He has contributed to over 50 scientific publications with approximately 4300 citations and an h-index of 22. In addition, he has also edited and authored several books with renowned international publishing houses.

Contributors

Jyoti Ajudiya Faculty of Science, Gujarat Vidyapith, Ahmedabad, Gujarat, India

Manoj Kumar Banjare Chemistry Division, State Forensic Science Laboratory, Raipur, Chhattisgarh, India

Kamalakanta Behera Department of Chemistry, University of Allahabad, Prayagraj, Uttar Pradesh, India

Suchandra Bhattacharjee University Center for Research and Development, Chandigarh University, Mohali, India

Natália Stram Bierhals Programa de Pós-Graduação em Biociências, Universidade Federal de Ciências da Saúde de Porto Alegre, Porto Alegre, RS, Brazil

Filipe Smith Buarque School of Chemistry, Federal University of Rio de Janeiro, Rio de Janeiro, Brazil

Vishwajit Chavda Department of Chemistry, Institute of Science, Humanities and Liberal Studies, Indus University, Shilaj, Ahmedabad, Gujarat, India

Department of Environmental Engineering, Kyungpook National University, Daegu, South Korea

Unnati Dani Department of Zoology, B.P. Baria Science Institute, Navsari, Gujarat, India

Cristiane Nunes da Silva Chemistry Institute, Federal University of Rio de Janeiro, Rio de Janeiro, Brazil

Lucas dos Santos Silva School of Chemistry, Federal University of Rio de Janeiro, Rio de Janeiro, Brazil

Welison Oliveira dos Santos Chemistry Institute, Federal University of Rio de Janeiro, Rio de Janeiro, Brazil

Kallol Kumar Ghosh School of Studies in Chemistry, Pt. Ravishankar Shukla University, Raipur, Chhattisgarh, India

Piyush Gupta Department of Chemistry, Faculty of Engineering and Technology, SRM Institute of Science and Technology, NCR Campus, Delhi-NCR Campus, Modinagar, Ghaziabad, Uttar Pradesh, India

Nor Munira Hashim Department of Toxicology, Advanced Medical and Dental Institute (AMDI), Universiti Sains Malaysia, Kepala Batas, Penang, Malaysia

Hikmat Research Center for Chemistry, National Research and Innovation Agency (BRIN), Tangerang Selatan, Indonesia

Darshna Hirpara Department of Applied Chemistry, Faculty of Technology and Engineering, The Maharaja Sayajirao University of Baroda, Vadodara, Gujarat, India

Pallavi Jain Department of Chemistry, Faculty of Engineering and Technology, SRM Institute of Science and Technology, NCR Campus, Delhi-NCR Campus, Modinagar, Ghaziabad, Uttar Pradesh, India

Ritu Jangid Department of Science, Biyani Girls College, Jaipur, Rajasthan, India

Anil Kumar Jangir Department of Chemistry, Shri Jagdishprasad Jhabarmal Tibrewala University, Vidyanagri, Jhunjhunu, Rajasthan, India

Nisha Jarwal Department of Chemistry, Indian Institute of Technology Roorkee, Roorkee, Uttarakhand, India

Tarnveer Kaur Department of Basic & Applied Sciences, Sri Guru Granth Sahib World University, Fatehgarh Sahib, Punjab, India

Sanjeev Kumar Department of Applied Chemistry, Faculty of Technology and Engineering, The Maharaja Sayajirao University of Baroda, Vadodara, Gujarat, India

Bong-Seop Lee Department of Chemical Engineering, Kangwon National University, Chuncheon, Kangwon, Republic of Korea

Kartika A. Madurani Laboratory of Instrumentation and Analytical Sciences, Chemistry Department, Faculty of Science and Data Analytics, Institut Teknologi Sepuluh Nopember, Arief Rahman Hakim, Surabaya, Indonesia

Gandhi Mansi Department of Chemistry, Gujarat Technological University, Chandkheda, Ahmedabad, Gujarat, India

Josias Merib Programa de Pós-Graduação em Biociências, Universidade Federal de Ciências da Saúde de Porto Alegre, Porto Alegre, RS, Brazil

Departamento de Farmacociências, Universidade Federal de Ciências da Saúde de Porto Alegre, Porto Alegre, RS, Brazil

Sofia Aquino Monteiro Programa de Pós-Graduação em Biociências, Universidade Federal de Ciências da Saúde de Porto Alegre, Porto Alegre, RS, Brazil

Siddharth Pandey Department of Chemistry, Indian Institute of Technology Delhi, Hauz Khas, New Delhi, India

Kushan Parikh Department of Chemical Science, Faculty of Applied Sciences, Parul University, Vadodara, India

Sapna Raghav Department of Chemistry, Shri Jagdishprasad Jhabarmal Tibbrewala University, Jhunjhunu, Rajasthan, India

Yuvrajsinh B. Rajput Department of Chemistry, Institute of Science, Humanities and Liberal Studies, Indus University, Shilaj, Ahmedabad, Gujarat, India

Bernardo Dias Ribeiro School of Chemistry, Federal University of Rio de Janeiro, Rio de Janeiro, Brazil

Chemistry Institute, Federal University of Rio de Janeiro, Rio de Janeiro, Brazil

Abhay Sawant Ramniranjan Jhunjhunwala College of Arts, Science, and Commerce, Ghatkopar, Mumbai, Maharashtra, India

Priyanka A. Shah School of Forensic Science, National Forensic Sciences University, Gandhinagar, Gujarat, India

Anshu Sharma Department of Chemical Engineering, Kangwon National University, Chuncheon, Kangwon, Republic of Korea

Priyanka Sharma Ramniranjan Jhunjhunwala College of Arts, Science, and Commerce, Ghatkopar, Mumbai, Maharashtra, India

Vaibhav Sharma Department of Chemistry, SRM Institute of Science and Technology, NCR Campus, Delhi-NCR Campus, Ghaziabad, Uttar Pradesh, India

Pranav S. Shrivastav Department of Chemistry, School of Sciences, Gujarat University, Ahmedabad, Gujarat, India

Yagnik Vora Department of Chemistry, Sardar Vallabhbhai National Institute of Technology (SVNIT), Ichchhanath, Surat, Gujarat, India

Noorfatimah Yahaya Department of Toxicology, Advanced Medical and Dental Institute (AMDI), Universiti Sains Malaysia, Kepala Batas, Penang, Malaysia

Nur Nadhirah Mohamad Zain Department of Toxicology, Advanced Medical and Dental Institute (AMDI), Universiti Sains Malaysia, Kepala Batas, Penang, Malaysia

Chapter 1
Deep Eutectic Solvents: A Sustainable Approach to Modern Chemistry "History, Classification, and Synthesis of DES"

Darshna Hirpara, Yuvrajsinh B. Rajput, Vishwajit Chavda, Jyoti Ajudiya, Sapna Raghav, and Pallavi Jain

1.1 Introduction

The increasing demand for green chemical techniques has driven the development of alternative solvents that adhere to the principles of green chemistry. Deep eutectic solvents (DESs) have emerged as a new type of solvent in green chemistry, providing a more sustainable and environmentally friendly alternative to conventional organic solvents and ionic liquids (ILs). These unique solvents are formed by combining a hydrogen bond acceptor (HBA), typically a quaternary ammonium salt

D. Hirpara
Department of Applied Chemistry, Faculty of Technology and Engineering, The Maharaja Sayajirao University of Baroda, Vadodara, Gujarat, India

Y. B. Rajput
Department of Chemistry, Institute of Science, Humanities and Liberal Studies, Indus University, Shilaj, Ahmedabad, Gujarat, India

V. Chavda
Department of Chemistry, Institute of Science, Humanities and Liberal Studies, Indus University, Shilaj, Ahmedabad, Gujarat, India

Department of Environmental Engineering, Kyungpook National University, Daegu, South Korea

J. Ajudiya
Faculty of Science, Gujarat Vidyapith, Ahmedabad, Gujarat, India

S. Raghav
Department of Chemistry, Shri Jagdishprasad Jhabarmal Tibbrewala University, Jhunjhunu, Rajasthan, India

P. Jain (✉)
Department of Chemistry, Faculty of Engineering and Technology, SRM Institute of Science and Technology, NCR Campus, Delhi-NCR Campus, Modinagar, Ghaziabad, Uttar Pradesh, India

K. Kuperkar et al. (eds.), *Deep Eutectic Solvents*, Environmental Footprints and Eco-design of Products and Processes,
https://doi.org/10.1007/978-3-032-03801-2_1

like choline chloride (ChCl), with a hydrogen bond donor (HBD), such as urea, glycerol, or carboxylic acids, which gives a liquid that has a lower melting point than its components [1]. This feature results from the formation of large hydrogen bond networks, which break the crystalline structures of pure chemicals and create stable liquid phases at normal temperatures. DESs are mixtures of pure compounds whose actual eutectic temperature falls below the theoretical (ideal) eutectic temperature at a specific molar ratio, owing to extra hydrogen-bonding interactions in the liquid phase. In contrast, when a mixture's real eutectic temperature matches or exceeds the ideal value—but remains below the melting points of the individual components—it is called a "eutectic solvent." Figure 1.1 illustrates how the real and ideal eutectic temperatures differ for DESs versus eutectic solvents.

Abbott et al. [3] discovered DESs, which demonstrated novel physicochemical features and the ability to resolve many of the environmental and safety concerns connected with existing solvents. This was the turning point in solvent chemistry. Theoretically, DESs share physicochemical properties with ILs, such as low volatility, higher thermal stability, and the ability to solubilize an extensive range of both nonpolar and polar compounds [4]. Compared to ILs, which are known to require intricate and expensive synthesis procedures, DES are readily prepared using the simple techniques of heating and mixing inexpensive, biodegradable materials. Prominent examples are the widely studied mixtures of choline chloride with urea and glycerol. These solvents are of specific interest because of their tunable properties, nontoxicity, and ability to be compatible with sustainable materials as raw materials and as viable candidates for use in the production of green solvents [1, 4].

The importance of DESs in green chemistry cannot be ignored, as they align perfectly with sustainability principles and reduce environmental impact. Traditional solvents, such as dichloromethane and dimethylformamide, have posed significant

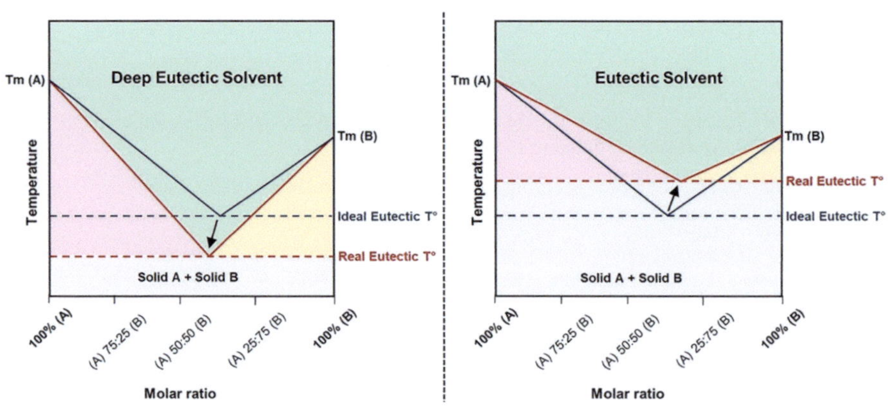

Fig. 1.1 The solid-liquid phase diagram (SLPD) depicts the difference in real and ideal eutectic temperatures between DESs (left panel) and eutectic solvents (right panel). The following colors describe the physical condition of the mixture at different temperatures and molar ratios: white (solid A + solid B), pink (liquid + solid A), yellow (liquid + solid B), and light green (DESs "left" and eutectic solvent "right"). Reproduced with the license granted by Wiley from ref. [2], Copyright 2023

ecological and health hazards, underscoring the need for alternatives that address the concerns associated with conventional solvents. DESs have several key properties, such as their components are often derived from renewable, nontoxic sources, their preparation requires minimum energy input, and their properties can be finely tuned by varying the HBA–HBD combination [5]. These characteristics have enabled their widespread adoption across diverse applications, such as biocatalysis, which enhances enzyme stability and activity, and extraction processes, where they demonstrate exceptional performance compared to conventional solvents in isolating valuable compounds from natural sources. Moreover, the development of natural deep eutectic solvents (NADESs) consisting entirely of primary metabolites such as sugars, organic acids, and amino acids has further expanded their green chemistry credentials by providing fully biodegradable options that minimize the solvation environments found in living organisms [6]. Although ILs were initially proposed as "green" alternatives because of their negligible vapor pressure and possible recyclability, concerns over their ecotoxicity, high price, and difficulties in biodegradability have recently been raised [7]. On the contrary, DES represents a simpler, sustainable, and generally more environmentally friendly alternative [8–14]. DES is relatively easy to prepare, economically feasible, and often consists of components presenting lower environmental and health risks [15–19]. Table 1.1 presents a comparison of the features of DESs, ILs, and conventional solvents from an environmentally sustainable perspective.

The importance of this chapter lies in its detailed discussion of DES from both fundamental and practical perspectives. With DES gaining more and more applications in fields as varied as organic synthesis and catalysis [20], electrochemistry [21], nanomaterials [22, 23], pharmaceuticals [24], and bio-separations/analysis [25]. It is crucial to know their history, classification, and synthesis in detail. The chapter explains how research on DES evolved, with a focus on key milestones and

Table 1.1 Comparison of the fundamental characteristics of DESs to organic solvents and ILs

Property	DESs	ILs	Organic solvent
Composition	A mixture of HBDs and HBAs	Pure cation–anion pairs	Organic compounds
Melting point	Generally low	Liquid at or below 100 °C	Varies widely, often low
Viscosity	High	Moderate to high	Low
Volatility	Very low	Very low	Often high
Thermal stablility	Moderate	High	Moderate
Flammability	Nonflammable	Nonflammable	Often flammable
Environmental impact	Often greener, biodegradable	Moderate to high, depending on the structure	Often toxic and nonbiodegradable
Cost	Less expensive	Expensive	Less expensive
Tunability	High	High	–
Solubility of chemicals	Organic and inorganic compounds	Organic and inorganic compounds	Organic compounds
Example	Choline chloride + urea, citric acid + glucose	1-Butyl-3-methylimidazolium chloride ([BMIM]Cl), [EMIM][BF_4]	Ethanol, hexane, acetone, and methanol

advances that contributed to the development of the topic. A comprehensive discussion of the classification methods and characterization techniques follows this. The chapter also explores various synthetic approaches, ranging from conventional heating techniques to emerging strategies, such as microwave and ultrasonic-assisted preparation. There is an emphasis on designing DES with particular properties for use in specialized chemical processes. By providing a structured and scientifically valid overview, this chapter emphasizes the paradigm-shifting influence of DES in modern-day chemistry and its potential to bring meaningful contributions toward a more sustainable and circular chemical economy.

1.2 Historical Development of DESs

The search for more eco-friendly and safer alternatives to conventional solvents has been at the forefront of modern chemistry, particularly in the context of green and sustainable development. DESs have been the key breakthrough in solvent chemistry of the last decades [1, 26]. The basis of DES is the classical eutectic method, where a mixture of two or more solid components is formed such that it possesses a melting point that is significantly lower than the individual components. A depression of the melting point due to a large amount of hydrogen bonding and interaction between the components brings about a stable liquid at, or near, room temperature.

The history of eutectic systems dates back more than a century, first in metallurgy and materials science, where salts or mixtures of metals with reduced melting points were utilized to manage phase behavior in alloys. The application of eutectic effects to the development of environmentally friendly solvents was not known until much later. It was only in the early 2000s that researchers began to consider the application of eutectic mixtures, particularly those composed of organic or biodegradable constituents, in chemistry. This was a shift from interest at the theoretical level to applied solvent design [27].

One of the milestones in DES history was accomplished in 2003 when Abbott [28] and others introduced the preparation of a liquid mixture by the mixing of choline chloride, a quaternary ammonium salt, and urea in a mole ratio of 1:2. Figure 1.2 illustrates the evolution of DES innovations. The system, the prototype of modern-day DES, had demonstrated that nonvolatile, stable, low-cost solvents could be prepared by the method of simple mixing and heating of solid precursors [29]. The significance of the work lies in its simplicity: without the need for chemical reactions or complicated synthesis, a new family of solvents was created with IL-like properties while addressing some of the shortcomings of ILs. ChCl–urea DES was used as a model, and therefore, other combinations of different HBDs, such as glycerol, ethylene glycol, organic acids, and amino acids, were rapidly studied.

Following the early discovery, DES research gained momentum, with scientists appreciating the tunability and flexibility of these systems. By adjusting the components and their ratios, scientists were able to modify the physicochemical properties, such as polarity, viscosity, conductivity, and solubility, of the solvent to meet

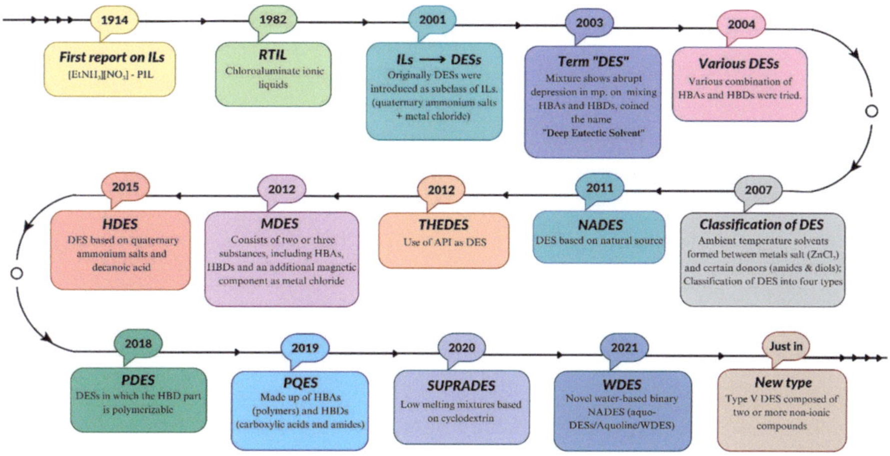

Fig. 1.2 Highlights and achievements in DES research. Reproduced with the license granted by Elsevier from ref. [24] Copyright 2023

specific demands. Such tunability made DES attractive for electrochemistry, separations, catalysis, and biomass processing. Moreover, the nontoxicity of much of the material, especially in naturally occurring substance-based systems, made DES a safer alternative to traditional organic solvents, which tend to be volatile, toxic, and of nonrenewable petrochemical origin.

As the subject evolved, efforts were made to formalize a category of DES in terms of composition and interaction nature. While there are several approaches in which DES may be classed, one standard classification into four major classes has been predominantly accepted. They range from mixtures of quaternary ammonium salts and metal salt or hydrate to more ecologically friendly systems in the presence of nonionic HBDs. Among these, DES composed of ChCl and HBDs such as urea or glycerol (Type III DES) have been studied extensively, as they are simple to prepare, less toxic, and precursors can be readily procured.

Including DES in green chemistry methods has further rooted them as vital in sustainable chemical engineering. Unlike most ILs, they are not expensive to produce and are mostly not persistent in the environment. Their creation typically requires no chemical reaction, no organic solvent, and minimal waste. These advantages align with the principles of green chemistry, particularly those related to waste minimization, the use of benign solvents, and the utilization of renewable feedstocks.

In the industrial context, DES has begun to emerge as a viable solvent for industrial-scale processes such as electrodeposition [30], carbon dioxide capture [31], and dissolution of cellulose and other biopolymers [32]. Their ability to function under mild conditions and engage suitably with polar and nonpolar compounds makes them suitable for various industries. The development of DESs represents a shift toward greener and multifunctional alternatives for traditional, often harmful solvent systems. From a theoretical background based on eutectic mixtures, deep eutectic solvents have become globally recognized as eco-friendly solvents, continuously revolutionizing the field of solvent chemistry. Their simple preparation, extensive functional range, and amenability for ecologically friendly and commercial purposes make them a fundamental tool for chemists. With the advancement of the chemical sciences toward a new generation of more responsible and efficient methods, deep eutectic solvents have the potential to become increasingly prominent features of the future of academic and industrial chemistry.

1.3 Classification of DESs

The term DES describes an aqueous phase created at room temperature through the interactions of polar constituents, which may be either solid or liquid, with hydrogen bonding as the primary interaction mechanism. Various combinations with distinct elements, origins, and physicochemical behaviors have been studied and systematically categorized. Abbott suggested a generalized structural formula for DES, which is $RN^+X^-\cdot zY$. Here, one of the cationic parts is RN^+, which may contain ions of sulfonium, ammonium, and phosphonium that are countered by another halide anion X^- or Lewis's base [33]. In addition, Lewis or Brønsted acidic part is also present, which is denoted as Y, and z is the number of molecules of Y associated with the halide.

The classification of DESs into five main types depends on the chemical characteristics of the acid and base constituents. Type I DESs consist of anhydrous metal chlorides serving as HBDs and quaternary ammonium halides functioning as HBAs, resulting in compounds with comparatively lower melting temperatures. When anhydrous metal halides are replaced with their hydrated analogues, Type II DESs are produced. These complexes have very low melting points due to the presence of water molecules through hydration, enhancing their economic efficiency and water resistance, features that favor their use in industrial applications. Type III DESs are produced when HBAs are mixed with a wide range of HBDs; these systems are often cost-effective, readily available, and biocompatible. The enormous range of HBDs available for this type confers great flexibility. Type IV DES consists of eutectics of lanthanide halides and transition metals with organic ligands that can endure ambient temperatures and exhibit features such as lower viscosity, density, and increased surface tension. A new type of DES, Type V DESs, has been created based on nonionic components (HBD and HBA), in which a significant hydrogen bonding network creates DES characteristics [1, 4, 34]. Figure 1.3 gives a list of some HBAs and HBDs. The literature includes classifications such as alkaline or acidic, natural or synthetic, and ionic or nonionic. DES can be hydrophobic or hydrophilic, medicinal, polymeric, or poly-quasi.

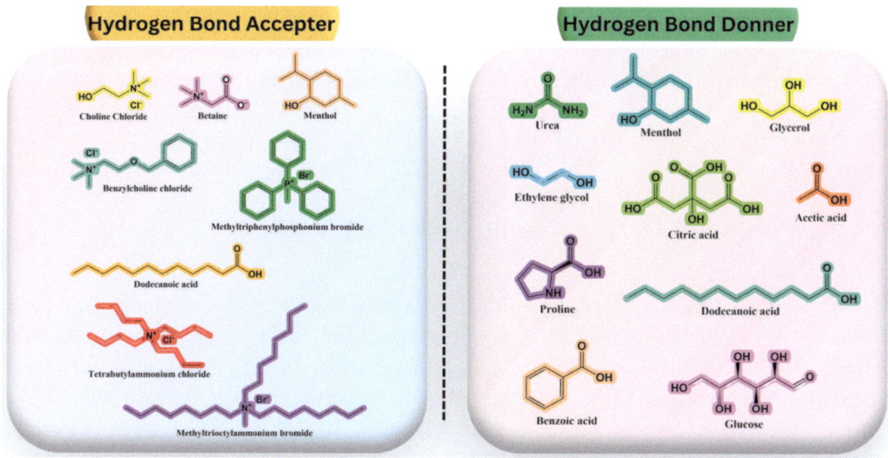

Fig. 1.3 Library of HBDs and HBAs

1.3.1 Alkaline or Acidic DESs

It can be classified according to its alkalinity or acidity. The acidic DESs can be divided into two main subgroups: Brønsted acidic deep eutectic solvents (BADESs) and Lewis acidic deep eutectic solvents (LADESs). Ionizable protons in the system are responsible for the Brønsted acidity displayed by the BADESs [35]. However, LADESs are described as Lewis acidic due to their electron deficiency [36]. Alkaline or basic DESs (BDESs) have shown promise for a wide range of industrial uses, with a particular emphasis on environmentally friendly solvents in rice straw processing. Alkaline DES is also produced by combining ethylene glycol, glycerol, and urea with choline chloride. This kind of alkaline DES has been examined as a suitable medium for the micellization of surfactants [37, 38].

1.3.2 Natural Deep Eutectic Solvents (NADESs)

NADES, as a unique form of DES, was identified by Choi et al. in 2011 [39]. These NADESs contain ChCl, sugars, and natural carboxylic acids in the presence or absence of water in minute quantities. NADESs are less hazardous and more environmentally benign than other DESs due to their naturally generated constituents. NADES mainly contains plant metabolism or eukaryotic cellular systems such as HBD and HBA. One advantage of NADESs is that they form liquid microenvironments, which reduces the physicochemical limits of metabolite movement and processes in cells [40]. ChCl and betaine are the most often used NADESs due to their inexpensive cost, minimal toxic effects, and biodegradable properties. Other ecologically benign and biocompatible compounds employed include glycine, alanine, proline, lidocaine, histidine, nicotinic acid, and acetylcholine chloride. NADESs are generated by the intermolecular interactions between hydrogen bonds; they do not yield substitute mixtures, resulting in fewer purification stages and less waste creation [41, 42].

1.3.3 Magnetic Deep Eutectic Solvents (MDESs)

In 2017, a new magnetic DES was developed by Khezeli et al., in which they combined ChCl: ethylene glycol: $FeCl_3$, ChCl: phenol: $FeCl_3$ [43]. These prepared DESs were employed for the extraction of thiophene from fuel via ultrasound-assisted liquid-liquid microextraction. The first MDESs, which included ChCl, $MgCl_2$:$6H_2O$, and ethylene glycol, were published in 2011 [44, 45]. It has been shown that an external magnetic field can affect MDESs [46]. Not only that, but the magnetic nature of MDEs makes them highly effective in separation techniques such as absorption, adsorption, extraction, and microextraction. They can speed up, simplify, and improve the productivity of these methods while lowering the number of necessary steps [47–49]. In most MDESs, quaternary salts, such as phosphonium or ammonium, act as HBAs.

1.3.4 Polymerized Deep Eutectic Solvents (PDESs)

Polymerizable HBDs produce PDESs [50]. Different ammonium salts and acrylamides or acrylic acids as HBDs were used by Mota-Morales et al. [51] to produce DESs. The utilization of ammonium salts determines the reactivity and process of frontal polymerization, resulting in fully converted polymers suitable for biomedical applications, such as drug delivery. Here, the monomer serves as a solvent for the polymerization medium and a functional filler, while DES has several uses [51]. Apart from that, few studies have demonstrated the application of DESs in environmentally benign free radical polymerization to generate valuable chemicals [52]. Deep eutectic monomers (DEMs) were synthesized by dissolving specific tetrabutylammonium salts, acting as HBAs, in methacrylic acid (MAA), which functions both as an HBD and as a vinyl monomer [53].

1.3.5 Poly-Quasi–Eutectic Solvents (PQESs)

PQESs are a unique kind of DES that has recently been suggested [54, 55]. These PQESs are prepared from polymers like HBAs (poly (ethylene glycol)-block-poly (propylene glycol)-block poly ethyleneglycol)(P123), polyethylene glycol (PEG), poly (ethylene glycol) dimethyl ether (DMPEG), poly (propylene glycol) bis

(2-aminopropyl ether) (PPG-NH_2), and HBDs (amides and carboxylic acids) [54]. PQESs have been used to treat metal oxides and advance oxygen operations. Among their many advantages are their intrinsic conductivity, low viscosity, excellent solvating qualities, environmental friendliness, and affordability. PQESs are a suitable solvent for metal oxide remediation because they possess the advantageous characteristics that set them apart from DESs [55].

1.3.6 Hydrophobic and Hydrophilic Deep Eutectic Solvents (HDESs)

HPBDES or HDES and hydrophilic deep eutectic (HPLDES) solvents are two forms of DESs classified according to how they interact with water. HPLDES refers to DESs containing hydrophilic groups, such as carboxyl, amino, or hydroxyl groups. However, HPLDESs become unstable when exposed to water, limiting their hands-on applicability. In contrast, HPLDESs significantly benefit from removing hydrophilic compounds from nonaqueous materials.

Van Osch et al. [56] proposed the first hydrophobic deep eutectic solvent (HPBDES), which consisted of decanoic acid in combination with a long-chain quaternary ammonium salt to enhance the use of DESs in aqueous media. Since the initial discovery, various formulations of HPBDES have been developed by mixing different quaternary ammonium salts, long-chain fatty acids, and fatty alcohols with extended alkyl chains [56]. Later, Ribeiro et al. [57] proposed a second generation of HPBDES, utilizing DL-menthol, a naturally sourced monoterpene, as the HBA with short organic acid HBDs, such as pyruvic, lactic, and acetic acids. More recently, other terpenes have been utilized for the synthesis of HPBDES, including thymol [58], camphor [59], and lidocaine [60]. Even with these developments, introducing hydrophilic additives in these systems may compromise their stability.

1.3.7 Supramolecular Deep Eutectic Solvents (SUPRADESs)

El Achkar and team achieved great strides in the field with the synthesis of a new type of solvent known as supramolecular deep eutectic solvents (SUPRADESs) [61]. Their vision was to blend levulinic acid (LevA) with a cyclodextrin (CD) derivative, forming a very low-temperature-melting system that can dissolve various substances well. SUPRADES are an unusual class of DESs characterized by an intense network of hydrogen bonds within their structure. This network enables them to dissolve all different compounds and capture molecules well-suited to the hydrophobic cavities of cyclodextrins. Cyclodextrins possess an outer segment that is water-attracting and an interior segment that does not attract water very much, which enables them to form complexes with compounds that do not prefer water very much and have an acceptable shape. This allows for noncovalent host-guest interactions [62, 63]. By merging the CD's molecular recognition and encapsulation capabilities with the beneficial characteristics of DESs, SUPRADESs become versatile solvents suitable for a wide range of compounds. This blend of characteristics makes SUPRADESs unique compared to normal DESs.

1.3.8 Therapeutic Deep Eutectic Solvents (THEDESs)

THEDESs have the potential to address a wide range of formulation issues in medicine, including manufacturing, bioavailability, solubility challenges, permeability, and handling. Stott and team [64] reported the synthesis of eutectic mixes of menthol and ibuprofen via hydrogen bond exchange, which could improve transdermal drug delivery by decreasing formulation melting points. Morrison and colleagues [46] also observed that the dissolution of different drugs, containing danazol, benzoic acid, itraconazole, and griseofulvin, increased by 5 to 22,000 times when dissolved in ChCl: urea and ChCl: malonic acid. Tuntarawongsa and Phaechamud [65] investigated THEDESs, a menthol- and camphor-based solution for dissolving ibuprofen. The results showed that ibuprofen is considerably more soluble in THEDES than in water.

1.3.9 Water-Based Deep Eutectic Solvents (WDESs)

The effect of water on DESs has been the focus of many studies, especially for systems in which ChCl is coupled with widely used HBDs like urea, glycols, sugars, and lactic acid. Later research, though, indicates that particular molar compositions of aqueous ChCl mixtures can possess DES-like behavior independently. One of the foremost examples is the so-called "aquoline" system, an officially recognized water-in-salt (WiS) type nonaqueous DES (NaDESs), constituted from ChCl and water in a 1:3.33 molar relationship [66]. Despite these promising findings, the solid-liquid phase diagram (SLPD) for the ChCl water system under these given conditions is not yet fully elucidated.

Follow-up investigations have focused on systems where water is the only HBD, with ChCl being the HBA. This system was studied by using techniques such as Brillouin spectroscopy, differential scanning calorimetry (DSC), and ^{1}H nuclear magnetic resonance (NMR) [67]. In addition, aquoline has been studied with the aim of its effect upon the micellization behavior of gemini and ionic surfactants in the presence of ChCl and salt-induced micellar morphologies, where water is the HBD [68, 69].

1.4 Synthesis of DESs

DESs have gained prominence as eco-friendly alternatives to conventional solvents owing to their low cost, biodegradability, and tunable properties. This section explores the synthesis of DESs, focusing on principles, preparation methods, the synthesis of hydrophobic DESs, and strategies for optimization and industrial scale-up by understanding the chemistry and practical considerations of DES.

1.4.1 General Synthesis Principles

DESs are formed by combining an HBA and an HBD in specific molar ratios, resulting in a eutectic mixture with a lower melting point than its constituents. This phenomenon, known as eutectic point depression, happens because of the formation of strong intermolecular interactions, mainly hydrogen bonding, that disrupt the crystalline structures of the individual components, stabilizing the liquid phase [4]. Common HBAs include quaternary ammonium salts, such as ChCl, while HBDs range from urea and glycerol to carboxylic acids and alcohols. For example, ChCl with urea in a 1:2 molar ratio yields a DES with a melting point as low as 12 °C [70]. This lowering of the melting point is caused by the formation of a hydrogen-bonded complex that prevents recrystallization. The choice of HBA and HBD, along with their molar ratio, determines key physicochemical properties such as viscosity, density, and conductivity, allowing DESs to be modified for specific applications ranging from catalysis to pharmaceuticals. The synthesis process of DESs is straightforward, but optimizing these properties for practical use requires careful consideration of synthesis conditions, as discussed later in this section.

1.4.2 Methods of Preparation

Several methods are available to prepare DESs. The following subsections detail the most common approaches, their advantages, and their limitations. Figure 1.4 illustrates the sustainable preparation of DESs using the various methods reported in the literature.

1.4.2.1 Heating and Stirring

Heating and stirring are the most widely used methods for synthesizing DES due to their simple procedure and accessibility. In this method, the HBA and HBD are combined in a fixed molar ratio and heated with continuous stirring until a clear, homogeneous liquid forms. The preparation of choline chloride urea DES involves

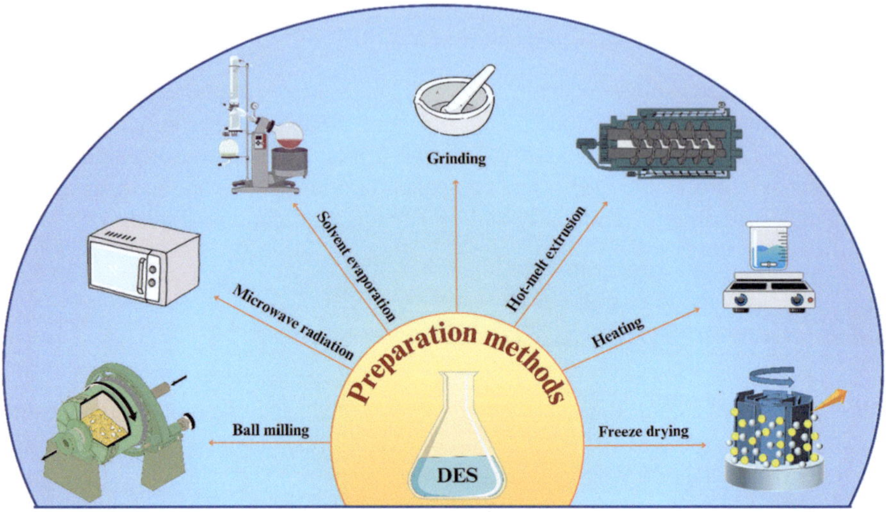

Fig. 1.4 Synthesis methods of DESs. Reproduced with the license granted by Elsevier from ref. [24], Copyright 2023

heating the mixture at 80 °C for 1–2 h. This method requires minimal equipment, making it ideal for laboratory-scale experiments. However, prolonged heating can degrade thermally sensitive components, and achieving homogeneity may be challenging for viscous mixtures or components with disparate melting points. For example, sugars or amino acids used as HBDs can undergo caramelization or the Maillard reaction if overheated. In addition, very viscous blends can require extended stirring time to reach uniformity, which then increases the energy cost. Careful monitoring of the temperature and mixing time is necessary to ensure uniformity [71].

1.4.2.2 Solvent-Assisted Synthesis

Solvent-driven synthesis promotes HBA–HBD contact by using a cosolvent, such as water, ethanol, or methanol, which reduces the viscosity of the blend. The ingredients are dissolved or suspended in the cosolvent and stirred at room or slightly elevated temperatures. Then the cosolvent is removed by evaporation or vacuum distillation to form the DES. This method is effective for high viscosity DESs or components with poor miscibility in their pure forms. For example, synthesizing DESs with high melting point HBDs such as citric acid can be done by using water as a cosolvent to facilitate mixing at lower temperatures and keeping the component identity [72]. In this method, control over molar ratios is achieved because the cosolvent ensures uniform dispersion before DES formation. Solvent-assisted synthesis has the advantage of preparing DESs with biologically active components, such as NADESs, used in food or cosmetic applications where low-temperature processing is important [73]. The only drawback is the additional step of cosolvent removal, which increases energy costs and processing time. Incomplete removal can also leave residual solvent in the DES that affects its properties for specific applications.

1.4.2.3 Freeze-Drying

Freeze-drying, also known as lyophilization, is another method for synthesizing DESs that contain thermally sensitive or biologically active components. The HBA and HBD are dissolved in water to form an aqueous solution, which is then frozen and subjected to sublimation under vacuum to remove the water, leaving behind the DES as a viscous liquid or amorphous solid [74]. This method is mainly crucial for pharmaceutical and food-grade DESs, where maintaining the properties of components like enzymes, vitamins, or polyphenols is important. DESs of lactic acid or glucose as HBDs can be prepared using the freeze-drying method to prevent thermal degradation, ensuring their suitability for biomedical applications. The method also allows the use of water-sensitive components by controlling the aqueous environment during the initial mixing phase. Freeze-drying can synthesize DESs with unique microstructures by enhancing their solubility or stability in specific applications. The freeze-drying method is time-consuming, requiring 24–48 h of processing.

1.4.2.4 Microwave-Assisted Synthesis

Microwave-assisted synthesis uses microwave irradiation to rapidly and uniformly heat the HBA and HBD, significantly minimizing the synthesis time as compared to conventional methods. The components are mixed in a microwave and irradiated for 1–10 minutes, often at 50–100 °C, until a homogeneous DES forms [75]. This method is effective for solid components with high melting points, as microwave energy promotes rapid molecular interactions and lowers the energy barrier for DES formation. For example, DESs based on ChCl and carboxylic acids, which may require extended heating in traditional methods, can be synthesized in minutes using microwave irradiation. Challenges include the need for specialized microwave reactors and precise control to prevent overheating, which could degrade components or produce inconsistent DESs. Uneven heating may also occur in large batches, making careful optimization of irradiation parameters crucial.

1.4.2.5 Grinding/Milling

Grinding or milling is a solvent-free and environmentally friendly method in which solid HBA and HBD components are mechanically mixed at room temperature until a liquid DES forms. This can be achieved by using a mortar and pestle for small-scale synthesis or ball mills for larger-scale synthesis. Grinding utilizes mechanical energy to enhance molecular-level interactions, facilitating the formation of hydrogen bonds and the evolution of eutectic systems—all without the need for external heat or solvents. This method is especially beneficial for creating DESs from solid components that have poor solubility in traditional solvents, such as betaine or proline. It is also ideal for green chemistry applications as it excludes solvent waste and reduces energy consumption compared to heating-based methods. For example, grinding ChCl with oxalic acid for 30–60 minutes can yield a stable DES without additional processing [76–78].

1.4.3 Optimization and Scale-Up

Optimizing DES synthesis involves adjusting parameters such as temperature, molar ratio, and mixing time to achieve desired properties like low viscosity, high stability, and low melting point [4]. Small changes in the molar ratio of choline chloride to urea can significantly alter the viscosity and conductivity of the resulting DES. The temperature must be carefully controlled to accelerate synthesis without causing degradation, while mixing time ensures homogeneity without excessive processing. Scaling up DES synthesis for industrial applications requires addressing cost, reproducibility, and environmental impact. ChCl-based DESs are cost-effective due to the affordability of their components, but energy-intensive methods like freeze-drying or microwave-assisted synthesis can increase production costs [79, 80].

1.5 Characterization Techniques for DESs

Numerous physicochemical aspects should be investigated before employing DESs in industrial applications. Melting point, density, pH, and viscosity are the most essential features of these solvents. However, additional features of eutectic systems employed in biomedical applications, such as toxicity, biodegradability, and biocompatibility, must also be investigated. Furthermore, the scientific community has investigated other features of these systems, such as surface tension, polarity, and ionic conductivity, as they have been utilized in various fields of study, including extraction, nanotechnology, and organic synthesis. As a result, to fully map the components and understand the behavior of DESs, a wealth of information is required.

1.5.1 Differential Scanning Calorimetry

Differential scanning calorimetry (DSC) is used to determine thermal properties such as melting and crystallization temperatures, glass transition temperature (Tg), reaction enthalpy, and heat capacity, as a sample physically transforms from one form of matter to another. Determining the thermal characteristics of DESs is critical for comprehending their behavior at various temperatures. DSC is a popular thermal analysis method because of its versatility in sample volume, superior sensitivity, and rapidity. Furthermore, it can operate at extremely low and high temperatures, ranging from −180 to 700 °C [81]. For the reasons stated above, the DSC approach is discussed in the majority of published works dealing with DESs. Martins et al. [58] used DSC to monitor phase transitions at different molar compositions and created the SLP diagram for several eutectic solvents containing the terpenes thymol or l-(−)-menthol and monocarboxylic acids. Craveiro et al. [82] also used DSC to assess the thermal stability of different NADESs at temperatures as high as 250 °C. They also demonstrated the influence of water on NADESs by measuring Tg. In summary, DSC is an essential technique for characterizing DESs and is the primary way utilized to confirm their development.

1.5.2 Nuclear Magnetic Resonance

Nuclear magnetic resonance (NMR) spectroscopy is an effective method of analysis used to study and evaluate the presence of various functional groups in a sample. Its applications range from purity validation to the recognition of functional groups; because this technique can detect a vast number of elements, it is utilized to establish a baseline understanding of what functional groups exist and how they rearrange and interact. As a result, this technique enables researchers to determine the types of associations among the HBD and the HBA that constitute the DES system, as well as the moieties of each counterpart participating, allowing them to gain insight into the hydrogen bond network. Noro et al. [83] employed NMR spectroscopy to demonstrate the formation of hydrogen bonds in three distinct DESs, where menthol served as the HBD and saturated fatty acids as the HBAs. The hydroxyl group of saturated fatty acids disappeared from the DES spectrum, whereas the spectrum of pure fatty acids revealed strong and defined signals. Furthermore, a well-defined doublet signal was discovered in the menthol spectrum, attributed to its hydroxyl group, while a bigger singlet was recorded in the DES spectrum. The results indicate that hydrogen bonds form between the carboxyl groups of saturated fatty acids and the hydroxyl groups of menthol. The use of NMR also allows investigators to evaluate the influence of water on DESs and detect any contaminants or byproducts produced during the production process.

1.5.3 Infrared Spectroscopy

Infrared (IR) spectroscopy is a crucial method for studying molecular interactions and detecting the presence of water or contaminants in DESs. The theory behind this technique is based on the idea that particles absorb particular wavelengths of light that are unique to their configuration due to the continual oscillation of their bonds. As a result, the creation of intermolecular hydrogen bonds can cause significant changes in the IR spectrum, such as shifts in frequency and/or intensity, for bands associated with vibrational modes of functional groups involved in the hydrogen bond bridges. Qu et al. [84] used Fourier transform infrared spectroscopy (FTIR) to study the molecular structure of five hydrophilic and three hydrophobic DES. The existence of a broad peak attributed to the stretching vibration of the hydroxyl groups (-OH) and substantially wider than the peak of the pure components demonstrated that the eight DESs were successfully prepared.

1.5.4 Other Experimental Techniques

Other equipment, such as a densimeter, viscometer, pH meter, and conductometer, is utilized to measure other qualities to investigate the characteristics of DESs thoroughly. Density is a key characteristic that governs the dispersion and dissolution of solvents in different liquids. The bulk of established DESs have densities higher than water, except HDESs [85]. Another essential and well-studied feature of DESs is their viscosity. Similar to density, the majority of the examined DESs had a higher viscosity compared to that of water, which is primarily due to the extensive hydrogen bond network formed between the DES components. Low viscosity would hinder API distribution via oral or intravenous routes, potentially leading to toxicity. In contrast, a highly viscous formulation would be advantageous for topical applications if it could be easily dispersed over the skin. In addition to density and viscosity, DES can be affected by ionic conductivity and pH levels [86]. Eutectic systems have limited ionic conductivity at RT. due to their high viscosity. Furthermore, the wide pH ranges that DESs may achieve make them suitable for use in a variety of pharmaceutical and industrial areas [87]. The properties above, as well as others such as polarity, surface tension, and refractive index, must be investigated for each eutectic system because they are affected by the molecular structure and molar ratio of the DES components. Furthermore, the water content of DESs influences their properties. Therefore, this element can be leveraged to manage and tailor its qualities for specific roles and applications.

1.6 Applications of DESs in Different Fields

DESs have multiple beneficial solvent characteristics, making them appropriate for dissolving various compounds, including metal oxides, carbon dioxide, dyes, and drugs [88–90]. Due to its strong polarity, DES can be utilized to remove glycerol during biodiesel refining. It can replace traditional reaction media used in acid–base reactions, transition metal-catalyzed reactions, and bioanalytical reactions. This is essential because volatile organic solvents (VOSs) are frequently used in organic synthesis; however, DES can partially or fully replace VOSs, giving a more sustainable and ecologically friendly option. DES is similar to ILs; therefore, it can be utilized in electrochemistry as solvents, electrolytes, or for the dissolution of metals [21, 30]. Furthermore, DES has been used in dye-sensitized solar cells. Recently, DES has been utilized as a solvent medium to produce various products, including porous materials, metal phosphates, carbon compounds, and nanoparticles. Additionally, DES is expected to be a substantially less hazardous and adaptable medium, allowing for new API delivery options. This conceptualization is motivated by the presence of naturally occurring DESs in living bodies. Pharmacy experts are increasingly interested in using DESs to administer lifesaving drugs and provide focused treatment. DESs have significantly better solubility for certain drugs or APIs than water, making them ideal for maintaining extended availability.

1.7 Limitations and Challenges

Despite their widespread advancement as green alternatives to traditional solvents, the industrial use of DESs is greatly hindered by various limitations and unresolved challenges. To maximize their potential in the context of green chemistry, it is necessary to consider these issues carefully. One of the most significant disadvantages of DESs is their relatively high viscosity at room temperature. This characteristic typically inhibits mass transfer, reduces reaction rates, and complicates product recovery in extraction and separation operations. While increased temperature can reduce viscosity, it may negate the environmental advantage by increasing energy consumption. Moreover, the viscosity of DESs is highly composition-dependent, in a way that it becomes challenging to generalize or predict their behavior in different systems. This is especially relevant in applications such as catalysis and electrochemistry, where mass transport is a crucial factor.

Another major challenge that relates to a lack of understanding of the physicochemical properties of DESs is that DESs often exhibit nonideal behavior due to complex hydrogen bonding and ionic interactions compared to classical solvents. Properties such as polarity, conductivity, thermal stability, and solvatochromic properties can vary drastically with even slight changes in composition or moisture content. This lack of standardization creates a hindrance to their systematic design for target applications. Moreover, the intricacy of predicting the effect of slight structural changes on the performance of HBA or HBD underscores the need for a unified and broader theory.

The issues of toxicity and biodegradability are of concern. Although many DESs are derived from ingredients that are widely considered to be safe, the combination of these materials does not always maintain the desirable characteristics of their constituents. Some DESs have already shown moderate to high toxicity to aquatic organisms and human cells when they contain halide salts or amides. Additionally, the long-term persistence of these chemicals in the environment is poorly understood. As DESs become increasingly explored in the pharmaceutical, agricultural, and food sectors, it is essential to perform an exhaustive toxicological analysis to determine the risks associated with their use and disposal.

Another critical issue is the hygroscopic nature of many DESs. The absorption of atmospheric moisture can significantly alter their physical and chemical properties, including viscosity, density, and solubility. This complicates their storage, handling, and reproducibility, particularly in industrial-scale operations where moisture control can be challenging. Moreover, water incorporation may enhance or diminish performance depending on the specific application, making the need for moisture sensitivity assessments crucial in process development.

From a practical point of view, a lack of knowledge on the long-term stability and recyclability of DESs is a hindrance to their application on an industrial scale. Although some research has shown that DESs have good thermal and chemical stability, their recyclability is often dependent on the properties of dissolved solutes and reaction conditions. On subsequent cycling, contaminants can accumulate, decreasing their stability or performance. Reliable regeneration processes are still at the development stage and need to be specially tailored for each DES composition.

Additionally, the lack of a unified classification system and nomenclature creates uncertainty within communication and documentation schemes. Different publications may use varying names or acronyms for DESs with similar components, making comparisons difficult. To improve reproducibility and facilitate meta-analyses, a unified method for reporting preparation methods, molar ratios, and physicochemical properties is needed.

1.8 Future Perspectives and Research Directions

DESs have immense potential as green alternatives to traditional solvents in various chemical processes. To reach their full potential, however, continued research is necessary to expand their applications, optimize their performance, and address existing challenges. A key aspect of this research regards the systematic design of task-specific DESs, where the intentional variation of HBDs and HBAs can produce tailored solvent systems for specific reactions or separations. Progress toward this goal will be significantly facilitated by computational modeling and quantum chemical calculations, which can predict physicochemical properties and interaction dynamics before experimental synthesis.

Another key area is in the elucidation of structure-property relationships, which is underdeveloped for DES compared to traditional solvents and ILs. Even with expanded datasets, predictive knowledge of molecular structure based on primary criteria such as viscosity, polarity, conductivity, and thermal stability remains limited. High-throughput experiments and machine learning methods have the potential to expedite the discovery of new DES systems and help correlate molecular properties with macroscopic properties.

Long-term use of NADESs has great potential, particularly for applications in pharmaceuticals, agriculture, and biomedicine. NADESs based on metabolites, amino acids, and organic acids have higher biocompatibility and often show better solvation for biologically relevant molecules. They are, however, in need of a thorough toxicological and ecotoxicological evaluation to enable regulatory approval and promote extensive applications in bio-relevant fields.

Another area of advancement is hybrid systems that involve DES–water mixtures or DES–polymer blends. These integrations could be designed to improve tunability, reduce viscosity, and fine-tune performance for specific applications. Integration with renewable feedstocks or nanostructured platforms is expected to yield multifunctional systems with novel properties for sensing, drug delivery, and environmentally friendly nanotechnology applications.

Finally, the integration of circular economy concepts will be necessary. The development of bio-based, recyclable, and low-impact DESs for use in closed-loop systems aligns with long-term sustainability goals. Efforts to assess the overall life cycle of DES applications, from raw material sourcing to end-of-consumption management, will be critical to establishing their actual environmental impacts.

1.9 Conclusion

DESs have developed as a widely applicable and sustainable class of solvents that provide a viable alternative to classical organic solvents and ionic liquids. The unique physicochemical properties of DESs, resulting from strong hydrogen-bonded interactions, enable a substantial lowering of the melting point, reduced volatility, and flexibility in chemical composition. These properties make DESs particularly attractive for applications in synthetic chemistry, extraction, electrochemical applications, catalysis, and biotechnological processes. The historical development of DESs illustrates the growing desire for cleaner and more affordable solvent systems. Their ease of preparation from inexpensive and often biodegradable components with low energy input predisposes them to alignment with green chemistry principles. The classification of DESs into types has also helped to delineate their chemical diversity and enable targeted design approaches. Despite their promise, DESs have some limitations. Problems such as high viscosity, sensitivity to moisture, a lack of toxicological data, and variability in recyclability hinder their large-scale application. The current research is directed toward addressing these issues through the use of structure-property modeling, improving analytical methods, and performing extensive safety assessments. Recent advancements, particularly within the area of NADES, have opened up new avenues of opportunity within biologically relevant applications, such as pharmaceuticals and bioprocessing. Larger-scale industrial implementation, however, depends on overcoming technical hurdles, refining predictability, and meeting regulatory criteria. Prospectively, DESs have an enormous potential for transforming solvent systems in sustainable chemistry. Their biodegradable nature, small ecological impact, and wide range of applications suggest a groundbreaking role for both research and industry purposes.

Interdisciplinary research is needed to unlock their full potential, improve their design, and ensure their safe and efficient integration into future chemical processes.

Acknowledgments The authors would like to acknowledge The Maharaja Sayajirao University of Baroda, Indus University, Gujarat Vidyapith, and Shri Jagdishprasad Jhabarmal Tibrewala (JJT) University, as well as SRM Institute of Science and Technology, for their support of this work.

Conflict of Interest Statement The authors declare that they have no conflict of interest.

Declaration of Generative AI and AI-Assisted Technologies in the Writing Process While preparing this work, the authors utilized the ChatGPT tool to enhance the language and readability of several statements in the manuscript. After using this tool/ service, the authors reviewed and edited the content as needed and take full responsibility for the publication content.

References

1. Hansen BB, Spittle S, Chen B et al (2021) Deep eutectic solvents: a review of fundamentals and applications. Chem Rev 121:1232–1285. https://doi.org/10.1021/acs.chemrev.0c00385
2. Oyoun F, Toncheva A, Henríquez LC et al (2023) Deep eutectic solvents: an eco-friendly design for drug engineering. ChemSusChem 16. https://doi.org/10.1002/cssc.202300669
3. Abbott AP, Boothby D, Capper G et al (2004) Deep eutectic solvents formed between choline chloride and carboxylic acids: versatile alternatives to ionic liquids. J Am Chem Soc 126:9142–9147. https://doi.org/10.1021/ja048266j
4. Smith EL, Abbott AP, Ryder KS (2014) Deep eutectic solvents (DESs) and their applications. Chem Rev 114:11060–11082. https://doi.org/10.1021/cr300162p
5. Ijardar SP, Singh V, Gardas RL (2022) Revisiting the physicochemical properties and applications of deep eutectic solvents. Molecules 27:1368. https://doi.org/10.3390/molecules27041368
6. Bragagnolo FS, Strieder MM, Pizani RS et al (2024) Revisiting natural deep eutectic solvents (NADES) as extraction media and ready-to-use purposes. TrAC Trends Anal Chem 175:117726. https://doi.org/10.1016/J.TRAC.2024.117726
7. de Jesus SS, Maciel Filho R (2022) Are ionic liquids eco-friendly? Renew Sust Energ Rev 157:112039. https://doi.org/10.1016/J.RSER.2021.112039
8. Jangir AK, Dani U, Chaudhary VK (2025) Exploring the influence of composition on physicochemical and biological properties of choline chloride-based deep eutectic solvents: an experimental and theoretical endorsement. J Mol Liq 427:127419. https://doi.org/10.1016/j.molliq.2025.127419
9. Jangir AK, Bhawna VG et al (2022) Design and thermophysical characterization of betaine hydrochloride-based deep eutectic solvents as a new platform for CO_2 capture. New J Chem 46:5332–5345. https://doi.org/10.1039/D1NJ05373F
10. Jangir AK, Nain AK, Kuperkar K (2021) Insight into structural properties and molecular interactions of maline (choline chloride + malonic acid) and 1, 4- butanediol based pseudo-binary mixture: a thermophysical, spectral, and simulation portrayal. J Mol Liq 334:116050. https://doi.org/10.1016/j.molliq.2021.116050
11. Jangir AK, Sethy P, Verma G et al (2021) An inclusive thermophysical and rheology portrayal of deep eutectic solvents (DES) for metal oxides dissolution enhancement. J Mol Liq 332:115909. https://doi.org/10.1016/j.molliq.2021.115909
12. Jangir AK, Mandviwala H, Patel P et al (2020) Acumen into the effect of alcohols on choline chloride: L-lactic acid-based natural deep eutectic solvent (NADES): a spectral investigation unified with theoretical and thermophysical characterization. J Mol Liq 317:113923. https://doi.org/10.1016/j.molliq.2020.113923

13. Jangir AK, Lad B, Dani U et al (2020) *In vitro* toxicity assessment and enhanced drug solubility profile of green deep eutectic solvent derivatives (DESDs) combined with theoretical validation. RSC Adv 10:24063–24072. https://doi.org/10.1039/C9RA10320A
14. Kumar JA, Patel D, More R et al (2019) New insight into experimental and computational studies of choline chloride-based 'green' ternary deep eutectic solvent (TDES). J Mol Struct 1181:295–299. https://doi.org/10.1016/j.molstruc.2018.12.106
15. Chavda V, Patel B, Singh S et al (2023) A sustainable approach for the adsorption of methylene blue from an aqueous background: an adsorbent based on DES/CGS modified GO@ZrO $_2$. RSC Sustain 1:2038–2057. https://doi.org/10.1039/D3SU00236E
16. Babu Singh M, Prajapat A, Jain P et al (2023) Investigate the ability of deep eutectic solvent (ChCl-glycerol) to sense the Sulphur dioxide using density functional theory calculations and molecular dynamics simulations. J Mol Liq 388:122720. https://doi.org/10.1016/j.molliq.2023.122720
17. Singh MB, Jain P, Mohammad F et al (2024) Significant increase in the dipole moment of graphene on functionalization: DFT calculations and molecular dynamics simulations. ACS Omega 9:16458–16468. https://doi.org/10.1021/acsomega.4c00173
18. Singh MB, Prajapat A, Jain P et al (2024) Investigate the significance of DES to enhance the solubility of noscapine: DFT calculations, MD simulations, and experimental approach. Ionics (Kiel) 30:1795–1813. https://doi.org/10.1007/s11581-023-05359-3
19. Raman APS, Singh MB, Vishvakarma VK et al (2024) Exploring bioinspired designed DES for their acetylene sensing capabilities via DFT calculations and molecular dynamics simulations. J Solut Chem 53:1685–1700. https://doi.org/10.1007/s10953-024-01407-2
20. Nayak VDD, Jude Martis G, Gaonkar SL (2025) Advancements in biocatalysis: a comprehensive review of applications in organic synthesis. Synth Commun 55:443–464. https://doi.org/10.1080/00397911.2024.2449258
21. Protsenko VS (2025) Electrochemical corrosion properties and protective performance of coatings electrodeposited from deep eutectic solvent-based electrolytes: a review. Materials 18:558. https://doi.org/10.3390/ma18030558
22. Shah PA, Chavda V, Hirpara D et al (2024) Advances in deep eutectic solvent-based synthesis of nanomaterials for environmental remediation. J Mol Liq 416:126482. https://doi.org/10.1016/J.MOLLIQ.2024.126482
23. Chavda V, Patel HD, Patel AK et al (2024) Deep eutectic solvent engineered <scp>GO</scp> / <scp> ZrO $_2$ </scp> nanofiller for mixed matrix membranes: an efficient hydrogen gas separation. J Appl Polym Sci 141. https://doi.org/10.1002/app.55938
24. Shah PA, Chavda V, Hirpara D et al (2023) Exploring the potential of deep eutectic solvents in pharmaceuticals: challenges and opportunities. J Mol Liq 390:123171. https://doi.org/10.1016/j.molliq.2023.123171
25. Shah PA, Shrivastav PS, Sharma VS, Chavda V (2023) Uncovering the green frontier: harnessing deep eutectic solvents for sustainable bioanalysis. Bioanalysis 15:815–821. https://doi.org/10.4155/bio-2023-0088
26. Meher AK, Zarouri A (2025) Green analytical chemistry—recent innovations. Analytica 6:10. https://doi.org/10.3390/analytica6010010
27. Zimmerman JB, Anastas PT, Erythropel HC, Leitner W (2000) Designing for a green chemistry future. Science 367:397–400. https://doi.org/10.1126/science.aay3060
28. Abbott AP, Capper G, Davies DL et al (2003) Novel solvent properties of choline chloride/urea mixtures. Chem Commun 70–71. https://doi.org/10.1039/b210714g
29. Yu D, Jiang D, Xue Z, Mu T (2024) Deep eutectic solvents as green solvents for materials preparation. Green Chem 26:7478–7507. https://doi.org/10.1039/D4GC00136B
30. Kityk A, Pavlik V, Hnatko M (2024) Breaking barriers in electrodeposition: novel eco-friendly approach based on utilization of deep eutectic solvents. Adv Colloid Interf Sci 334:103310. https://doi.org/10.1016/J.CIS.2024.103310

31. Guo T, Zhu D, Zhao C et al (2024) Efficient solar-driven carbon dioxide capture system for greenhouse using graphene-contained deep eutectic solvents. Sep Purif Technol 332:125754. https://doi.org/10.1016/J.SEPPUR.2023.125754
32. Zhu J, Shao C, Hao S et al (2025) Recent progress on the dissolution of cellulose in deep eutectic solvents. Ind Crop Prod 228:120844. https://doi.org/10.1016/J.INDCROP.2025.120844
33. Abbott AP, Capper G, Davies DL et al (2001) Preparation of novel, moisture-stable, Lewis-acidic ionic liquids containing quaternary ammonium salts with functional side chains. Chem Commun 2010–2011. https://doi.org/10.1039/b106357j
34. Abranches DO, Coutinho JAP (2022) Type V deep eutectic solvents: design and applications. Curr Opin Green Sustain Chem 35:100612. https://doi.org/10.1016/j.cogsc.2022.100612
35. Zhu Y, Yang TX, Qi BK et al (2023) Acidic and alkaline deep eutectic solvents (DESs) pretreatment of grapevine: component analysis, characterization, lignin structural analysis, and antioxidant properties. Int J Biol Macromol 236:123977. https://doi.org/10.1016/J.IJBIOMAC.2023.123977
36. Guo Y, Xu L, Shen F et al (2022) Insights into lignocellulosic waste fractionation for lignin nanospheres fabrication using acidic/alkaline deep eutectic solvents. Chemosphere 286:131798. https://doi.org/10.1016/j.chemosphere.2021.131798
37. Banjare RK, Banjare MK, Behera K et al (2020) Micellization behavior of conventional cationic surfactants within glycerol-based deep eutectic solvent. ACS Omega 5:19350–19362. https://doi.org/10.1021/acsomega.0c00866
38. Hirpara D, Patel B, Chavda V et al (2022) Micellization and clouding behaviour of an ionic surfactant in a deep eutectic solvent: a case of the reline-water mixture. J Mol Liq 364:119991. https://doi.org/10.1016/J.MOLLIQ.2022.119991
39. Choi YH, van Spronsen J, Dai Y et al (2011) Are natural deep eutectic solvents the missing link in understanding cellular metabolism and physiology? Plant Physiol 156:1701–1705. https://doi.org/10.1104/pp.111.178426
40. Durand E, Villeneuve P, Bourlieu-lacanal C, Carrière F (2021) Natural deep eutectic solvents: hypothesis for their possible roles in cellular functions and interaction with membranes and other organized biological systems. Adv. Bot. Res 97:133–158. https://doi.org/10.1016/bs.abr.2020.09.005
41. Cao J, Chen L, Li M et al (2018) Two-phase systems developed with hydrophilic and hydrophobic deep eutectic solvents for simultaneously extracting various bioactive compounds with different polarities. Green Chem 20:1879–1886. https://doi.org/10.1039/C7GC03820H
42. Nam MW, Zhao J, Lee MS et al (2015) Enhanced extraction of bioactive natural products using tailor-made deep eutectic solvents: application to flavonoid extraction from Flos sophorae. Green Chem 17:1718–1727. https://doi.org/10.1039/C4GC01556H
43. Khezeli T, Daneshfar A (2017) Synthesis and application of magnetic deep eutectic solvents: novel solvents for ultrasound assisted liquid-liquid microextraction of thiophene. Ultrason Sonochem 38:590–597. https://doi.org/10.1016/J.ULTSONCH.2016.08.023
44. Wang H, Jing Y, Wang X et al (2011) Ionic liquid analogous formed from magnesium chloride hexahydrate and its physico-chemical properties. J Mol Liq 163:77–82. https://doi.org/10.1016/j.molliq.2011.08.004
45. Yue D, Jing Y, Sun J et al (2011) Structure and ion transport behavior analysis of ionic liquid analogues based on magnesium chloride. J Mol Liq 158:124–130. https://doi.org/10.1016/J.MOLLIQ.2010.11.005
46. Morrison HG, Sun CC, Neervannan S (2009) Characterization of thermal behavior of deep eutectic solvents and their potential as drug solubilization vehicles. Int J Pharm 378:136–139. https://doi.org/10.1016/j.ijpharm.2009.05.039
47. Wang M, Zhao L, Niu Y et al (2023) Magnetic deep eutectic solvent-based dispersive liquid–liquid microextraction for determination of strobilurin fungicides in water, juice, and vinegar by high-performance liquid chromatography. Food Chem X 18:100711. https://doi.org/10.1016/j.fochx.2023.100711

48. Zainal-Abidin MH, Hayyan M, Matmin J et al (2023) Greening industrial applications with magnetic-based deep eutectic solvents: a promising future. J Ind Eng Chem 124:1–16. https://doi.org/10.1016/j.jiec.2023.04.011
49. Sajid M (2019) Magnetic ionic liquids in analytical sample preparation: a literature review. TrAC Trends Anal Chem 113:210–223. https://doi.org/10.1016/j.trac.2019.02.007
50. Ren'ai L, Zhang K, Chen G et al (2018) Green polymerizable deep eutectic solvent (PDES) type conductive paper for origami 3D circuits. Chem Commun 54:2304–2307. https://doi.org/10.1039/C7CC09209A
51. Mota-Morales JD, Gutiérrez MC, Ferrer ML et al (2013) Deep eutectic solvents as both active fillers and monomers for frontal polymerization. J Polym Sci A Polym Chem 51:1767–1773. https://doi.org/10.1002/pola.26555
52. Mota-Morales JD, Sánchez-Leija RJ, Carranza A et al (2018) Free-radical polymerizations of and in deep eutectic solvents: green synthesis of functional materials. Prog Polym Sci 78:139–153. https://doi.org/10.1016/j.progpolymsci.2017.09.005
53. Wierzbicki S, Mielczarek K, Topa-Skwarczyńska M et al (2021) Visible light-induced photopolymerization of deep eutectic monomers, based on methacrylic acid and tetrabutylammonium salts with different anion structures. Eur Polym J 161:110836. https://doi.org/10.1016/J.EURPOLYMJ.2021.110836
54. Jiang J, Yan C, Zhao X et al (2017) A PEGylated deep eutectic solvent for controllable solvothermal synthesis of porous $NiCo_2S_4$ for efficient oxygen evolution reaction. Green Chem 19:3023–3031. https://doi.org/10.1039/C7GC01012E
55. Jiang J, Bai X, Zhao X et al (2019) Poly-quasi-eutectic solvents (PQESs): versatile solvents for dissolving metal oxides. Green Chem 21:5571–5578. https://doi.org/10.1039/C9GC02604E
56. van Osch DJGP, Zubeir LF, van den Bruinhorst A et al (2015) Hydrophobic deep eutectic solvents as water-immiscible extractants. Green Chem 17:4518–4521. https://doi.org/10.1039/C5GC01451D
57. Ribeiro BD, Florindo C, Iff LC et al (2015) Menthol-based eutectic mixtures: hydrophobic low viscosity solvents. ACS Sustain Chem Eng 3:2469–2477. https://doi.org/10.1021/acssuschemeng.5b00532
58. Martins MAR, Crespo EA, Pontes PVA et al (2018) Tunable hydrophobic eutectic solvents based on terpenes and monocarboxylic acids. ACS Sustain Chem Eng 6:8836–8846. https://doi.org/10.1021/acssuschemeng.8b01203
59. Makoś P, Przyjazny A, Boczkaj G (2018) Hydrophobic deep eutectic solvents as "green" extraction media for polycyclic aromatic hydrocarbons in aqueous samples. J Chromatogr A 1570:28–37. https://doi.org/10.1016/j.chroma.2018.07.070
60. van Osch DJGP, Parmentier D, Dietz CHJT et al (2016) Removal of alkali and transition metal ions from water with hydrophobic deep eutectic solvents. Chem Commun 52:11987–11990. https://doi.org/10.1039/C6CC06105B
61. El Achkar T, Moufawad T, Ruellan S et al (2020) Cyclodextrins: from solute to solvent. Chem Commun 56:3385–3388. https://doi.org/10.1039/D0CC00460J
62. Moufawad T, Moura L, Ferreira M et al (2019) First evidence of cyclodextrin inclusion complexes in a deep eutectic solvent. ACS Sustain Chem Eng 7:6345–6351. https://doi.org/10.1021/acssuschemeng.9b00044
63. Di Pietro ME, Ferro M, Mele A (2020) Drug encapsulation and chiral recognition in deep eutectic solvents/β-cyclodextrin mixtures. J Mol Liq 311:113279. https://doi.org/10.1016/j.molliq.2020.113279
64. Stott P (1998) Transdermal delivery from eutectic systems: enhanced permeation of a model drug, ibuprofen. J Control Release 50:297–308. https://doi.org/10.1016/S0168-3659(97)00153-3
65. Phaechamud T, Tuntarawongsa S, Charoensuksai P (2016) Evaporation behavior and characterization of eutectic solvent and ibuprofen eutectic solution. AAPS PharmSciTech 17:1213–1220. https://doi.org/10.1208/s12249-015-0459-x

66. Triolo A, Lo Celso F, Brehm M et al (2021) Liquid structure of a choline chloride-water natural deep eutectic solvent: a molecular dynamics characterization. J Mol Liq 331:115750. https://doi.org/10.1016/j.molliq.2021.115750
67. Zhang H, Ferrer ML, Roldán-Ruiz MJ et al (2020) Brillouin spectroscopy as a suitable technique for the determination of the eutectic composition in mixtures of choline chloride and water. J Phys Chem B 124:4002–4009. https://doi.org/10.1021/acs.jpcb.0c01919
68. Hirpara D, Patel B, Chavda V, Kumar S (2022) Micellization of conventional and gemini surfactants in aquoline: a case of exclusively water based deep eutectic solvent. J Mol Liq 362:119672. https://doi.org/10.1016/J.MOLLIQ.2022.119672
69. Hirpara D, Chavda V, Hirapara N, Kumar S (2024) Inorganic salt-induced micellar morphologies in deep eutectic solvent: structure and curcumin solubilization. J Mol Liq 411:125761. https://doi.org/10.1016/j.molliq.2024.125761
70. Wang J, Zhang S, Ma Z, Yan L (2021) Deep eutectic solvents eutectogels: progress and challenges. Green Chem Eng 2:359–367. https://doi.org/10.1016/j.gce.2021.06.001
71. Martín MI, García-Díaz I, Rodríguez ML et al (2024) Synthesis and properties of hydrophilic and hydrophobic deep eutectic solvents via heating-stirring and ultrasound. Molecules 29:3089. https://doi.org/10.3390/molecules29133089
72. Malaiyarasan V, Devakumar Vethakkan JD, Ramalingam A, Vijayan Nithya V (2024) Experimental and computational studies on isolation of cardanol from cashew nut shell liquid using diethanolamine-based deep eutectic solvent at different molar ratios. J Chem Technol Biotechnol 99:1083–1100. https://doi.org/10.1002/jctb.7612
73. Tien NNT, Le NL, Khoi TT, Richel A (2022) Optimization of microwave-ultrasound-assisted extraction (MUAE) of pectin from dragon fruit peels using natural deep eutectic solvents (NADES). J Food Process Preserv 46. https://doi.org/10.1111/jfpp.16117
74. Gutiérrez MC, Ferrer ML, Mateo CR, del Monte F (2009) Freeze-drying of aqueous solutions of deep eutectic solvents: a suitable approach to deep eutectic suspensions of self-assembled structures. Langmuir 25:5509–5515. https://doi.org/10.1021/la900552b
75. Santana APR, Mora-Vargas JA, Guimarães TGS et al (2019) Sustainable synthesis of natural deep eutectic solvents (NADES) by different methods. J Mol Liq 293:111452. https://doi.org/10.1016/j.molliq.2019.111452
76. Farooq MQ, Abbasi NM, Anderson JL (2020) Deep eutectic solvents in separations: methods of preparation, polarity, and applications in extractions and capillary electrochromatography. J Chromatogr A 1633:461613. https://doi.org/10.1016/J.CHROMA.2020.461613
77. Wang L, Guo J, Fang Z (2025) Lower temperature pretreatment of wheat straw for high production of fermentable sugars using ball-milling combined with deep eutectic solvent. Renew Energy 241:122240. https://doi.org/10.1016/j.renene.2024.122240
78. Lee CBTL, Wu TY, Yong KJ et al (2022) Investigation into Lewis and Brønsted acid interactions between metal chloride and aqueous choline chloride-oxalic acid for enhanced furfural production from lignocellulosic biomass. Sci Total Environ 827:154049. https://doi.org/10.1016/j.scitotenv.2022.154049
79. Vargas-Serna CL, Ochoa-Martínez CI, Vélez-Pasos C (2022) Microwave-assisted extraction of phenolic compounds from pineapple peel using deep eutectic solvents. Horticulturae 8:791. https://doi.org/10.3390/horticulturae8090791
80. Chen Y, Xu F, Pang M et al (2022) Microwave-assisted hydrodistillation extraction based on microwave-assisted preparation of deep eutectic solvents coupled with GC-MS for analysis of essential oils from clove buds. Sustain Chem Pharm 27:100695. https://doi.org/10.1016/j.scp.2022.100695
81. De Las HI, Dufour J, Coto B (2021) A novel method to obtain solid–liquid equilibrium and eutectic points for hydrocarbon mixtures by using differential scanning calorimetry and numerical integration. Fuel 297:120788. https://doi.org/10.1016/J.FUEL.2021.120788
82. Craveiro R, Aroso I, Flammia V et al (2016) Properties and thermal behavior of natural deep eutectic solvents. J Mol Liq 215:534–540. https://doi.org/10.1016/J.MOLLIQ.2016.01.038

83. Noro J, Cabo J, Freitas DS et al (2023) Deep eutectic solvents as suitable solvents for lipase-catalyzed transesterification reactions. ChemSusChem 16. https://doi.org/10.1002/cssc.202300615
84. Qu Q, Lv Y, Liu L et al (2019) Synthesis and characterization of deep eutectic solvents (five hydrophilic and three hydrophobic), and hydrophobic application for microextraction of environmental water samples. Anal Bioanal Chem 411:7489–7498. https://doi.org/10.1007/s00216-019-02143-z
85. Omar KA, Sadeghi R (2022) Physicochemical properties of deep eutectic solvents: a review. J Mol Liq 360:119524. https://doi.org/10.1016/j.molliq.2022.119524
86. Agieienko V, Buchner R (2019) Densities, viscosities, and electrical conductivities of pure anhydrous reline and its mixtures with water in the temperature range (293.15 to 338.15) K. J Chem Eng Data 64:4763–4774. https://doi.org/10.1021/acs.jced.9b00145
87. Pedro SN, Freire CSR, Silvestre AJD, Freire MG (2021) Deep eutectic solvents and pharmaceuticals. Encyclopedia 1:942–963. https://doi.org/10.3390/encyclopedia1030072
88. Chavda V, Hirpara D, Rao V, Kumar S (2024) Deep eutectic solvent/surfactant-engineered GO@TiO_2 advanced material for sustained and ultrafast adsorption of hazardous dye from contaminated aqueous environment. ACS Appl Eng Mater 2:604–617. https://doi.org/10.1021/acsaenm.3c00696
89. Chavda V, Hirpara D, Kumar S (2022) GO/ionic surfactant inspired photophysical modulation of rhodamine B in reline with or without additives. J Mol Liq 368:120614. https://doi.org/10.1016/j.molliq.2022.120614
90. Zhu J, Wei Y, Zhang J et al (2023) Are all poorly soluble drugs dissolved in deep eutectic solvents true solutions? J Colloid Interface Sci 645:813–822. https://doi.org/10.1016/j.jcis.2023.04.164

Chapter 2
Exploration of the Physicochemical Properties of Deep Eutectic Solvents

Priyanka Sharma, Yagnik Vora, and Abhay Sawant

2.1 Introduction

Over the past 20 years, ionic liquids (ILs) that remain stable at ambient temperatures have garnered significant interest, particularly in catalysis, electrochemistry, materials science, and, more recently, biomass pretreatment. Since the emergence of "green chemistry," there has been a growing interest in developing metal-free ILs. Owing to their high boiling points and low vapor pressure, ILs are considered environmentally friendly solvents, as these properties facilitate easier recovery and reuse. Despite their promising properties, ILs have raised concerns due to reports of toxicity, environmental hazards, and limited biodegradability. Their application demands high purity, as even minimal impurities can significantly influence their physical behavior [1]. Moreover, the synthesis of ILs often involves complex procedures requiring substantial amounts of salts and solvents, especially for anion exchange, making the process less environmentally favorable. The elevated cost of most ILs has also hindered their large-scale industrial application, emphasizing the need for innovative alternatives that offer greater practicality and sustainability. In response to these challenges, the early twenty-first century witnessed the emergence of a new class of solvents known as deep eutectic solvents (DESs) [2]. These solvents were introduced as cost-effective and less toxic substitutes for ILs. DESs are chemically stable in the presence of water, allowing for

P. Sharma · A. Sawant (✉)
Ramniranjan Jhunjhunwala College of Arts, Science, and Commerce,
Ghatkopar, Mumbai, Maharashtra, India
e-mail: abhaysawant@rjcollege.edu.in

Y. Vora
Department of Chemistry, Sardar Vallabhbhai National Institute of Technology (SVNIT),
Ichchhanath, Surat, Gujarat, India

K. Kuperkar et al. (eds.), *Deep Eutectic Solvents*, Environmental Footprints and Eco-design of Products and Processes,
https://doi.org/10.1007/978-3-032-03801-2_2

easier storage, and can be prepared by simply mixing two components, avoiding the complications associated with IL synthesis. Unlike ILs, DESs bypass purification and waste disposal issues, enhancing their environmental credentials. DESs are recognized for being biodegradable, nontoxic, and biocompatible, aligning well with the concepts of green chemistry. Their physicochemical properties, such as density, viscosity, ionic conductivity, surface tension, and freezing point, resemble traditional ILs [3, 4]. Thanks to their lower environmental impact and affordability, DESs have garnered increasing interest in both academic and industrial spheres. The growing body of literature reflects the rapid expansion of research into DES applications. This chapter further explores the detailed physicochemical characteristics of DESs.

2.2 Physicochemical Properties of DESs

DESs are versatile and tunable solvents that resemble ILs. They can be custom-designed by pairing a variety of quaternary ammonium salts, such as choline chloride (ChCl), which act as hydrogen bond acceptors (HBAs), with a wide range of hydrogen bond donors (HBDs). This flexibility allows for the formulation of DESs with specific physicochemical characteristics, including but not limited to density, viscosity, conductivity, and freezing point, tailored to suit particular applications. Owing to their promising range of applications, considerable attention has been given to the physicochemical characterization of DESs. This section provides an overview and discussion of the key physicochemical properties associated with DESs (Fig. 2.1).

2.2.1 Density

The density of the solvent, one of the fundamental physical properties, is essential for designing, operating, and optimizing various applications. Many DESs generally exhibit densities greater than water. The variation in density across different DESs primarily arises from the nature of their individual components and the specific ratios in which they are combined.

Density is a key physical property of both ILs and, more specifically, DESs. While it is commonly measured using experimental methods, this approach can be impractical or unfeasible in situations where density needs to be determined across a range of temperatures [5, 6]. As a result, there is a strong demand for reliable estimation techniques to accurately predict the density, especially when experimental data is unavailable or difficult to obtain. In the literature, two primary approaches have been proposed for estimating the densities of ILs. One approach utilizes the principle of corresponding states, while the other relies on group contribution

Fig. 2.1 Some common thermophysical properties of DESs, which are discussed in this chapter

methods [7, 8]. Direct application of these methods often leads to significant prediction errors due to the unique ionic nature of both ILs and DESs. In many prediction models, parameters such as critical properties, normal boiling point, and more are commonly used to estimate thermophysical and transport properties. However, for DESs, these critical properties are typically unavailable. This is because DES components tend to decompose before reaching their boiling points, making it impossible to determine these values experimentally. Moreover, the observed boiling point of a DES mixture does not accurately reflect the behavior of the entire system. Given the similarities in behavior between ILs and DESs, it is reasonable to adapt estimation models developed for ILs to DESs. Additionally, the variation in DES density under different conditions can be analyzed through the lens of the Hole Theory [9]. The Hole Theory establishes a relationship between the properties of DESs and the presence of voids or "holes" within the solvent structure that are roughly comparable in size to the mobile species. DESs are conceptualized as being composed of these free volume elements. When HBD is introduced to HBA, the normal size of these holes is altered, leading to a corresponding change in the overall density of the DES. In essence, variations in DES density can be attributed to shifts in molecular interactions and changes in available free volume. For instance, in a DES system such as $ZnCl_2$: urea, a decrease in the average hole radius was observed, leading to a marginal gain in density compared to a DES formed from ChCl: urea [10, 11].

2.2.1.1 Nature of HBD

The density of DESs is clearly impacted by the type of HBD used. For instance, Florindo et al. investigated DESs formed by combining ChCl with various carboxylic acids, which differ in their alkyl chain lengths. It has been shown that DESs prepared with levulinic acid and glutaric acid typically had lower densities than those prepared with the other carboxylic acids. The longer C5 alkyl chains in these acids primarily cause this density reduction. Furthermore, DESs containing levulinic acid show even lower densities than those with glutaric acid, which can be linked to the higher acid-to-HBA molar ratio (1:2) used in their formulation. Generally, the density of DESs follows the trend: oxalic/glycolic acid (C2) > malonic acid (C3) > glutaric/levulinic acid (C5). As the chain length of the alkyl group increases, the molar volume of the mixture also increases, leading to a corresponding decrease in overall density [12]. In order for DESs to form, hydrogen bonding between the HBA and HBD is essential. More hydrogen bonding interactions result from the increased number of –OH groups of HBD. These additional hydrogen bonds reduce the available free volume within the mixture, resulting in higher DES densities. For example, DESs formed using benzyl trialkylammonium chloride salts with ethylene glycol as HBAs and HBD, respectively, exhibit lower densities compared to those formulated with diethylene glycol, triethylene glycol, or glycerol, which contain more –OH groups and, therefore, support greater hydrogen bonding [13]. With an increase in the molecular weight of polyethylene glycol (PEG) from 200 to 600, the density of the resulting DES with tetraalkylammonium bromide (TBAB) also increases. This is attributed to the enhanced interactions between the higher-molecular-weight PEG and TBAB [14]. The densities of carbohydrate-based DESs formulated using three different HBAs, such as ChCl, acetylcholine chloride, and benzylcholine chloride, have been reported in the literature. The carbohydrates used to formulate these DESs included D-fructose, D-glucose, ribose, xylose, and mannose. The density of the resultant DESs showed the following pattern while employing the usual carbohydrate-based HBA: mannose < glucose = fructose < ribose < xylose. Among them, DESs containing xylose exhibited the lowest density. A comparable pattern was also observed in DESs prepared with ethylammonium chloride combined with urea, 1-(trifluoromethyl) urea, and acetamide. The density of the resultant DESs was found to decrease when N, N-diethyl ethanolammonium chloride or methyltriphenylphosphonium bromide were employed as the HBA when mixed with glycerol or ethyl glycol. When ChCl is used as the HBA, the following descending order may be seen in the densities of DESs formed using various HBDs in a 1:2 molar ratio: urea > glycerol > ethylene glycol > p-cresol [15–17].

DESs formed using tetraethylammonium chloride or bromide salts as HBAs, combined with levulinic acid, ethylene glycol, and glycol in a 1:4 molar ratio, exhibited a reduction in density depending on the type of HBD used [18]. For salt hydrate-based DESs, a change in the HBD leads to a variation in density values. Among the various systems investigated, the DES formulated with glycine exhibited the highest density (1.677 $g \cdot cm^{-3}$), whereas the one prepared with ethylene glycol showed the lowest density (1.605 $g \cdot cm^{-3}$) [19]. A comparable trend was noted for the studied carbohydrates when different HBAs were used. The densities of the resulting DESs followed the order: ChCl: carbohydrate > AcChCl: carbohydrate > BzChCl: carbohydrate [20]. In both scenarios, the variation in DES density was primarily influenced by the nature and strength of hydrogen bonding and the amount of available free volume within the system. Additionally, when the length of alkyl groups of tetraalkylammonium salts in PEG-based DESs increased from ethyl to butyl, a noticeable decrease in density was observed [14]. Kroon et al. also observed a similar result for DESs formed from tetraalkylammonium chloride and decanoic acid [21]. The density of DESs is strongly influenced by the halide ion in the HBA. Typically, DESs formulated with tetraalkylammonium bromide salts display higher densities than those containing their chloride counterparts. Additionally, when the chain length of the alkyl group of the HBD increases from ethyl to butyl, DESs formulated with Tetraethylammonium bromide (TEABr) or Tetraethylammonium chloride (TEACl) show the highest density, which can be attributed to the compact structure of these systems.

2.2.1.2 Molar Ratio of HBA to HBD

The molar ratio or composition of the HBA and HBD can alter the density of DESs. Shafie et al. investigated ChCl: citric acid-based DESs at three different molar ratios and found that increasing the proportion of ChCl relative to citric acid led to decreased density [22]. Similarly, for TBABr: PEG-based DESs, a slight increase in density was observed with a higher molar ratio of PEG. DESs prepared with a 1:3 molar ratio generally exhibit higher densities compared to those with a 1:2 ratio. For instance, the densities of TBABr: PEG 200 were reported as 1.11077 $g{\cdot}cm^{-3}$ for the 1:2 ratio and 1.11360 $g{\cdot}cm^{-3}$ for the 1:3 ratio. A similar slight increase in density was observed for DESs based on N,N-diethylethanolammonium chloride (DEACl), Methyltriphenylphosphonium bromide (MTPBr), and Tetrahexylammonium bromide (THABr) when combined with common HBDs such as glycerol or ethylene glycol at a 1:2 molar ratio [14].

2.2.1.3 Temperature

As the temperature increases, the densities of DESs tend to decrease due to thermal expansion. The isobaric thermal expansion coefficient (α_p) describes this temperature-dependent change in density, which reflects the amount of free volume within the DES. These coefficients are also useful in characterizing the compressibility of DESs. The isobaric thermal expansion coefficient is calculated by evaluating density measurements obtained at various temperatures.

$$\alpha_p = -\frac{1}{\rho}\left(\frac{\delta\rho}{\delta T}\right)_p \tag{2.1}$$

A linear decline in the density of all DESs was detected as the temperature increased. This reduction in density is attributed to the greater availability of free volume between the HBA and HBD at elevated temperatures. The increase in free space within DESs correlates with changes in both density and the isobaric thermal expansion coefficient (α_p). Notably, the α_p values for DESs are relatively low compared to those of conventional solvents, but they are comparable to the α_p values observed for imidazolium-based ILs. Temperature had a minimal impact on both density and the isobaric thermal expansion coefficient (α_p), indicating that DESs undergo less expansion or compression compared to ILs and conventional organic solvents [23, 24].

The change in density with temperature was modeled using a linear relationship.

$$\rho = a + bT \tag{2.2}$$

Here, ρ represents the density in g·cm^{-3}, a and b are fitting constants, and T denotes the temperature in Kelvin.

The density of all examined DESs declines as the temperature rises. This behavior is due to increased free volume within the DES, allowing molecules to move more freely, resulting in a lower overall density. The α_p values for the studied DESs range from 0.5 to 0.7, more than those observed for typical organic solvents. Shah and Mjalli et al. studied the density of a ChCl: urea DES at two different temperatures to investigate the effect of water content on it. Furthermore, the intermolecular interactions in both pure and water-containing DES systems have been explored using molecular dynamics (MD) simulations [25]. The findings demonstrated that interactions between urea and chloride anions contribute to the lower melting point of the ChCl: urea DES. Among the constituents, chloride ions demonstrated a stronger affinity for hydration than both urea and the choline cation. Molecular simulations also indicated that interactions between urea and the anion diminished the self-association of urea molecules and choline ions, leading to a decrease in the melting point of the DES. The influence of water on density was categorized into three stages. At first, the number of hydrogen bonds decreased as the water content rose. In the final stage, the anion associated with the HBD exhibited more extensive hydration under higher water concentrations than the cation.

The density of ChCl-based DESs prepared with HBDs such as carboxylic acid was affected by water absorption. After being left in ambient conditions for a month, these DESs absorbed between 14% and 20% water by weight. It was observed that the density of DESs containing 5% water was found to be lower than that of their anhydrous counterparts. This suggests that even minimal water content can influence the density of DESs, indicating that moisture has a noticeable, though slight, impact on their physical properties [12]. Dai et al. investigated how varying water amounts affect the physicochemical properties and structural characteristics of several Natural Deep Eutectic Solvents (NADES) [26]. A DES prepared from 1,2-propanediol, ChCl, and water in a 1:1:1 molar ratio was diluted with deuterium oxide and analyzed using NMR spectroscopy at 40 °C. The Nuclear magnetic resonance (NMR) analysis revealed that the structural integrity of the NADES was preserved when the water content remained below 50% (v/v). However, when the dilution exceeded this threshold, the DES structure broke down, separating the HBD and HBA into individual components within the solution. Adding water to a ChCl: phenol DES was found to lower its density [27]. The introduction of water into pure DESs led to a significant disruption of the hydrogen bonding between their components, resulting in a continuous decline in the density of the aqueous DES solutions. A comparable disruption in hydrogen bonding was observed with increasing temperature. This led to an expansion in the free volume within the DESs, displacement of their constituent components, and a subsequent decrease in their density.

The molecular volume (V_m), standard entropy (S^o), and lattice potential energy (U_{POT}) at a specific temperature (K) can also be determined using density data, based on established empirical equations (Eqs. 2.3–2.5).

2.2.1.4 Molecular Volume (V_m)

Typically, the molecular volume of solvents is highly sensitive to temperature changes. At lower temperatures, the kinetic energy of fluid molecules is reduced due to intermolecular van der Waals forces, which restrict molecular movement to a confined region. As the temperature rises, the kinetic energy increases, allowing molecules to move farther apart. This results in the expansion of the solvent and an increase in its molecular volume. The findings suggest that strong hydrogen bonding interactions between the HBA and HBD components of DESs maintain molecular cohesion and contribute to the system behaving as a unified entity. The molar volume at 303.15 K can be calculated using Eq. 2.3, based on the molar mass (M) of the DES in g mol^{-1} and its corresponding density in g·cm^{-3} [28].

$$V_m = \frac{M}{\rho} \tag{2.3}$$

2.2.1.5 Standard Entropy (S°)

The standard molar entropy of a DES, composed of a binary mixture of HBA and HBD, can be estimated from its density. An increase in entropy is generally associated with a greater number of hydrogen bonds formed in the DES compared to its individual components. Based on Glasser's empirical relationship (Eq. 2.4), the standard entropy value can be determined from the known density data [29].

$$S^o \approx 1246.5 V_m + 29.5 \tag{2.4}$$

2.2.1.6 Lattice Potential Energy (U_{POT})

DESs typically exhibit low melting points, which can be attributed to their composition of highly asymmetrical molecules and consequently low lattice energies. The reduced lattice energy leads to stronger interactions between the individual components, contributing to the depression of the melting point. To achieve a lower melting point in DESs, a combination of weak crystalline organization, large ionic species, and robust intermolecular interactions is essential. The U_{POT} can be estimated at 303.15 K using Eq. 2.5 [29].

$$U_{POT} = 1981.2(\rho / M)^{1/3} + 103.8 \tag{2.5}$$

2.2.2 *Viscosity*

Viscosity is a crucial transport property of DESs, representing the resistance of a fluid to deformation under shear. It provides key insights into molecular-level interactions within the liquid phase and significantly influences the practical applications of DESs [30]. DESs generally exhibit high viscosity at room temperature due to extensive hydrogen bonding networks, van der Waals forces, and electrostatic interactions between the HBD and HBA components. These interactions contribute to reduced ionic mobility and limited free volume within the liquid structure. The viscosity of DESs is also directly linked to their conductivity, mass transfer properties, and reaction kinetics, making it a critical parameter in process design. Due to their high viscosity, DESs are often used in mixtures with less viscous solvents to enhance their applicability in industrial processes. The viscosity of these mixtures plays a fundamental role in determining their suitability for specific applications. For instance, DESs are widely employed as solvents in chemical and physical processes, where viscosity directly affects process efficiency, equipment design, and energy requirements. Experimentally measured viscosity data and its temperature dependence are essential for designing equipment such as pipelines, pumps, and reactors. The viscosity range of DESs is broad, typically between 10 and 10^4 mPa·s at room temperature, significantly higher than conventional organic solvents (0.2–10 mPa·s). For example, ChCl-based DESs exhibit viscosities varying from 0.6 to 15,678 mPa·s, emphasizing the variability and tunability of these solvents for different applications. Moreover, several factors affect the viscosity of DESs, including the chemical nature of HBAs and HBDs, their molar ratio, temperature, molar mass, and the size of ions relative to the available void space.

2.2.2.1 Effect of Temperature

Temperature plays a crucial role in determining DES viscosity. As temperature increases, the hydrogen bond network between the HBD and HBA weakens, decreasing viscosity. This behavior follows the Hole Theory, which explains that rising temperature adds energy to the system, increasing molecular mobility and creating larger voids that facilitate fluid movement. Consequently, DES viscosities decrease exponentially with increasing temperature. Abbott et al. reported that adding organic salts to ChCl: glycerol-based DESs significantly reduces viscosity. This occurs because of the partial disordering of the hydrogen bond network in glycerol, enhancing molecular mobility. The rise in temperature further increases fluidity and reduces resistance to flow, making low-viscosity DESs more attractive for green technology applications [17].

2.2.2.2 Effect of HBA–HBD Molar Ratio

The molar ratio of the HBA and HBD significantly affects viscosity. Increasing the ratio of one of the components generally decreases viscosity, as the salt may act as a bridging agent, connecting ionic groups and forming a less dense network. However, in some systems, such as ChCl: Glycerol and ChCl: Py, decreasing the ChCl ratio increases viscosity. This is attributed to the ability of ChCl to disorder the three-dimensional hydrogen bond network of glycerol, leading to greater molecular freedom and lower viscosity. The viscosity of these systems often exhibits minima or maxima depending on the HBA–HBD molar ratio [31].

2.2.2.3 Effect of Component Size and Structure

The molecular size and structure of DES components also impact viscosity. In systems like alkylammonium bromide: glycerol, tetraalkylammonium halides: levulinic acid, and tetraalkylammonium halides: ethylene glycol, viscosity tends to rise with increasing chain length of the alkyl group of the salt (HBA) due to enhanced van der Waals interactions that limit molecular movement. However, in certain high-viscosity DESs, such as tetraalkylammonium bromide: Pyrogallol (PY), tetraalkylammonium bromide: Barbituric acid (BA), and tetraalkylammonium chloride: decanoic acid, the opposite behavior is observed, with viscosity decreasing as the alkyl chain length grows. This variation suggests that steric effects and molecular packing influence viscosity differently, depending on the DES composition. For TBAB-based DESs, viscosity follows the order: TBAB: EG < TBAB: 1,3-propanediol < TBAB: 1,5-pentanediol < TBAB: glycerol. The TBAB: glycerol system has the highest viscosity due to multiple hydroxyl groups, which enhance hydrogen bonding and intermolecular attractions. Additionally, the longer carbon chain in TBAB:1,5-pentanediol results in entanglement, further increasing the viscosity [32].

2.2.2.4 Effect of HBA Anion Type

The type of anion in the HBA also influences viscosity. In general, viscosity follows the trend: RCl: HBD < RBr: HBD < RI: HBD. This suggests that larger halide anions lead to higher viscosity due to their greater polarizability and ability to form stronger interactions [32].

2.2.2.5 Entropy and Liquid Structure Effects

Viscosity is strongly influenced by the entropy and molecular structure of the liquid. The deviation of viscosity from ideal behavior depends on molecular interactions, size, and shape. For example, the negative viscosity deviation observed in ethaline-Dimethyl Sulfoxide (DMSO) mixtures suggests a lack of strong specific interactions, a conclusion supported by excess volume measurements [32].

Here, some studies have highlighted the factors influencing viscosity. Florindo et al. suggested that the discrepancies in viscosity data among different studies could be attributed to variations in preparation techniques, experimental design, and the presence of contaminants. Numerous studies have explored the impact of both temperature and compositional variation on the viscosity of DESs [20]. Interestingly, the presence of water in the DES system did not show a statistically significant impact on viscosity. Since low-viscosity liquids offer broader applicability in industrial processes, designing DESs with reduced viscosity remains a key objective in this field. Most reported DESs exhibit high viscosities at room temperature (>100 cP), primarily due to the extensive hydrogen bond network among their components. However, hydrophobic DESs based on DL-menthol have demonstrated significantly lower viscosities, such as 7.61 cP at 298.15 K for a 1:3 DL-menthol: octanoic acid system, highlighting the potential for viscosity reduction through the careful selection of components. Crespo et al. highlighted the importance of accurate density data for modeling DES properties [33]. Zhu et al. examined how HBD substitution influences hydrogen bonding using Fourier-transform infrared (FT-IR) spectroscopy [34]. Lapeña et al. found that while water affects transport properties, it does not significantly alter the internal structure of anhydrous DESs [35]. Molecular dynamics (MD) simulations revealed that while temperature reduces viscosity, increases diffusivity, and enhances ion conductivity in DES–water mixtures, it has minimal structural effects. Balaraman and Rathnasamy et al. observed that diol-based DESs exhibit decreasing viscosity and density with rising temperature, while the refractive index remains unaffected. Increasing temperature significantly lowers viscosity, as seen in ChCl: urea and Benzyltrimethylammoniumchloride (BTMA): Gly DESs. Water addition drastically reduces the viscosity, while molar ratio changes, and increasing molecular weight generally raises viscosity. A study on TBAB- and tetraethylammonium p-toluenesulfonate (TEAPTS)-based DESs revealed viscosity variations (17.4–138.0 mPa·s), strongly influenced by the PEG chain length [36].

Viscosity data for DESs can be acquired through experimental measurements or predictive modeling. However, conducting experiments is often labor-intensive and expensive, particularly given the wide range of DESs that can be formed by combining various HBDs and HBAs in different molar ratios. These factors result in different physical properties, further influenced by temperature and pressure variations. Moreover, DESs are generally highly viscous, making their viscosity measurements more challenging than conventional solvents. Considering these complexities, it is essential to develop a reliable global model for accurately estimating the viscosities of diverse DESs. As far as we are aware, only a limited number of studies in the literature have focused explicitly on predicting the viscosity of DESs. Roosta et al. applied the group contribution (GC) approach alongside two machine learning (ML) models, the multilayer perceptron artificial neural network (MLPANN) and the least-squares support vector machine (LSSVM) to estimate the viscosities of various DESs. These predictive models incorporated temperature, the molar ratio of HBA to HBD, and the contributions of specific functional group weights as input variables to improve the accuracy of viscosity predictions [37]. The GC method operates on the theoretical foundation that each functional group in a molecule contributes a defined value to a specific property, enhancing the ability to make accurate predictions. An extensive dataset comprising 2533 data points across 305 different DESs was compiled from existing literature to develop the models grounded in GC theory. Statistical evaluations demonstrated that these models effectively replicated experimental results and successfully predicted the viscosities of DES systems lacking direct experimental measurements [38, 39].

The MLPANN–GC model attained an R^2 value of 0.98 and an average absolute relative deviation (AARD%) of 9.43%. The LSSVM-GC model outperformed it, showing an R^2 of 0.99 and an AARD% of 6.88%. Additionally, the evaluation of various MLP training algorithms revealed the following performance hierarchy: Bayesian Regularization (BR) was the most effective, followed by the Levenberg–Marquardt (LM) algorithm, and then Resilient Propagation (R_{prop}). Compared to previously published models, the methodology used in this study offered several key advantages. In addition to high calculation speed and relative accuracy, incorporating the GC method significantly enhanced model predictability, allowing for screening DESs that were yet to be synthesized. A combination of coupled simulated annealing and grid search optimization was employed to optimize the two crucial tuning parameters (γ and σ^2) of the LSSVM method. For chemists and chemical engineers, these developed models, particularly the LSSVM-GC model, provide straightforward predictive tools for estimating DES viscosities, thereby addressing the challenges posed by complex thermodynamic techniques. Additionally, when compared to correlations reported in the literature, the developed models demonstrated acceptable precision, further validating their practical utility in viscosity prediction [40]. Lloret et al. utilized the soft-Statistical Associating Fluid Theory (SAFT) equation of state (EoS) to model viscosity and other physicochemical characteristics of DESs, though their investigation covered only four specific DES systems. The soft-SAFT EoS was combined with free volume theory to effectively describe the viscosity trends, resulting in predictions that closely aligned with experimental findings. Although additional research has applied associative EoS models to DESs, these efforts have primarily concentrated on properties other than viscosity [41]. Verevkin et al. applied the Perturbed-Chain (PC)-SAFT equation of state (EoS) to evaluate the separation capabilities of DESs. Their study involved determining activity coefficients at infinite dilution for a set of 23 solutes, which included hydrocarbons, alcohols, ketones, ethers, and esters [42]. Pontes et al. examined the solid–liquid equilibria (SLE) of 15 binary systems consisting of quaternary ammonium chlorides and fatty acids. The phase behavior of these mixtures was successfully modeled using the PC-SAFT equation of state, showing strong consistency with the experimental data. Given the absence of a universal viscosity model for DESs, this study aimed to develop a broadly applicable approach rooted in robust theoretical principles. Two advanced associating mixture models, PC-SAFT and Cubic Plus Association (CPA) EoS, were coupled with free volume theory to account for molecular interactions in DESs. These models were applied to an extensive dataset of DESs with diverse HBDs, HBAs, and molar ratios to establish a comprehensive viscosity prediction framework [43].

Reza et al. applied the free volume theory, a widely recognized theoretical framework, to model the viscosities of a range of DESs. They incorporated this approach with the Cubic Plus Association (CPA) and PC-SAFT equations of state, both suited for associating systems. They employed the modified Lydersen–Joback–Reid method along with Lee–Kesler mixing rules to determine the critical properties and acentric factors. The pure component parameters for the CPA and PC-SAFT models were derived using an extensive density dataset comprising 27 DESs under various compositions, temperatures, and pressures. The models demonstrated high accuracy (AARD% = 2.7%) and strong agreement with experimental viscosity values, effectively capturing temperature-dependent trends and HBD ratio variations [44].

Several DES-specific viscosity models and correlations have been proposed in the literature. Most studies present system-specific correlations with unique parameter sets for each DES. For instance, Mjalli et al. developed a viscosity model for ChCl-based DESs, incorporating the molar compositions of HBAs and HBDs. Two model variations were introduced. More recently, the viscosities of 27 DESs with diverse HBAs and HBDs were predicted using a theoretical framework combining free volume theory with CPA and PC-SAFT equations of state. Another approach integrated the friction theory with these equations. While EoS-based models offer accurate viscosity estimations, they involve complex calculations and require extensive parameterization, limiting their applicability to a small set of DESs [45]. Srijan et al. combined polarization-selective pump-probe spectroscopy, two-dimensional infrared (2D IR) spectroscopy, and MD simulations to investigate viscosity effects in the alcohol-based DES, ethaline. Comparing solvent fluctuation and solute reorientation dynamics in ethaline and ethylene glycol, they found that while solute reorientation scales with bulk viscosity, solvation dynamics do not. Instead, the solute experiences domain-specific local viscosity governed by hydrogen bond dynamics. Spectral diffusion was slowed by a factor of 3 with a twofold viscosity increase, highlighting microheterogeneity in DESs. This study demonstrates the power of 2D IR, PSPP, and MD simulations in revealing localized viscosity effects, paving the way for future investigations into domain-specific interactions in DESs [46–48].

Mohan et al. also developed machine learning (ML) models for accurate and rapid viscosity prediction across a diverse range of DESs to facilitate DES design using one of the most comprehensive datasets, covering 672 DESs with 4949 viscosity data points over a temperature range of 278.15 – 385.25 K and varying molar ratios (1:19 – 19:1). Several machine learning models were developed and tested, including support vector regression (SVR), feedforward neural networks (FFNNs), and categorical boosting (CatBoost). These were benchmarked against basic multilinear and polynomial regression approaches. The input features for all models were derived from Conductor-like Screening Model for Real Solvents (COSMO-RS)-based sigma profiles (σ-profiles) obtained via quantum chemical calculations. Among all models, CatBoost demonstrated exceptional accuracy, with an R^2 value of 0.99, a low root-mean-square error (RMSE), and an average absolute relative deviation (AARD) of 5.22%. Notably, 98% of the predictions were within a 15% AARD threshold. Shapley Additive Explanation (SHAP) analysis was used to enhance model interpretability, offering valuable insights into feature contributions and the viscosity behavior of DESs. This ML approach enhances viscosity prediction accuracy and accelerates DES selection for industrial applications. By assembling an extensive training dataset and leveraging quantum chemistry-derived features, this study establishes a robust framework for transferable ML models. The findings underscore the potential of data-driven approaches in understanding DES viscosity and optimizing their use in various scientific and industrial domains [49]. This study compiled a significantly larger and structurally diverse dataset of DES viscosities than previous literature, covering a wide range of DES structures and compositions. The developed ML models demonstrate high predictive accuracy and broad applicability, making them valuable for various industrial applications, including sustainable materials and energy. A key innovation of this work is the use of the SHAP analysis to interpret viscosity predictions, revealing that polar interactions are more dominant than nonpolar interactions. Another significant advantage is the efficient feature selection approach, which achieves high prediction accuracy without explicitly modeling complex intermolecular geometries of HBA/HBD pairs. Instead, the σ-profile features derived from isolated HBA and HBD molecules effectively capture intermolecular interactions while significantly reducing computational costs. The CatBoost model accurately predicts DES viscosity across different research applications. Lower-viscosity DESs exhibit greater affinity and capability for enhancing the solubility and absorption of polymers and gases. The superior performance of CatBoost over traditional models, such as multilinear regression, two-factor polynomial regression, FFNN, and SVR, suggests that viscosity follows a complex, nonlinear relationship with molecular features. The ensemble-based approach of CatBoost prevents overfitting, making it more robust compared to single-function models. Ultimately, the developed CatBoost model streamlines the screening of DESs, minimizing the need for time-consuming experiments. Additionally, the SHAP analysis provides valuable insights to guide the rational design of DESs with tailored viscosities for industrial applications [50].

2.2.3 Surface Tension

Until now, research on the surface tension of DESs has been quite limited. Surface tension is a key physical property that reflects a liquid's tendency to minimize its surface area [51]. In industrial applications, surface tension is crucial, as it significantly influences properties like permeability and bubbling. These factors are essential in the design and operation of equipment such as distillation columns, thermal systems, and heat exchangers [52]. Surface tension arises from intermolecular forces, and as the temperature of a liquid enhances, the kinetic energy of its molecules also rises. This increase weakens the intermolecular attractions, leading to a decrease in surface tension. For DESs, in addition to temperature, the molar ratios of the HBD and HBA components play a key role in determining their surface tension [53]. At the interface between a liquid and a gas, the surface molecules experience unbalanced and directional forces. These intermolecular attractions draw the surface molecules inward, toward the bulk of the liquid, resulting in surface tension. In this manner, the surface layer of a liquid experiences tension, which naturally acts to reduce the surface area under equilibrium conditions. This molecular-level tension present at the liquid's surface is referred to as surface or interfacial tension [54]. Surface tension in DESs is anticipated to exhibit a trend similar to that of viscosity, as both properties are strongly influenced by the intermolecular forces that drive DES formation. Abbott and colleagues provided surface tension data for ChCl- and $ZnCl_2$-based DESs. For example, the surface tensions measured for ChCl: malonic acid (1:1) and ChCl: phenylacetic acid (1:2) were approximately 65.68 and 41.86 mN/m, respectively [16]. Moreover, the surface tension of the ChCl: glycerol DES was found to decrease with an increase in ChCl concentration. This observation aligns with earlier findings related to viscosity, suggesting that the incorporation of ChCl disrupts the robust hydrogen bonding network present in pure glycerol [52, 55–58].

2.2.4 Refractive Index

Refractive index is an essential property of DESs, playing a vital role in numerous applications, such as optical characterization of materials, evaluating their purity, and measuring solute concentrations within solutions. A linear relationship between refractive index and temperature is typically expressed using the following equation.

$$\eta_{\mathrm{D}} = d + d_0 T \tag{2.6}$$

where η_D is the refractive index, T is the temperature in K, and d and d_0 are constants.

A comprehensive study on the refractive index of pure DESs and their mixtures with water demonstrated the influence of temperature, HBAs, and HBDs. It was observed that the refractive index of both neat DESs and their aqueous mixtures decreases linearly as the temperature rises. Additionally, the choice of HBA and HBD significantly impacts the refractive index values [8, 59–61].

2.2.5 *Freezing Point* (T_f)

The low freezing point of DESs is fundamental to their practical applications and versatility. This depression in freezing point arises from the strength of interactions between the HBD and HBA, their specific molar ratios, and the change in entropy upon DES formation. Specifically, combinations such as ChCl with glycerol, polyalcohols, or urea demonstrate significantly lower freezing points compared to those of the individual components [62]. The freezing point of the DES formed by ethylene glycol and N, N-diethylethanolammonium chloride rises as the HBA: HBD molar ratio increases from 1:2 to 1:4. In contrast, for the DES composed of N, N-diethylethanolammonium chloride and glycerol, the freezing point decreases as the HBA: HBD molar ratio changes from 1:2 to 1:4 [63, 64].

2.2.6 *Ionic Conductivity* (κ) *and Ionicity*

Ionic conductivity is primarily explained by the ionic conduction mechanism, which reflects how readily a substance permits the movement of ions and the flow of electric current. In applications such as redox flow batteries and power systems, the conductivity characteristics of DESs are critically important. The ionic conductivities of DESs are relatively low and are correlated with the viscosity of DESs. Since viscosity is the main conductivity controller, most DESs typically exhibit low ionic conductivities ($\kappa < 2$ mS/cm at ambient temperature). However, raising the temperature generally lowers viscosity, enhancing the ionic conductivity. This behavior is influenced by various factors, including the molar ratio of the HBA to HBD, the specific nature of the organic salt and HBD used, the identity of the salt's anion, and the presence of water. Notably, DESs formed using ethylene glycol or imidazole with ChCl demonstrate relatively high conductivities, 7.61 mS/cm at 20 °C and 12 mS/cm at 60 °C, respectively, primarily due to their low viscosities. The relationship between conductivity and viscosity in DESs can be more deeply understood by analyzing their ionicity through a Walden plot. Evaluating the ionicity is crucial, as it reflects the proportion of free, mobile ions that actively contribute to electrical conductivity, an essential factor when assessing the potential of DESs in energy storage technologies [65–69].

2.3 Conclusion and Future Potential

The physicochemical properties of DESs are important fundamental properties for which large-scale substantiation and details are necessary. DES physicochemical properties are intensively regulated by HBD/HBA combinations, their molar ratio, the types of functional groups involved, and the alkyl chain lengths of the HBD and HBA components. Exhaustive cognizance of the effect of the above-mentioned factors on the DES physicochemical properties would be beneficial for the future development of novel DESs and their use in industry. Conducting repeated laboratory experiments is often constrained by both time and financial resources. Hence, documentation of the physicochemical properties of DESs is to make predictive models and correlations easier, which can be useful for screening the viability of HBDs and HBAs as DES constituents. With the rise of AI tools, new routes in the study of DES physicochemical properties are foreseen for discovery.

Acknowledgments The authors acknowledge the Director/Principal, Ramniranjan Jhunjhunwala College of Arts, Science and Commerce (Empowered Autonomous), Ghatkopar (W), 400086, Mumbai, Maharashtra, and the Department of Chemistry, Sardar Vallabhbhai National Institute of Technology (SVNIT), Gujarat, India, for providing infrastructure facilities and support.

Conflict of Interest Statement The authors declare no conflict of interest.

Declaration of Generative AI and AI-Assisted Technologies in the Writing Process While preparing this work, the authors utilized the AI tool to enhance the language and readability of several statements in the manuscript. After using this tool/service, the authors reviewed and edited the content as needed, and they take full responsibility for the publication content.

References

1. Blanchard LA, Hancu D, Beckman EJ, Brennecke (1999) Green processing using ionic liquids and CO_2. Nature 399(6731):28–29. https://doi.org/10.1038/19887
2. Abbott AP, Capper G, Davies DL, Rasheed RK, Tambyrajah VJ (2003) Novel solvent properties of choline chloride/urea mixtures. Chem Commun 1:70–71. https://doi.org/10.1039/B210714G
3. Jangir AK, Dani U, Chaudhary VK (2025) Exploring the influence of composition on physicochemical and biological properties of choline chloride-based deep eutectic solvents: an experimental and theoretical endorsement. J Mol Liq 427:127419. https://doi.org/10.1016/j.molliq.2025.127419
4. Jangir AK, Verma G, Pandey S, Kuperkar K (2022) Design and thermophysical characterization of betaine hydrochloride-based deep eutectic solvents as a new platform for CO_2 capture. New J Chem 46(11):5332–5345. https://doi.org/10.1039/D1NJ05373F
5. Wang Y, Hou Y, Wu W, Liu D, Ji Y, Ren SJ (2016) Roles of a hydrogen bond donor and a hydrogen bond acceptor in the extraction of toluene from n-heptane using deep eutectic solvents. Green Chem 18(10):3089–3097. https://doi.org/10.1039/C5GC02909K
6. Jibril B, Mjalli F, Naser J, Gano ZJ (2014) New tetrapropylammonium bromide-based deep eutectic solvents: synthesis and characterizations. J Mol Liq 199:462–469. https://doi.org/10.1016/j.molliq.2014.08.004

7. Mjalli FS, Naser J, Jibril B, Alizadeh V, Gano ZJ (2014) Tetrabutylammonium chloride based ionic liquid analogues and their physical properties. J Chem Eng Data 59(7):2242–2251. https://doi.org/10.1021/je5002126
8. Su HZ, Yin JM, Liu QS, Li CP (2015) Properties of four deep eutectic solvents: density, electrical conductivity, dynamic viscosity and refractive index. Acta Phys -Chim Sin 31(8):1468–1473. https://doi.org/10.3866/PKU.WHXB201506111
9. Abbott (2004) Application of hole theory to the viscosity of ionic and molecular liquids. ChemPhysChem 5(8):1242–1246. https://doi.org/10.1002/cphc.200400190
10. Abbott AP, Capper G, Davies DL, Rasheed R (2004) Ionic liquids based upon metal halide/substituted quaternary ammonium salt mixtures. Chem Eur J 43(11):3447–3452. https://doi.org/10.1021/ic049931s
11. Freitas DS, Cavaco-Paulo A, Silva C (2024) Enhancing insights into the phenomena of deep eutectic solvents. Sustain Mater Technol:e01039. https://doi.org/10.1016/j.susmat.2024.e01039
12. Florindo C, Oliveira FS, Rebelo LPN, Fernandes AM, Marrucho IM (2014) Insights into the synthesis and properties of deep eutectic solvents based on cholinium chloride and carboxylic acids. ACS Sustain Chem Eng 2(10):2416–2425. https://doi.org/10.1021/sc500439w
13. Basaiahgari A, Panda S, Gardas RL (2018) Effect of ethylene, diethylene, and triethylene glycols and glycerol on the physicochemical properties and phase behavior of benzyltrimethyl and benzyltributylammonium chloride based deep eutectic solvents at 283.15–343.15 K. J Chem Eng Data 63(7):2613–2627. https://doi.org/10.1021/acs.jced.8b00213
14. Ijardar SP (2020) Deep eutectic solvents composed of tetrabutylammonium bromide and PEG: density, speed of sound and viscosity as a function of temperature. J Chem Thermodyn 140:105897. https://doi.org/10.1016/j.jct.2019.105897
15. Ijardar SP, Singh V, Gardas RL (2022) Revisiting the physicochemical properties and applications of deep eutectic solvents. Molecules 27(4):1368. https://doi.org/10.3390/molecules27041368
16. Abbott AP, Harris RC, Ryder KS, D'Agostino C, Gladden LF, Mantle MD (2011) Glycerol eutectics as sustainable solvent systems. Green Chem 13(1):82–90. https://doi.org/10.1039/C0GC00395F
17. Shahbaz K, Baroutian S, Mjalli F, Hashim M, AlNashef I (2012) Densities of ammonium and phosphonium based deep eutectic solvents: prediction using artificial intelligence and group contribution techniques. Thermochim Acta 527:59–66. https://doi.org/10.1016/j.tca.2011.10.010
18. Deng D, Jiang Y, Liu X, Zhang Z, Ai NJT (2016) Investigation of solubilities of carbon dioxide in five levulinic acid-based deep eutectic solvents and their thermodynamic properties. J Chem Thermodyn 103:212–217. https://doi.org/10.1016/j.jct.2016.08.015
19. Liu F, Xue Z, Zhao X, Mou H, He J, Mu T (2018) Catalytic deep eutectic solvents for highly efficient conversion of cellulose to gluconic acid with gluconic acid self-precipitation separation. Chem Commun 54(48):6140–6143. https://doi.org/10.1039/C8CC03798A
20. Florindo C, Oliveira MM, Branco LC, Marrucho IM (2017) Carbohydrates-based deep eutectic solvents: Thermophysical properties and rice straw dissolution. J Mol Liq 247:441–447. https://doi.org/10.1016/j.molliq.2017.09.026
21. van Osch DJ, Zubeir LF, van den Bruinhorst A, Rocha MA, Kroon MC (2015) Hydrophobic deep eutectic solvents as water-immiscible extractants. Green Chem 17(9):4518–4521. https://doi.org/10.1039/C5GC01451D
22. Shafie MH, Yusof R, Gan C-Y (2019) Synthesis of citric acid monohydrate-choline chloride based deep eutectic solvents (DES) and characterization of their physicochemical properties. J Mol Liq 288:111081. https://doi.org/10.1016/j.molliq.2019.111081
23. Jangir AK, Sethy P, Verma G, Bahadur P, Kuperkar K (2021) An inclusive thermophysical and rheology portrayal of deep eutectic solvents (DES) for metal oxides dissolution enhancement. J Mol Liq 332:115909. https://doi.org/10.1016/j.molliq.2021.115909
24. Jangir AK, Nain AK, Kuperkar K (2021) Insight into structural properties and molecular interactions of maline (choline chloride+ malonic acid) and 1, 4-butanediol based pseudo-binary

mixture: a thermophysical, spectral, and simulation portrayal. J Mol Liq 334:116050. https://doi.org/10.1016/j.molliq.2021.116050
25. Shah D, Mjalli FS (2014) Effect of water on the thermo-physical properties of reline: an experimental and molecular simulation based approach. Phys Chem Chem Phys 16(43):23900–23907. https://doi.org/10.1039/C4CP02600D
26. Dai Y, Witkamp G-J, Verpoorte R, Choi YH (2015) Tailoring properties of natural deep eutectic solvents with water to facilitate their applications. Food Chem 187:14–19. https://doi.org/10.1016/j.foodchem.2015.03.123
27. Haghbakhsh R, Duarte ARC, Raeissi SJT (2021) Volumetric investigation of aqueous mixtures of the {choline chloride+ phenol (1: 4)} deep eutectic solvent. J Chem Thermodyn 158:106440. https://doi.org/10.1016/j.jct.2021.106440
28. Jangir AK, Mandviwala H, Patel P, Sharma S, Kuperkar K (2020) Acumen into the effect of alcohols on choline chloride: L-lactic acid-based natural deep eutectic solvent (NADES): a spectral investigation unified with theoretical and thermophysical characterization. J Mol Liq 317:113923. https://doi.org/10.1016/j.molliq.2020.113923
29. El Achkar T, Greige-Gerges H, Fourmentin S (2021) Basics and properties of deep eutectic solvents: a review. Environ Chem Lett 19:3397–3408. https://doi.org/10.1007/s10311-021-01225-8
30. Kumar Jangir A, Patel D, More R, Parmar A, Kuperkar K (2019) New insight into experimental and computational studies of choline chloride-based 'green'ternary deep eutectic solvent (TDES). J Mol Struct 1181:295–299. https://doi.org/10.1016/j.molstruc.2018.12.106
31. Abbott AP, Boothby D, Capper G, Davies DL, Rasheed RKJ (2004) Deep eutectic solvents formed between choline chloride and carboxylic acids: versatile alternatives to ionic liquids. J Am Chem Soc 126(29):9142–9147. https://doi.org/10.1021/ja048266j
32. Omar KA, Sadeghi R (2023) Database of deep eutectic solvents and their physical properties: a review. J Mol Liq:121899. https://doi.org/10.1016/j.molliq.2023.121899
33. Crespo EA, Silva LP, Martins MNA, Fernandez L, Ortega J, Ferreira O, Sadowski G, Held C, Pinho SOP, Coutinho JOA (2017) Characterization and modeling of the liquid phase of deep eutectic solvents based on fatty acids/alcohols and choline chloride. Ind Eng Chem Res 56(42):12192–12202. https://doi.org/10.1021/acs.iecr.7b02382
34. Zhu J, Yu K, Zhu Y, Zhu R, Ye F, Song N, Xu Y (2017) Physicochemical properties of deep eutectic solvents formed by choline chloride and phenolic compounds at T=(293.15 to 333.15) K: the influence of electronic effect of substitution group. J Mol Liq 232:182–187. https://doi.org/10.1016/j.molliq.2017.02.071
35. Lapeña D, Lomba L, Artal M, Lafuente C, Giner B (2019) Thermophysical characterization of the deep eutectic solvent choline chloride: ethylene glycol and one of its mixtures with water. Fluid Phase Equilib 492:1–9. https://doi.org/10.1016/j.fluid.2019.03.018
36. Balaraman HB, Rathnasamy SK (2019) Thermophysical and molar volume aberration of amphiphilic eutectic mix of bivalent diols and ammonium-ionic liquid. J Chem Eng Data 64(8):3307–3315. https://doi.org/10.1021/acs.jced.9b00134
37. Roosta A, Haghbakhsh R, Duarte ARC, Raeissi S (2023) Deep eutectic solvent viscosity prediction by hybrid machine learning and group contribution. J Mol Liq 388:122747. https://doi.org/10.1016/j.molliq.2023.122747
38. Duarte T, de Alencar LV, Rodríguez-Reartes SB, Tavares FW, Llovell F (2024) Assessing viscosity in sustainable deep eutectic solvents and cosolvent mixtures: an artificial neural network-based molecular approach. ACS Sustain Chem Eng 12(21):7987–8000. https://doi.org/10.1021/acssuschemeng.3c07219
39. Liu H, Xu H, Zhu W, Zhou Y, Xue K, Zhu Z, Wang Y, Qi (2025) Prediction of the viscosity of green deep eutectic solvents by constructing ensemble model based on machine learning. Chem Eng Sci 304:120987. https://doi.org/10.1016/j.ces.2024.120987
40. Haghbakhsh R, Parvaneh K, Raeissi S, Shariati A (2018) A general viscosity model for deep eutectic solvents: the free volume theory coupled with association equations of state. Fluid Phase Equilib 470:193–202. https://doi.org/10.1016/j.fluid.2017.08.024

41. Lloret J, Vega L, Llovell F (2017) Accurate description of thermophysical properties of tetraalkylammonium chloride deep eutectic solvents with the soft-SAFT equation of state. Fluid Phase Equilib 448:81–93. https://doi.org/10.1016/j.fluid.2017.04.013
42. Verevkin SP, Sazonova AY, Frolkova AK, Zaitsau DH, Prikhodko IV, Held C (2015) Separation performance of BioRenewable deep eutectic solvents. Ind Eng Chem Res 54(13):3498–3504. https://doi.org/10.1021/acs.iecr.5b00357
43. Pontes PV, Crespo EA, Martins MA, Silva LP, Neves CM, Maximo GJ, Hubinger MD, Batista EA, Pinho SP, Coutinho JA (2017) Measurement and PC-SAFT modeling of solid-liquid equilibrium of deep eutectic solvents of quaternary ammonium chlorides and carboxylic acids. Fluid Phase Equilib 448:69–80. https://doi.org/10.1016/j.fluid.2017.04.007
44. Haghbakhsh R, Raeissi S, Parvaneh K, Shariati A (2018) The friction theory for modeling the viscosities of deep eutectic solvents using the CPA and PC-SAFT equations of state. J Mol Liq 249:554–561. https://doi.org/10.1016/j.molliq.2017.11.054
45. Mjalli FS, Mousa HJC (2017) Viscosity of aqueous ionic liquids analogues as a function of water content and temperature. Chin J Chem Eng 25(12):1877–1883. https://doi.org/10.1016/j.cjche.2017.09.008
46. Chatterjee S, Chowdhury T, Bagchi SJT (2024) Solvation dynamics and microheterogeneity in deep eutectic solvents. J Phys Chem B 128(51):12669–12684. https://doi.org/10.1021/acs.jpcb.4c06295
47. Sakpal SS, Deshmukh SH, Chatterjee S, Ghosh D, Bagchi S (2021) Transition of a deep eutectic solution to aqueous solution: a dynamical perspective of the dissolved solute. J Phys Chem Lett 12(36):8784–8789. https://doi.org/10.1021/acs.jpclett.1c02118
48. Chatterjee S, Deshmukh SH, Bagchi SJT (2022) Does viscosity drive the dynamics in an alcohol-based deep eutectic solvent? J Phys Chem B 126(41):8331–8337. https://doi.org/10.1021/acs.jpcb.2c06521
49. Mohan M, Jetti KD, Smith MD, Demerdash ON, Kidder MK, Smith JC (2024) Accurate machine learning for predicting the viscosities of deep eutectic solvents. J Chem Theory Comput 20(9):3911–3926. https://doi.org/10.1021/acs.jctc.3c01163
50. Andrade JD, Smith LM, Gregonis DE (1985) The contact angle and interface energetics. In: Surface and interfacial aspects of biomedical polymers: volume 1 surface chemistry and physics. Springer, pp 249–292. https://doi.org/10.1007/978-1-4684-8610-0_7
51. Kumar V, Pal N, Jangir AK, Manyala DL, Varade D, Mandal A, Kuperkar K (2020) Dynamic interfacial properties and tuning aqueous foamability stabilized by cationic surfactants in terms of their structural hydrophobicity, free drainage and bubble extent. Colloids Surf A Physicochem Eng Asp 588:124362. https://doi.org/10.1016/j.colsurfa.2019.124362
52. Ghaedi H, Ayoub M, Sufian S, Shariff AM, Lal B (2017) The study on temperature dependence of viscosity and surface tension of several Phosphonium-based deep eutectic solvents. J Mol Liq 241:500–510. https://doi.org/10.1016/j.molliq.2017.06.024
53. Chen Y, Chen W, Fu L, Yang Y, Wang Y, Hu X, Wang F, Mu T (2019) Surface tension of 50 deep eutectic solvents: effect of hydrogen-bonding donors, hydrogen-bonding acceptors, other solvents, and temperature. Ind Eng Chem Res 58(28):12741–12750. https://doi.org/10.1021/acs.iecr.9b00867
54. Gharagheizi F, Eslamimanesh A, Sattari M, Mohammadi AH, Richon D (2013) Development of corresponding states model for estimation of the surface tension of chemical compounds. AICHE J 59(2):613–621. https://doi.org/10.1002/aic.13824
55. Li X, Hou M, Han B, Wang X, Zou L (2008) Solubility of CO_2 in a choline chloride+ urea eutectic mixture. J Chem Eng Data 53(2):548–550. https://doi.org/10.1021/je700638u
56. Gajardo-Parra NF, Lubben MJ, Winnert JM, Leiva Á, Brennecke JF, Canales RI (2019) Physicochemical properties of choline chloride-based deep eutectic solvents and excess properties of their pseudo-binary mixtures with 1-butanol. J Chem Thermodyn 133:272–284. https://doi.org/10.1016/j.jct.2019.02.010
57. Khanbabaeimoghri N, Lashkarbolooki M, Bakhshi H, Farzam S (2025) Measurement and thermodynamic modeling of the surface tension of binary mixtures containing

deep eutectic solvent. Chem Eng Commun 212(7):1013–1030. https://doi.org/10.1080/00986445.2024.2445229
58. Fronduti M, Del Giacco T, Rossi E, Tiecco M, Germani R (2023) Insights into the structural features of deep eutectic solvents: the eutectic point as an unicum in their physical properties and the surface tension as a method for its determination. J Mol Liq 379:121679. https://doi.org/10.1016/j.molliq.2023.121679
59. Luan J, Cheng Y, Xue F, Cui L, Wang D (2023) Refractive index of 48 neat deep eutectic solvents and of selected mixtures: effect of temperature, hydrogen-bonding donors, hydrogen-bonding acceptors, mole ratio, and water. ACS Omega 8(28):25582–25591. https://doi.org/10.1021/acsomega.3c03502
60. Shahbaz K, Bagh FG, Mjalli F, AlNashef I, Hashim M (2013) Prediction of refractive index and density of deep eutectic solvents using atomic contributions. Fluid Phase Equilib 354:304–311. https://doi.org/10.1016/j.fluid.2013.06.050
61. Ferreira E, Ramos-Ortiz G, Vazquez A, Trejo-Durán M (2023) Third-order nonlinear optical properties of choline chloride based deep eutectic solvents: theoretical and experimental studies. J Mol Liq 384:122253. https://doi.org/10.1016/j.molliq.2023.122253
62. Piemontese L, Perna FM, Logrieco A, Capriati V, Solfrizzo M (2017) Deep eutectic solvents as novel and effective extraction media for quantitative determination of ochratoxin A in wheat and derived products. Molecules 22(1):121. https://doi.org/10.3390/molecules22010121
63. Khajeh A, Shakourian-Fard M, Parvaneh K (2021) Quantitative structure-property relationship for melting and freezing points of deep eutectic solvents. J Mol Liq 321:114744. https://doi.org/10.1016/j.molliq.2020.114744
64. Jiang Y, Liao Y, Chen J, Ma K (2025) Advancing anti-freezing hydrogel electrolyte based on deep eutectic solvent for wide temperature range aqueous zinc-ion battery. J Environ Sci 156:596–605. https://doi.org/10.1016/j.jes.2025.02.022
65. de Lacalle JL, Gallastegui A, Olmedo-Martínez JL, Moya M, Lopez-Larrea N, Picchio ML, Mecerreyes D (2023) Multifunctional ionic polymers from deep eutectic monomers based on polyphenols. ACS Macro Lett:125–132. https://doi.org/10.1021/acsmacrolett.2c00657
66. Huddleston JG, Visser AE, Reichert WM, Willauer HD, Broker GA, Rogers RD (2001) Characterization and comparison of hydrophilic and hydrophobic room temperature ionic liquids incorporating the imidazolium cation. Green Chem 3(4):156–164. https://doi.org/10.1039/B103275P
67. Kudłak B, Owczarek K, Namieśnik (2015) Selected issues related to the toxicity of ionic liquids and deep eutectic solvents—a review. Environ Sci Pollut Res 22:11975–11992. https://doi.org/10.1007/s11356-015-4794-y
68. van Osch DJ, Kollau LJ, van den Bruinhorst A, Asikainen S, Rocha MA, Kroon MC (2017) Ionic liquids and deep eutectic solvents for lignocellulosic biomass fractionation. Phys Chem Chem Phys 19(4):2636–2665. https://doi.org/10.1039/C6CP07499E
69. Cao P, Feng J, Yang T, Ao H, Shang T, Xing B (2023) MXene-enhanced deep eutectic solvent-based flexible strain sensor with high conductivity and anti-freezing using electrohydrodynamic direct-writing method. Colloids Surf A Physicochem Eng Asp 677:132349. https://doi.org/10.1016/j.colsurfa.2023.132349

Chapter 3
Deep Eutectic Solvents: Revolutionizing Biotechnology Applications

Anil Kumar Jangir, Ritu Jangid, Unnati Dani, and Nisha Jarwal

3.1 Introduction

A eutectic was named in 1884 for the mixture of metal alloys that melt together at a lower point than their separate parts would. At the beginning of the twentieth century, deep eutectic solvents (DESs) were introduced as a suitable alternative to ionic liquids (ILs). DESs are also characterized as a eutectic mixture with a lower melting point, formed by mixing two or more components that act as hydrogen bond donors (HBDs) and hydrogen bond acceptors (HBAs) in a particular molar ratio. A strong hydrogen bonding network is established between the HBA and the HBD [1]. A large number of DESs are nontoxic, biodegradable, and environmentally benign. Distinctive features of DESs are their low volatility, high heat resistance, ability to conduct electricity, low melting point, and great capacity to turn insoluble materials in water into dissolved compounds [2]. In addition, the process involves strengthening enzymes, carrying out biocatalytic reactions, and recycling useful substances from plant substances, microorganisms, and sea life components. Regenerative medicine and advanced methods to administer drugs with more biocompatibility were also witnessed [3].

A. K. Jangir
Department of Chemistry, Shri Jagdishprasad Jhabarmal Tibrewala University, Vidyanagri, Jhunjhunu, Rajasthan, India

R. Jangid (✉)
Department of Science, Biyani Girls College, Jaipur, Rajasthan, India

U. Dani
Department of Zoology, B.P. Baria Science Institute, Navsari, Gujarat, India

N. Jarwal
Department of Chemistry, Indian Institute of Technology Roorkee, Roorkee, Uttarakhand, India

K. Kuperkar et al. (eds.), *Deep Eutectic Solvents*, Environmental Footprints and Eco-design of Products and Processes,
https://doi.org/10.1007/978-3-032-03801-2_3

3.2 Deep Eutectic Solvents (DESs)

3.2.1 Definition of DESs

Deep Eutectic Solvents (DESs), also known as green solvents, are composed of hydrogen bond donors (HBDs) and hydrogen bond acceptors (HBAs) by forming a hydrogen bonding network. Consequently, the charges spread through the molecule, which makes the material melt at a lower temperature and requires less lattice energy. Combining HBDs and HBAs led to a wide range of different material types. In general, ethylene glycol (EG), glycerol (GLY), urea (U), 1,2-propanediol (PD), citric acid (CA), lactic acid (LA), succinic acid (SA), tartaric acid (TA), xylitol (XY), sorbitol (SOR), and fructose (FR) are the typical hydrogen bond donors (HBDs) used. Included among the common HBAs are metal halides, choline chloride (ChCl), betaine (BH), "tetraalkylammonium salts (e.g., tetrabutylammonium bromide [TBAB], tetraethylammonium bromide [TEAB], and tetrapropylammonium bromide [TPAB]), and many others, as demonstrated in Fig. 3.1. DESs usually align well with the main ideas of green chemistry by being nonflammable, biodegradable, recyclable, cost-effective, usable within a wide liquid range, suitable for a variety of applications, and produced without loss and not requiring any further purification [4].

3.2.2 Physicochemical Properties of DESs

DESs can be tailored similarly to ILs with different combinations of a quaternary ammonium salt (QAS) (ChCl) and an HBD group. Combining changes in freezing point (T_f), viscosity (η), surface tension (γ), refractive index (n_D), and conductivity (σ) makes it possible to create DESs, which are suited for specific purposes. Literature studies showed that the researchers have reported the comprehensive

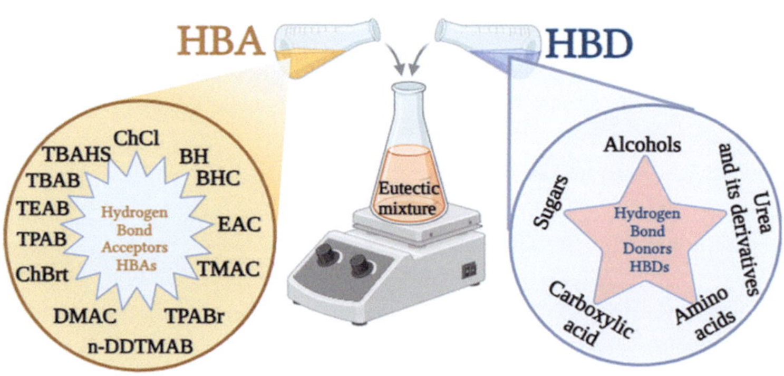

Fig. 3.1 A representation of DESs' preparation by varying the nature of the molar ratios of HBAs and HBDs in the correct amounts

examination on the physicochemical properties of DESs and emphasized their potential applications in academic and industrial sectors [5–18].

3.2.2.1 Freezing Point (T_f)

The term eutectic point in DESs means that the lowest melting point is reached when two solid materials mix using hydrogen bonding. For instance, ChCl:urea (1:2) can show this happening [19]. Table 3.1 lists the T_f of some DESs, in which the ChCl is strongly combined with a particular HBD, all having $\boldsymbol{T_f}$ less than 100 °C. The smaller the molecule's mass, the greater its depression of the freezing point. Abbott

Table 3.1 Illustrations of different types of DESs prepared through diverse methods

Sr. No.	DES composition	Molar ratio	Preparation method	Ref.
1.	ChCl:urea	1:2	Thermal mixing	[38]
			Mortar and pestle (grinding)	[38]
			Vacuum evaporation	[39]
			Freeze-drying	[40]
			Twin-screw extrusion	[41]
2.	ChCl:glycerol	1:2	Thermal mixing	[42]
			Mortar and pestle (grinding)	[43]
			Vacuum evaporation	[39]
			Freeze-drying	[40]
			Twin-screw extrusion	[41]
			Microwave irradiation	[44]
			Ultrasonication	[45]
3.	ChCl:lactic acid	1:2	Vacuum evaporation	[39]
		1:10	Ultrafiltration	[46]
		1:1	Thermal mixing	[47]
4.	ChCl:ethylene glycol	1:2	Ultrafiltration and electrodialysis	[39]
			Thermal mixing	[48]
			Grinding	[43]
			Vacuum evaporation	[39]
			Twin-screw extrusion	[41]
			Microwave irradiation	[44]
			Freeze-drying	[48]
5.	ChCl:1,4-butanediol	1:4	Microwave irradiation	[44]
6.	ChCl:$ZnCl_2$	1:3	Ultrasonication	[49]
			Vacuum evaporation	[50]
			Freeze-drying	[39]
7.	Menthol:DecA	1:1	Thermal mixing	[51]
			Grinding	[52]
			Freeze-drying	[53]
			Twin-screw extrusion	[54]

(continued)

Table 3.1 (continued)

Sr. No.	DES composition	Molar ratio	Preparation method	Ref.
8.	ChCl:citric acid	1:1	Thermal mixing	[55]
			Vacuum evaporation	[43]
			Microwave irradiation	[56]
		1:2	Twin-screw extrusion	[54]
			Ultrasonication	[57]
9.	ChCl:malic acid	1:1	Microwave irradiation	[44]
			Thermal mixing	[58]
			Freeze-drying	[59]
			Vacuum evaporation	[59]
			Grinding	[59]
			Ultrasonication	[59]
10.	ChCl:levulinic acid	1:1	Microwave irradiation	[44]
			Thermal mixing	[60]
			Freeze-drying	[61]
			Vacuum evaporation	[61]
			Twin-screw extrusion	[62]
11.	ChCl:oxalic acid	1:1	Freeze-drying	[63]
			Microwave irradiation	[64]
			Grinding	[65]
			Vacuum evaporation	[54]
			Thermal mixing (heating and stirring)	[66]
			Thermal mixing (heating and stirring)	[55]
12.	ChCl:glutaric acid	1:3	Grinding	[65]

et al. explain that the T_f of DESs is affected by the structure of the DES lattice, bonds between the DES ions, and the variation in entropy as the DES becomes a liquid [20].

3.2.2.2 Density (ρ)

Density is a vital physical parameter that helps recognize the nature of the intermolecular bonds in DESs using different solvents. The eutectic mixtures consisting of Type IV $ZnCl_2$-HBD have ρ above 1.3 g.cm $^{-3}$. On the other hand, molar solvent ratios gave $ZnCl_2$:urea (1:3.5) a significantly higher ρ value than the lower reading for $ZnCl_2$-acetamide (1:4) [21]. The way molecules are grouped in DES samples may lead to the ρ difference observed in these structures. Florindo et al. examined various mixtures of ChCl and citric acid used in DES preparation. The findings show that a higher ratio of ChCl in DES makes the DES less dense. An increase in the quantity of citric acid was associated with an increase in the ρ value [12]. According to Basaiahgari et al., greater binding between HBD molecules raises ρ due to factors such as increasing the level or number of HBD substituents [22].

3.2.2.3 Viscosity (η)

DESs share a viscosity behavior similar to ILs. The three factors that play a major role in regulating the η of DESs are the type and proportion of chemical ingredients, temperature, and amount of water. The η of ChCl:urea (1:2) at room temperature is greater than 100 cP, while all other DESs have higher η values due to the significant interaction that occurs between the HBD and the HBA. DESs have higher viscosity due to less movement of free species. DES solutions have two characteristics: large-sized ions and limited volume for water [23].

3.2.2.4 pH

As cationic and anionic species in a DES have acidic traits, the pH log-scale is a major parameter when measuring and studying DES solutions [4]. A direct connection between both pH and temperature was found when measurements were taken over a temperature range of 25–85 °C. The researchers studied this issue and found that as the temperature increased, the pH value decreased linearly. Similar pH relationships were found among different DES mixtures, but variations were observed in the acidic value depending on the nature of HBD– HBA [24].

3.2.2.5 Surface Tension (γ)

Abbott et al. reported the values of γ for DESs formed from ChCl-$ZnCl_2$-based electrolytes. Besides, the change in γ value is significantly attributed to the change in nature of HBDs; for instance, ChCl as HBA with malonic acid in the ratio 1:1 and phenylacetic acid in the ratio 1:2 were reported, along with interesting changes in the composition of the DES [20].

3.2.2.6 Ionic Conductivity (σ)

This property is used to discover the ease with which a material allows ions to move and carry a current. According to Abbott et al., both lowering the cation size and using fluorinated replacements for the HBD make the free volume of the DES larger and its ionic conductivity higher [20]. Increasing the free volume of DESs lowers intermolecular friction, which reduces viscosity, assisting molecules to move faster, thereby increasing ionic conductivity. Since DESs have moderately high acidity, they typically have poor conductivity through ions [25]. Several researchers reported the σ of DES at several temperatures. Ionic conductivity increases strongly as the temperature increases due to a lowering of viscosity. Furthermore, changes in the organic salt/HBD ratio result in large changes in the DES viscosity, which greatly affects the ionic conductivity of the DES [12].

3.2.2.7 Refractive Index (n_D)

Refractive index is the science that measures how the rate of light traveling through materials compares to the light in vacuum. Examining the electronic charge density of molecules helps identify the interactions and open space available in each DES molecule maker. The group of Luan et al. examined the n_D of DES made from ChCl and glycerol in different amounts (1:1.5, 1:2, and 1:3) as the temperature varied (15–55 °C) and found that readings on the peak index are found only in the eutectic mixture. In another investigation, it was shown that HBDs significantly changed the n_D value of DESs [26]. Getie et al. observed n_D of various prepared DESs composed of different HBDs with HBA (tetrabutylammonium chloride (TBAC)). Among the two, the DES with equal quantities of TBAC and phenylacetic acid (PAA) had the n_D for 1:2 molar ratios [27]. Sanchez et al. reported the thermophysical properties of L-proline- and betaine-based DESs. [28].

3.2.3 Classification of DESs

Four types of DESs can be identified by their components and the affinity between HBAs and HBDs. In the case of Type I DESs, quaternary ammonium salts (QASs) were combined with metal chlorides. The structure of Type II DES had metal chloride hydrates in addition to the QAS compounds. Type III DESs have QASs, together with organic HBDs, while Type IV DESs have a chloride hydrate of a metal as the HBA and a Type III HBD [29]. Fig. 3.2 illustrates the DES classes and explains the main ingredients needed to make them.

3.2.3.1 Classification Based on Composition

DES Type I

Type I DESs are unique because they combine metal chlorides as HBAs with organic compounds as HBDs through strong intermolecular hydrogen bonding between metal salts and changed HBDs. This class holds a unique characteristic in that these solvents have been used for both research and industrial tasks, as they bring beneficial results in electrochemistry, metallurgy, and catalysis. Type I

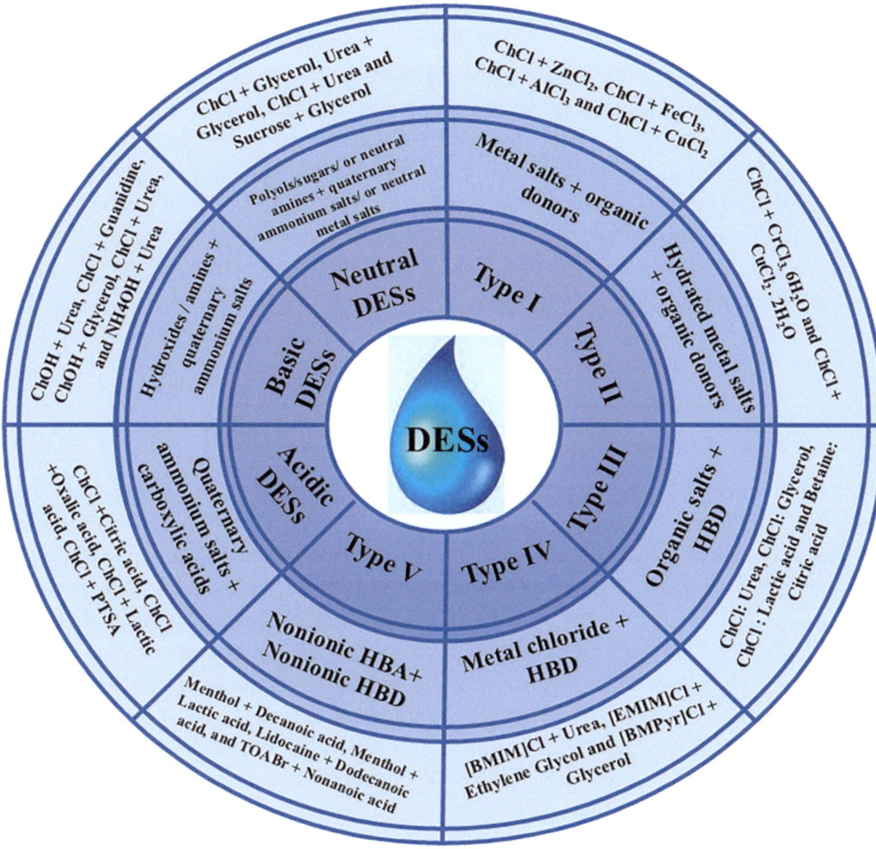

Fig. 3.2 A systematic representation of the various types of DESs based on the nature of HBAs and HBDs

DESs help save money and the environment by being flexible for different purposes. The appearance and function of DESs depend on which metal salts (HBAs) and which types of organic donors (HBDs) are used, which determine their conductivity levels, solubility, viscosity, and hydrophilic–hydrophobic ratios. To make the DES formulation, the HBAs are metal salts like cobalt chloride ($CoCl_2$), zinc chloride ($ZnCl_2$), tin chloride ($SnCl_2$), aluminum chloride ($AlCl_3$), and ferric chloride ($FeCl_3$), and the HBD is ChCl, alongside urea, ethylene glycol, and glycerol. Fig. 3.2 shows some important examples of DESs, which fall in the Type I category [30].

DES Type II

These DESs consist of metal salts for the HBA element, unlike Type I, which mostly incorporates metal halides, as they are hydrated in nature. Certain special properties arise when hydrated metal salts in Type II DESs are included in a system. New properties, phases, and solubility can be achieved in the solution simply by changing the HBA components of the metal salt complex. Type II DESs are impressive because they can be used in place of traditional solvents to help achieve what is needed for an application. Various studies and industries collectively use many famous DES compositions derived from ChCl + $CrCl_3{\cdot}6H_2O$ and ChCl + $CuCl_2{\cdot}2H_2O$ mixtures [30].

DES Type III

This crucial class of DESs comprises mainly HBDs, which exhibit an organic functionality [16]. These DESs contain both organic HBAs and organic HBDs to shape eutectic solvents with lower melting points than the individual HBAs and HBDs [2]. Because they are biodegradable, nontoxic, and adjustable, Type III DESs are becoming popular in pharmaceutical, biochemical, and environmental areas. DESs that belong to Type III are made possible by the strong hydrogen bonds between organic substances acting as both HBAs and HBDs. Selecting particular HBAs and HBDs in Type III DES structures controls important qualities such as polarity, viscosity, stability at different temperatures, and solubility. These systems are built using HBAs such as ChCl, betaine, proline, and *L*-carnitine, in combination with organic HBD compounds, which include urea, glycerol, various carboxylic acids, sugars, and amino acids. Using Type III DES compositions has been highlighted by numerous applications, and four examples are ChCl:glycerol, ChCl:urea in a 1:2 molar ratio, ChCl:lactic acid, and betaine:citric acid in a 1:1 molar ratio [31].

DES Type IV

Type IV DESs are a subclass of DESs that share similar features with ILs and can be applied in electrochemistry, pollution remediation, and energy storage systems. This group of DESs is becoming more popular since these DESs can function in place of traditional solvents and ILs, thanks to their excellent conduction of ions, strong ability to withstand heat, and low vapor pressure [2]. Type IV DESs are formed by adding urea, EG, GLY, and amine-based HBDs to metal chloride salts of pyridinium and imidazolium ILs. These DESs contain, as their ionic liquid part, a metal chloride salt combined with an organic cation-based IL. The most common ILs being used at present are 1-methyl-1-propylpyrrolidinium chloride ([MPPy]Cl), 1-butyl-3-methylimidazolium chloride ([BMIM]Cl), 1-butyl-1-methylpyrrolidinium chloride ([BMPyr]Cl), 1-ethyl-3-methylimidazolium chloride ([EMIM]Cl) 1-ethyl-3-methylimidazolium chloride ([EMIM]Cl), 1-butyl-3-methylimidazolium chloride ([BMIM]Cl), and 1-butyl-1-methylpyrrolidinium chloride ([BMPyr]Cl). The selection of HBDs alters the physicochemical properties of these DESs [29, 32].

DES Type V

These solvents are preferred by the industry since they can be changed as needed, are harmless to the environment, and work well in many processes. The water-repelling behavior and good stability in water make Type V DESs suitable to be called hydrophobic DESs (HDESs). Typically, DESs use compounds with low polarity and the ability to repel water. In Type V DESs, lidocaine, thymol, menthol, and tetraoctylammonium bromide (TOABr) play the biggest roles. Many people use octanoic acid, lauric acid, and nonanoic acid as the most common HBDs. Thymol:camphor, menthol:decanoic acid, menthol:lactic acid, TOABr: nonanoic acid, and lidocaine:dodecanoic acid are actual instances of Type V DES variations [33].

3.2.3.2 Classification Based on Functionality

Acidic DESs

Drawing from the reduced risks they carry and the ability to evolve depending on situations, DESs are now considered eco-friendly solvents in various fields. This type of DES is known as acidic DESs because their HBDs consist of organic acids and are acidic. The fact that acidic DESs are rich in protons, have impressive solubility, and serve as excellent catalysts makes them useful for acidic catalysis, as well as for pretreating biomass and using them in green processes. Acidic DESs are made up of two main types: some with just one type of acid and others with a mix of acids. HBAs include ChCl and betaine mixed with tetraalkylammonium salts, and also some acids, such as lactic acid, citric acid, p-toluenesulfonic acid (PTSA), and oxalic acid. The combination of ChCl with oxalic/citric/lactic and PTSA makes up acidic DESs [34].

Basic DESs

Basic DESs represent a subclass of DESs, and their basic nature is mainly due to the hydroxide and amine groups present in HBA and HBDs. Due to this property, these DESs showed remarkable solvating ability, promoted faster chemical reactions, and demonstrated high efficiency in processes such as biomass delignification and CO_2 capture. Thanks to their unique properties, basic DESs are important for both green research and environmentally responsible industrial work. Basic DESs have two parts in their design, only using direct and reverse parasitic components. Basic DESs currently generally use mixtures of two substances, such as ChOH with urea,

ChCl with guanidine, choline hydroxide with glycerol, ChCl with urea, or ammonium hydroxide with urea [34].

Natural Deep Eutectic Solvents (NADES)

When DESs are composed with natural HBAs such as choline chloride (ChCl) and HBDs such as sucrose, they exhibit very low melting temperatures and excellent biodegradability and non-toxicity. thereby termed as NADESs [35]. NADESs maintain pH balance in a particular range, facilitate enzymatic reactions, drug production, and other biochemical processes by supporting and stabilizing enzyme and drug activities. A few common examples of NADESs are such as ChCl: 1,2-propanediol/ glycerol/ sugars and amino acids. ChCl: citric acid/ malic acid/ oxalic acid/ glucose/ fructose, and Proline: glycerol [36].

3.2.4 Preparation Methods of DESs

A close perusal of Fig. 3.3 explains that mixing DESs at high temperatures is still considered among other techniques such as freeze-drying, vacuum evaporation, grinding, twin-screw extrusion, ultrasonication, and microwave irradiation. Information about numerous features is compulsory to select the precise technique for a project based on the needs and target properties.

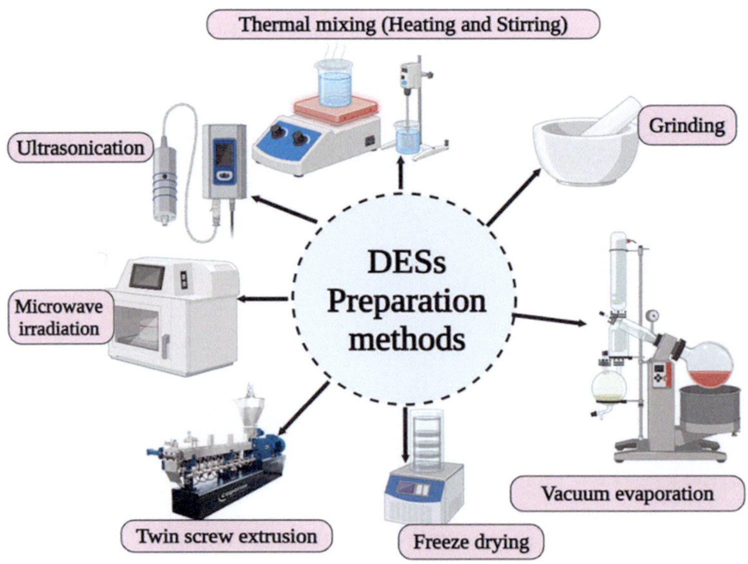

Fig. 3.3 Various preparation methods of DESs

3.2.4.1 Thermal Mixing (Heating and Stirring)

Adding and mixing a hydrogen bond acceptor and a donor in the same molar proportion is the preferred way to obtain a stable eutectic mixture. Mixing an HBA and an HBD while heating them and stirring them continuously allows a stable, transparent, and homogeneous liquid to appear [37]. The heat source used makes the parts of the sample mixture move apart from each other and mix in a way that lowers how cold it takes for the mixture to freeze and gives it new physical properties. The simplicity, along with being inexpensive and easy to redo, makes this method the one often picked by researchers. Thermal mixing relies on the idea that when different solids are mixed and exposed to heat, the molecules inside them start to separate, letting them mix until a steady alloy is reached. Table 3.1 reveals the DESs prepared by thermal mixing [38].

3.2.4.2 Grinding

A mixture of the components in the solid state using grinding creates DESs without using solvents or causing pollution. Unlike most other methods, grinding allows for mixing HBDs and HBAs together to make a eutectic mixture without having to warm or stir the material. More people prefer this approach because it promotes sustainability, uses a lot less energy, and is fairly easy to set up. Preparation of thick or solid DESs is better done using grinding, which is more effective than using liquid-based methods. The grinding method works well for DESs that need to form hydrogen bonds, as it helps make the right mixes by allowing you to slowly add and mix the chemicals while keeping the water level and temperature low. The high-energy grinding approach called ball milling mixes the solid components by grinding them intensely, and twin-screw extrusion grinding uses twin-screw machines to mix, grind, and heat the components all at once (some examples are mentioned in Table 3.1) [38].

3.2.4.3 Vacuum Evaporation

Many people use vacuum evaporation to create DESs because it helps get rid of extra water and preserve the purity of the product. Since it reduces degradation and fosters improved formation of a eutectic mixture, the method is ideal for temperature-sensitive parts. This way, the mixture can attain a eutectic state without suffering from any heat damage. Using the vacuum evaporation technique, a lower pressure helps reduce the boiling points of the solvents or water inside the eutectic mixture. The evaporation of solvents can happen more easily under low air pressure, without affecting the DES ingredients. It allows HBAs to join with the HBDs properly to ensure that the mixture keeps a proper balance and has the least contamination from the solvent. Table 3.1 presents examples of DESs that were synthesized through the vacuum evaporation process [39].

3.2.4.4 Freeze-Drying

Freeze-drying has become a popular way to make DESs since it lets scientists create pure samples, and the DESs end up being less likely to break down or degrade. Using low temperatures and a special drying method, lyophilization (freeze-drying), scientists can achieve completely stable and pure DESs free from all solvents. Freeze-drying should be used for hygroscopic and temperature-sensitive DESs, as standard high-heat methods will damage them. The way the freeze-drying method works is by sublimation, made up of three basic phases. This includes freezing, primary drying using sublimation, and then secondary drying through desorption. With this method, thermal degradation is reduced, yet the HBA–HBD bonds between molecules in DES hold, which results in pure end products [40].

3.2.4.5 Twin-Screw Extrusion

The twin-screw extrusion method is now seen as a common way to make DESs that are eco-friendly. Traditional DES preparation techniques usually let the ingredients mix either in a solvent or by melting, but with twin-screw extrusion, the components are mixed together by a machine using heat and pressure. The several benefits of this technique include getting very accurate results, fast reactions, using fewer solvents, and better control over the mixture of DES used. When two screws are rotating together in an interlocking manner, they create pressure that helps mix and warm the different materials. Having temperature sensors on the extruder screws allows operators to set and stick to precise timings and temperatures during the whole process. It is a good way to make solid DESs and eutectic mixtures that include both solids and liquids. Producing DESs in the industry is more desirable when it includes the ability to scale processing and control mixing and the makeup of products. Refer to Table 3.1, which lists several DES illustrations, with their common formulas and the details of application, as well as processing times and temperatures [38].

3.2.4.6 Microwave Irradiation

Today, microwave irradiation is recognized as an effective and green process for preparing DESs. Usually, DESs are made by melting ingredients in a solvent or using radiation. Among its advantages are accurate results, fast reactions, using less solvent, and precisely controlling the DES composition. Industrial production of DESs is more attractive once the process includes ways to scale up, mix solutions efficiently, and manage how the DES is built [44]. Table 3.1 contains different examples of DES synthesis through microwave irradiation.

3.2.4.7 Ultrasonication

In DES synthesis, ultrasonication is useful due to its quick and eco-friendly means of preparation. Sound waves produced by ultrasonication have frequencies between 20 kHz and 1 MHz, which leads to the formation of cavitation bubbles that result in localized episodes of high pressure and high temperature before they collapse and implode. As a result of such conditions, molecules separate to allow the eutectic mixture of HBA and HBD to form. Rapid composition of DESs occurs during ultrasonication, thanks to the acoustic energy, with little temperature change and stable results for safe and efficient mixing. Blending of DES is more effective and is accomplished more rapidly because of the ultrasonication performed [46].

3.2.5 Comparison of DESs with Ionic Liquids (ILs)

A close perusal of Fig. 3.4 shows a comparison between DESs and ILs. DES generation requires only a couple of materials that need to be put together and heated until they create a uniform solution. Because of this simplicity, DESs can be easily and economically produced, since additional purification is not needed. In contrast, to remove impurities, ILs often require lengthy synthesis and must undergo chromatography before they can be produced in large amounts [67–71].

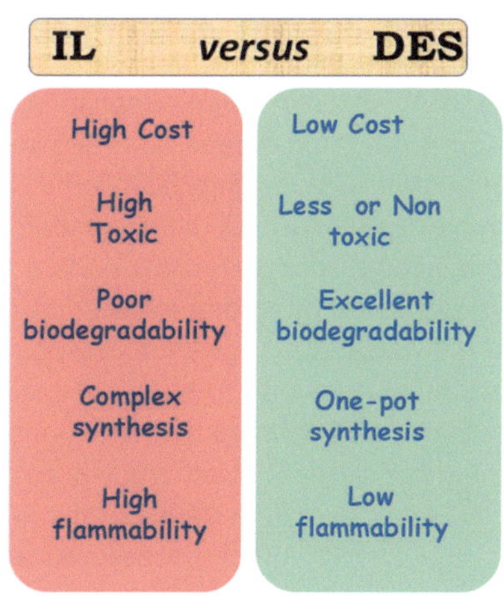

Fig. 3.4 A systematic comparison between DESs and ILs

3.3 Biotechnological Applications of DESs

3.3.1 Enzyme Stabilization and Biocatalysis

The application of DESs as biocatalytic solvents has gained recognition over recent years among researchers who work either with isolated enzymes or with whole cells. The solvent properties of DESs include both hydrogen bond abundance and low freezing temperature, while showing a possible inhibitory behavior toward enzymes, as reported. Hydrolytic processes and transesterification reactions, along with the Henry reaction and aldol reactions, occurred when certain hydrolases exhibited a superior activity in DES as a solvent compared to ILs. Lipases remain among the most popular enzymes for enzymatic reactions in DESs because researchers frequently study them in organic solutions. The first attempt to perform lipase-catalyzed biotransformation reactions in DESs took place in 2008 when Kazlauskas et al. conducted their study. Researchers evaluated lipases such as immobilized Candida antarctica lipase B (iCALB), Candida antarctica lipase A (CALA), Candida antarctica lipase B (CALB), and Pseudomonas cepacia lipase (PCL) for their ability to perform transesterification in different DES media. Lipase-assisted chemical transformations take place through four DESs. The potential uses of proteases have been demonstrated within systems that work with DESs [72]. Zhao and colleagues studied the enzymatic transesterification activity of glycerol-based choline salt (acetate or chloride) DESs with protease to determine if DESs could be applied in biocatalytic systems [73]. The epoxide hydrolysis and the subsequent production of diols represent an important chiral synthesis catalyzed by epoxide hydrolase (EH) enzymes. To date, the use of DESs in EH-catalyzed reactions has been limited. For instance, EHAD1 showed a significant improvement in styrene oxide conversion, achieving 92% conversion compared to 4.6% in the presence of 25% (v/v) ChCl: Glycerol in the buffer system. Similarly, other enzymatic processes, including those catalyzed by phospholipase D, benzaldehyde lyase, and haloalkane dehalogenases, exhibit enhanced reaction efficiency when DESs are incorporated. Whole-cell biocatalysis using natural or recombinant microorganisms, such as recombinant Saccharomyces cerevisiae and Acetobacter sp., offers advantages over isolated enzymes for specific chemical reactions. Whole-cell biocatalysts are cost-effective, provide an efficient cofactor regeneration system suitable for unconventional media, and can perform multistep cascade reactions within a single cell. [74]. As catalysts, whole-cell biocatalysts are differentiated from isolated enzymes by their affordability, together with a perfect cofactor regeneration system capacity for unconventional media use and the ability to conduct multistep cascade reactions within a single cell [75].

3.3.1.1 Enhancing Enzyme Activity and Stability

Enzymatic structures within DESs serve as enhancing factors for their stability and activity by adjusting enzyme structures to improve or reduce specific functionalities. The horseradish peroxidase (HRP) enzyme in ChCl-based DESs maintained a higher α-helix concentration when the solution contained greater molar ratios of the HBD. Enzyme activity, along with stability, shows an enhancement through elevated α-helix content and diminished β-sheet content in the enzymatic design. The ChCl portion disturbed α-helix elements in the enzyme, leading to decreased enzymatic activity based on recent findings. The enzyme stability could increase because the glycerol component promoted substantial α-helix formation and β-sheet depletion. The HBD and HBA have an inverse relationship with each other, which leads to activation or deactivation effects for a particular DES composition [76].

DESs show the capacity to modify enzyme three-dimensional enzyme structures when present in aqueous solutions. Enzyme functionality depends on their folding and unfolding patterns, which shows direct dependency. The stability and activity of proteins can be affected by a combination of solvent polarity, hydrophobicity, and hydrogen bond properties of ILs [77]. The refolding and unfolding rate of the enzyme experienced substantial changes based on the DES concentration in an aqueous medium through Circular Dichroism (CD) spectroscopy and intrinsic fluorescence evaluations. Table 3.2 includes the enzymatic performances of different enzymes in the presence of DESs.

3.3.2 Extraction and Purification of Bioactive Compounds

The literature demonstrates that DESs act as suitable extraction platforms for obtaining natural bioactive compounds (NBCs). Recently, various sources witnessed the use of DESs in the extraction of compounds more efficiently than traditional solvents. The use of natural products contains secondary metabolites that enable both natural defense protection and stress reaction processes in living organisms but do not participate in developmental or growth processes. Severe metabolic health issues emerge when vital substances or their necessary amounts are missing [34]. Several living organisms, including fungi, together with algae, microbes, and plants, possess NBCs. The industries of agrochemicals, cosmetics, pharmaceuticals, and nutraceuticals, together with the food-manufacturing sector, show interest in NBCs for their application in developing physiologically active compounds. Joana Gil-Chávez et al. report that NBCs serve as crucial synthetic precursors for pharmaceuticals, which have commercial value since anticancer drug production contains at least 60% NBCs. The chemical is widely used in food additives and insecticides, together with taste enhancers, nutritional supplements, and surfactants, while also functioning as preservatives and fragrances [83].

3.3.2.1 Plant-Derived Compounds

A specific approach exists in the form of the mechanical method used to extract bioactive plant constituents. The extraction methods encompass pressured liquid extraction as well as subcritical fluid extraction (SFE), supercritical extraction (SE),

Table 3.2 The role of different enzymes in the presence of DESs

S. no.	DES	Enzyme	Enzymatic performance	Applications	Ref.
1	ChCl:glycerol, ChCl:1,2-propanediol, ChCl:ethylene glycol, ChCl:urea	Cellulase	DESs generally have enhanced thermal stability compared to aqueous solutions	Biomass conversion (e.g., cell wall degradation for extraction), biofuel production	[78]
2	ChCl:ethylene glycol, ChCl:1,2-propanediol, ChCl:glycerol, ChCl:urea	Pectinase	Thermal stability: Similar to cellulase, DESs have improved thermal stability	Plant extraction processes (cell wall disruption), fruit juice clarification, and textile processing	[78]
3	ChCl:urea:glycerol, ChCl:glycerol, glycol:formamide, ChCl:ethylene	Lipase (*Candida rugosa* lipase)	Show high stability	Biocatalysis for synthesis (e.g., esterification and transesterification), fine chemical production	[79]
4	ChCl:urea, ChCl:glycerol	α-chymotrypsin	Enhanced thermal stability and preserved enzymatic activity	Protease-catalyzed reactions, peptide synthesis, protein modification	[80]
5	Betaine:sorbitol:water, betaine:ethylene glycol	Horseradish peroxidase	Enhanced the enzyme's stability and showed high enzymatic activity	Biosensors, diagnostics, wastewater treatment, organic synthesis (e.g., oxidation reactions)	[81]
6	Betaine:glycerol, betaine:sorbitol	Formate dehydrogenase	Increase in catalytic efficiency and also increased protection against thermal stress	CO_2 fixation, nicotinamide adenine dinucleotide (NADH) regeneration systems (crucial for many enzymatic reductions in the industry), chiral synthesis	[82]

ultrasound-assisted extraction (UAE), microwave-assisted extraction (MAE), and enzyme-assisted extraction (EAE). Solvent extraction, together with Soxhlet extraction and maceration and percolation, represent traditional extraction methods. Research by Hu Jinxia et al. showed that DESs provided better astaxanthin extraction from shrimp waste by producing a 47.3% higher maximum production rate than ethanol extraction [84].

The extraction of chlorogenic and caffeic acids from the herb *Artemisiae Scopariae* resulted in a 177% increase and a 138% increase, respectively [85]. DESs find applications in environmental sciences for separating materials while analyzing compounds and extracting metal ions as well as NBCs from environmental sources. Table 3.3 summarizes research on bioactive compounds extraction using DES-based mixtures.

Table 3.3 The role of DESs in the extraction and purification of bioactive compounds

DES	Bioactive compound	Source	Extraction efficiency	Purification role	Applications	Ref.
ChCl:1,2-propanediol	Polyphenols	Spent coffee ground	Good extraction yields at milder temperature conditions, without flammable solvents	Enhances the solubility of polyphenols	Food ingredients and/or nutraceuticals, antioxidants, antimicrobial agents, formulation aids for food ingredients	[86]
ChCl:lactic acid	Vanillin (4-hydroxy-3-methoxybenzaldehyde)	Vanilla pods (*Vanilla planifolia*)	High extraction rate, good efficiency, and potential for waste reduction	Enhances the purity of vanillin	Flavoring agents, antimicrobial agents, and antioxidants	[87]
ChCl:L-lactic acid	Volatile oil	Cumin	Pretreatment of the samples was done with low eutectic solvents. The essential (extracted) oils contained a higher amount of volatile components before pretreatment	Enhances the volatility of desired components	Flavoring, fragrance, medicinal	[88]
ChCl:citric acid	Volatile oil	*Angelica sinensis* radix	The extraction yield was improved	–	Traditional medicine, possibly food/fragrance	[89]
Methyltrioctylammonium chloride:1-butane alcohol	Terpene lactones	Artemisia	The extraction rate was significantly higher	–	Antimalarial drug	[90]
Betaine:triethylene glycol	Terpene lactones	*Ginkgo biloba*	The total extraction rate was 1.94 ± 0.03 mg/g	High recovery in a single run suggests good separation from the matrix	Cognitive enhancement, medicinal	[91]
ChCl:Triethylene glycol	Flavonoids	*Sophora japonica*	The extraction rate of rutin reached 194.17 ± 2.31 mg/g	–	Antioxidant, medicinal	[92]

(continued)

Table 3.3 (continued)

DES	Bioactive compound	Source	Extraction efficiency	Purification role	Applications	Ref.
ChCl:Citric acid	Isoflavone	Kudzu roots	More effective in the extraction of bioactive compounds, high antioxidant activity	Enhancing bioactivity implies a desirable "purity" for functional use	Medicinal (e.g., menopausal symptoms and cardiovascular health), antioxidant	[93]
ChCl:Malic acid	Anthocyanin	Sour cherry pomace	The extraction rate was higher	Enhances the stability of anthocyanins	Natural food colorants, antioxidants, and functional foods	[94]
ChCl:Ethylene glycol	Phenolics	*Prunella vulgaris*	The extraction rate was higher	–	Antioxidant, medicinal	[95]

3.3.2.2 Microbial Metabolites

The green chemistry era finds DESs to be promising antimicrobial solutions as alternatives to traditional antibiotics because they exhibit strong capabilities against resistant bacteria and demonstrate significant antifungal properties. Lab investigation of antimicrobial activity by DESs used disc diffusion tests alongside good diffusion, microtox assay, broth dilution, drop plate, and Fourier-transform infrared spectroscopy (FT-IR) bioassay [96]. Several NADESs were evaluated against gram-positive cocci bacteria, which included *Staphylococcus aureus* and *Streptococcus pneumoniae,* while two gram-positive rod-shaped bacteria, spore-forming *Clostridium perfringens* and *Bacillus subtilis,* were tested to assess NADESs' antibacterial properties. Studies showed that these NADESs based on acids produced the largest antibacterial inhibition zones among all tested formulations [97]. Fungal infections stand as major disease threats in public health. Research has tested several NADESs against fungi to assess their antifungal activity while highlighting *Candida* spp. as one of the major pathogens that cause serious fungal infections. ChCl:oxalic acid (1:1) molar ratio demonstrated a potent zone of inhibition against the fungal species *Candida albicans* [98].

The antimicrobial effects of NADESs extend to the biotreatment of biofilms and show cytotoxic action against tumor cells [99]. Research papers show that certain NADESs enable extraction processes through the purification of virus-like particles. Antibacterial photodynamic therapy strongly relies on the use of these compounds for effective treatment. Research has shown that NADESs function as a superior alternative medium to established substances for usage as antimicrobial agents and scientific solutions in different fields.

3.3.2.3 Marine Biomolecules

The green chemistry industry adopts DESs as versatile extraction and processing solvents for marine biomolecules. The first DESs were announced in 2012 when scientists synthesized marine-origin agarose polymer gels based on DES plasticizers. The most frequently employed DES systems contain choline as the HBA. In 2013, Sharma et al. described using DESs to break down alpha-chitin during marine biomass production by carrying out heating and ultrasonic heating, respectively. The yield of chitin was reported as 9% (w/w) when it was extracted using ChCl: thiourea (in a molar ratio of 1:2) with continuous heating of the mixture for 6 h at 100 °C [100]. Powerful attributes of DESs allow researchers to obtain a wide range of marine compounds from polysaccharides to minerals. Various DES systems are employed for screening the optimal system that best achieves both material processing and target transformation. The modification of HBA and HBD quantity using quaternary ammonium compounds, such as ChCl, betaine hydrochloride, and carboxylic acids, in combination with simple polyols, along with alkali, stands as another significant area of investigation. Early DES investigations required researchers to optimize the viscosity, which has remained crucial due to these liquids' extreme thickness. DESs show significant potential as multifunctional additives for improving polymeric marine materials. Marine biomaterials receive beneficial plasticization treatment from DESs, while bio-composites achieve better compatibility because of their DES-based compatibilization. Three high-value by-products resulting from DES-based downstream processing became accessible for extraction, while collagen, chondroitin sulfate, and hydroxyapatite successfully underwent recovery procedures [101]. DESs function as solvents when processing new bio-composites derived from raw materials. DESs are used in the processing of bio-composites derived from raw materials. So far, their use has mostly been limited to marine biomass. Research has primarily focused on crustacean shells, which are a major source of chitin and chitosan polymers. However, studies have also demonstrated the successful processing of fish, as well as crustaceans and seaweed, and initial efforts have explored the use of side-streams from primary marine resources.

DES technology establishes new opportunities to extract various algal-derived polysaccharides and carotenoids, together with proteins. The downscaled extraction process benefits from DES usage by making chitin deacetylation and chitosan modification more efficient. DNA storage media will be an important feature of the coming biotechnology revolution because DESs can perform this function. Research and industry will benefit significantly from the promising path presented by DESs [101].

3.3.3 Applications of DESs in Drug Delivery and Pharmaceuticals

Pharmaceutical applications, together with drug delivery systems, benefit from DESs' multifunctional physico-biochemical properties and characteristics that allow their potential development in drug systems. Before the utilization of DESs in the pharmaceutical developments, it is an essential step to determine the cytotoxicity profiles of DESs. The drug candidates need preliminary registration before they proceed into the preclinical and clinical trial phases [102, 103]. Drug solubility is a major concern encountered by the pharmaceutical industry. Researchers are examining the combinations of DESs and their performance with insoluble pharmaceutical compounds. Therapeutic DESs currently exist as new active pharmaceutical ingredients (APIs) in the pharmaceutical market. Drug delivery systems employ DESs for transporter roles through the skin. Various terms refer to DESs in pharmacology, where NADESs and Therapeutic Deep Eutectic Solvents (THEDESs) correspond to deproteinizing solvents, and both categories are classified as pharmaceutical DESs and hydrophobic DESs, and drug-based polymerized DESs exist separately. Using DESs represents a known method for delivering drugs in nano-sized forms. DESs act as functionalizing agents for developing drug delivery systems. The therapeutic properties of DESs as drug excipients have been evaluated extensively regarding their antiseptic, anticancer, anti-inflammatory, antifungal, antibacterial, antiviral, and antioxidant effects [3, 67, 103].

3.3.3.1 API-DESs in Drug Formulation

DESs function as room-temperature liquids that form when HBAs and HBDs mix in specific proportions and exhibit various physicochemical qualities, which make them valuable in medical chemistry for drug manufacturing as an effective drug formulation ingredient. APIs of DESs, also known as therapeutic DESs, provide a substitute for crystalline salts for designing natural, semisynthetic, and synthetic polymeric carriers within drug delivery systems, as well as acting as an alternative medium for functional material synthesis and solving polymorphic issues [67, 102]. API-DESs serve as crucial components during drug research and development because they enhance solubility as well as function as delivery system components. Several API-DESs include ibuprofen:menthol, lidocaine:1,8-octane diol, and lidocaine:urea, together with aspirin:choline chloride [103].

Medical chemistry research has focused increasingly on DESs due to their room-temperature liquid state as HBA–HBD mixtures at certain ratios, which has enabled numerous physicochemical properties useful in drug synthesis because they show promise as effective drug formulation components. API-DES operates as a therapeutically active DES while serving as a substitute for crystalline salts to develop natural, synthetic, and semisynthetic polymeric drug delivery carriers through DES-mediated designs and functions as both a synthesis reaction medium and a solution for polymorphism problems [103]. API-DESs function as essential components during drug development research since they contribute to drug formulation enhancement through solubility improvement and drug delivery improvement. Experimental studies have revealed that ibuprofen–menthol, lidocaine–1,8-octane diol, lidocaine–urea, and aspirin–ChCl represent examples of API-DESs [103].

3.3.3.2 Controlled Drug Release Systems

DESs represent emerging, promising excipients that act as transporters for developing controlled drug delivery systems. In drug delivery systems, these features, together with low volatility and biodegradability, along with viscosity control, make DESs an excellent choice. What makes DESs stand out as powerful solvents is their ability to dissolve drugs with low water solubility through drug–DES hydrogen bond formation, which improves the availability of drugs [104]. DESs protect drugs subject to oxidation or hydrolysis by reducing their contact with moisture and oxygen-based factors in the environment. Controlled drug release over extended periods becomes possible through DES's high viscosity, which acts as an important factor for slowing drug diffusion rates. The pharmaceutical use of DESs becomes possible because biocompatible components in ChCl and amino acids produce DESs that maintain low toxicity levels [96].

Table 3.4 Role of some DES compositions in drug delivery systems

DES	Drug(s)	Role in drug delivery	Ref.
ChCl:1,2-propanediol	Aspirin	The stability and solubility of the target drug are enhanced	[108]
ChCl:ethylene glycol	Acetaminophen	Increases the solubility of acetaminophen with an increase in the amount of the DES	[109]
Camphor:menthol	Ibuprofen	Higher water solubility and can be used for nanosuspension preparation	[110]
ChCl:glycolic acid:oxalic acid	Itraconazole	The solubility of itraconazole increases	[111]
ChCl:glycolic acid oxalic acid	Lidocaine	Enhances the solubility of the drug	[112]
ChCl:levulinic acid	Ketoprofen	Drug solubility increases in the DES as compared to water	[108]
ChCl:1,2-propanediol	Naproxen	The solubility of the drug increases in the DES as compared to its solubility in water	[108]
Proline:glutamic acid	Rutin	Improves bioavailability, increases rutin solubility, and is used as a formulating and solubilizing agent for polyphenol intake	[113]

Drug delivery platforms make use of DESs because they show flexibility across multiple platforms. The application of DESs in oral drugs helps enhance the solubility rates and absorption capabilities of drugs with low solubility to guarantee improved drug efficacy. The combination of ChCl:citric acid-based DESs leads to superior drug release patterns of anti-inflammatory medicines. Menthol and fatty acid mixture DESs serve as outstanding penetration enhancers because they help drugs pass through the skin better during transdermal drug delivery [105]. These DESs present minimal toxicity, which makes them appropriate for injectable delivery systems where their controlled medication release mechanism ensures drug persistence. The thick consistency of DESs makes them effective in delivery systems that use the mouth and under the tongue because improved tissue stickiness enhances drug delivery [106]. Table 3.4 represents the DES composition targeting the drugs, showing possible potential in the drug delivery system.

Several pharmaceutical compounds have been developed utilizing the Design of Experiment Systems to build drugs with defined release patterns. Through the combination of ibuprofen with choline chloride and glycerol, patients achieve better solubility, alongside sustained drug release capabilities. Similarly, diclofenac sodium creates an effective pain relief system through its combination with ChCl:citric acid. The combination of curcumin with a DES based on menthol produces a formulation that increases bioavailability and provides slow and controlled substance release. ChCl:urea (DES) stabilizes metformin while releasing the medication in a controlled manner to support better glucose control in diabetic patients. The combination of paracetamol with lactic acid-based drug-excipient systems enhances drug dissolution speed and expands therapeutic duration [107].

DESs present several important benefits when used in controlled drug release systems. Higher drug-loading capacity, along with reduced side effects, becomes achievable through DESs because they support stable, sustained drug release profiles, particularly for hydrophobic drugs. Using DESs provides an environmentally friendly solution that reduces operational costs when replacing conventional solvents. The wide range of formulation types, from solids to semisolids and liquids, makes DESs extremely suitable for pharmaceutical uses.

3.3.4 Bioplastics

Petroleum plastics cannot decompose, leading to major ecological destruction, which results in severe environmental contamination. Several renewable biomaterials, including lignin, starch, cellulose, lactic acid, chitosan, proteins, and more, are employed today to substitute traditional plastic products. The market demand for environmentally friendly, biomass-derived, biodegradable plastics shows high growth despite their higher price compared to conventional products, scarce raw materials, and restricted production methods. DESs have emerged as the most suitable solution to address this production challenge. Starch-based bioplastics by using ChCl:urea (1:2) DESs have been considered for starch plasticization, as reported by Abbott et al. Biomedical research has also analyzed that starch-based DESs can be used as solvents and modifiers and also act as plasticizers and, hence, can be used as a suitable substitute to starch plasticizers [114]. The combination of ChCl:oxalic acid in a 1:1 molar ratio DES serves as the primary treatment agent for lignocellulosic materials dedicated to their fractionation steps and value-added processes. The combination of 22 wt% gelatin with ChCl:EG in a 1:2 molar ratio produces an extensible gel that could serve as a flexible biocompatible element for energy-independent sensor applications, according to Qui et al. [115]. Thermoplastic chitosan semitransparent films were produced by Sanchez et al. through the use of NADESs based on ChCl with citric acid/malic acid/lactic acid and glycerol as HBDs [28]. Table 3.5 includes other examples of DESs in bioplastic processing.

Table 3.5 Examples of DESs in the bioplastic-processing system

DES	Bioplastic app.	Role in bioplastic processing	Ref.
ChCl:glycerol	Polyethylene terephthalate (PET) plastic waste	Acts as a pretreatment agent to modify the PET structure, increases amorphous regions and hydrophilicity, enhances enzymatic and microbial degradation, and improves the efficiency of biological plastic waste treatment	[116]
ChCl:ethylene glycol:oxalic acid	Cellulose	Acts as a pretreatment system to efficiently extract cellulose	[117]
ChCl:anhydrous oxalic acid	Lignocellulosic bioplastic	Enhances the crystallinity of the bioplastics	[118]
ChCl:malic acid, ChCl:citric acid, ChCl:lactic acid, ChCl:glycerol, ChCl:malonic acid, ChCl:urea	Chitosan-based films	Used as plasticizers, improve elasticity, influence water permeability, and increase surface homogeneity	[41]

(continued)

Table 3.5 (continued)

DES	Bioplastic app.	Role in bioplastic processing	Ref.
ChCl:malonic acid	Pectin-based film	Used as a plasticizer, decreases the tensile strength, increases the flexibility of the bioplastic, and increases the degradation rate	[119]
ChCl:urea, ChCl:glycerol	Starch-based plastic	Acts as efficient functional additives; used as plasticizer, water uptake inhibitor, and compatibilizer; and prevents the recrystallization of starch	[120]
ChCl:urea, ChCl:glycerol, ChCl EG, ChCl:succinic acid, ChCl:citric acid, ChCl:sorbitol	Starch-based films	Used as efficient plasticizers and water uptake inhibitors, act as compatibilizers, prevent the recrystallization of starch, act as solvents for starch, influence film appearance and morphology, and enhance ionic conductivity	[41]
ChCl:urea, ChCl:glucose, ChCl:glycerol, ChCl:citric acid	Cellulose	Effectively used as a plasticizer, suppresses crystallinity, enhances ionic conductivity, and improves heat stability	[41]
ChCl:urea	Agar	Used as plasticizers, influence on film-forming ability, plays a role in water resistance	[41]
ChCl:glycerol, ChCl:citric acid monohydrate	Pectin	Used as plasticizer, enabling thermoplastic behavior, influences the mechanical properties of pectin films	[41]
ChCl:urea	Agarose	Used as a plasticizer, it enhances mechanical properties and reduces the glass transition temperature	[41]

3.4 Environmental and Economic Perspectives

ChCl-based DES has recently gained major attention from biotech scientists because of its environmentally favorable properties. The environmental and economic aspects of six ChCl-based mixtures become clearer through a critical evaluation process. DES production relies on basic and efficient methods that mix ChCl with numerous HBDs under heating conditions between 60 and 100 °C. ChCl synthesis via antisolvents stands apart from the intricate production of ionic liquids, which need multiple solvents, together with monitored environmental conditions. DES manufacturing operations generate environmental consequences, mostly from the acquisition and production techniques of its structural components. Life cycle assessment (LCA) studies reveal that ChCl and HBD synthesis procedures lead to major environmental impacts.

The production of ChCl: urea (1:2) DES has a higher environmental impact than typical Volatile Organic Compounds (VOCs), such as methanol and ethanol, due to the upstream manufacturing processes of ChCl and urea [121]. DES sustainability depends directly on the renewable source of its constituent elements. The renewable resource-based synthesis of choline chloride improves the environmental aspects of this sustainable material. The renewable classification also extends to HBDs, including glycerol, which emerges from biodiesel manufacturing. Other HBDs display different renewability levels, while their production methods affect the sustainability outcome of the DES.

The environmental acceptability of DESs is contingent on two factors: the toxic potential of DESs as well as their degradation capabilities. Various studies suggest that ChCl-based DESs demonstrate low toxic levels and fast biodegradation properties, yet some DES products present environmental hazards to the environment. EG-based DESs result in environmental problems because this substance is known to be toxic. To apply DES at an industrial level, each solvent system needs a comprehensive evaluation of its environmental behavior. DESs demonstrate cost-effective properties, which make them superior to conventional solvents in applications. This widespread notion about DESs stems from their affordability, combined with basic preparation methods, along with the low prices of both ChCl and specific HBDs. The expense of a product depends on its purity needs and production size requirements.

The economic aspects are affected by challenges related to storing and using DESs when processing viscous liquids. The selection of DES applications for biotechnological programs requires consideration of environmental factors as well as efficiency requirements for production processes. The sustainability attributes of DESs depend on their composition and sourcing methods because they provide better solubility for biomolecules and support enzyme conservation [122].

DESs attract interest as green alternative solvents because they possess several advantageous attributes that include low volatility as well as biodegradability, matched by straightforward preparation methods. The performance evaluation of environmental sustainability for DES-based technologies requires an in-depth LCA analysis for complete evaluation. The LCA methodology serves as a structured method to evaluate environmental burdens across the complete product life cycle, beginning from extraction and extending to manufacture, product utilization, and eventually, discarding. Examining DESs through LCA starts with reviewing the environmental effects of each contributing substance and the entire solvent framework [123].

This study performed a representative LCA on ChCl and urea DES, which researchers refer to as "reline." This research examined the environmental profiles between reline and four standard organic solvents, which included methanol, ethanol, dichloromethane (DCM), and ethyl acetate. The alkane-to-ketone transition of acetophenone served as an example for solvent comparison [124]. Detailed research examined the environmental sustainability of ChCl-based DESs through investigations involving six mixtures. The sustainability of DES systems depends strongly on selecting the right hydrogen bond donors and acquiring suitable raw materials, according to the latest research findings.

3.5 Challenges for Advancements in DES Research

DESs have become an attractive solvent category due to their environmental sustainability, economic benefits, and simple preparatory methods compared to regular solvents and ILs. Despite their potential, several challenges impede the widespread adoption and advancement of DESs in various applications. This chapter discusses the various issues associated with these challenges, drawing on current and past research studies. is revised as "DESs exhibit a large variety of applications in biotechnology. This chapter provided a comprehensive insight into the possible utilizations and also the challenges associated with DESs while using these solvents on a large scale. DESs possess properties of high viscosity and density as their core limitations, which negatively impact mass transfer and fluidity operations, especially during continuous modes. Procedures that need optimal solvent flow, together with mixing, face major difficulties due to DESs' dense nature. Researchers studied different methods to address the DES challenges through DES water dilution and the usage of external microwave and temperature enhancements. These methods tend to alter the distinctive features of solvents and present supplementary technological challenges [34]. Low vapor pressures of DESs help reduce environmental emissions yet create difficulties in the recovery and separation of both solvents and products. Strengthening these alternatives needs further research to achieve maximum operational performance and reduced expenses [34]. The use of DESs depends on their essential influence on thermodynamic behavior and transport properties for industrial applications to become effective. The absence of detailed insights about DESs prevents the development of accurate predictive models while hampering the process design that adopts DESs as components. Molecular dynamics simulations need advancements, in addition to dependable force fields, to overcome present-day knowledge gaps [124]. DESs represent environmentally friendly solvents, according to typical assessments, but experts still need to fully understand their impact on the environment. Determination of environmental implications, in addition to designing standards for safe DES component selection, is needed to sustain DES-based processes [34]. Successful implementation of DESs in membrane separation methods requires the solution of unique operational difficulties. Scientists continuously work on improving membranes that can function with DESs while understanding solute behavior in this medium [125]. The hindrance to precise modeling and simulation of DESs is caused by their complex structural elements and interaction patterns. The accurate development of computational tools for DES predictions needs improvement because it helps researchers conduct experimental work more efficiently [126]. Research into nonstandard DESs brings promising opportunities but creates difficulties in determining their characteristics and practical uses. Additional research must examine the behavior patterns of these alternative DESs, according to Shishov et al. [127]. Widespread adoption of DESs as sustainable solutions requires a complete evaluation through life cycle assessments while performing economic studies. A thorough assessment that includes the study of raw material acquisition and energy utilization, as well as waste management at the product's conclusion, will determine whether DESs provide superior sustainability than traditional solvents, according to Cabezas R. et al. [122].

3.6 Future Opportunities for Advancements in DES Research

Multiple new opportunities will help scientists advance the applications of these methods across different fields. Recent research papers and review articles analyze these prospects, in addition to the challenges presented in this chapter.

(a) ***Design of Task-Specific DESs***: Researchers actively pursue the creation of designed DESs for particular uses as a growing field. Controlling DES properties requires choosing specific HBDs, together with HBAs. Through this method, DESs achieve desired properties, including proper viscosity and polarity and thermal stability. A well-designed, rational approach represents a key method to increase DES performance in specific applications [4].
(b) ***Integration with Advanced Materials***: DESs, when combined with nanoparticle or polymer components, enable the formation of exclusive composite materials. Extensive research on DES-based hybrid materials facilitates their deployment in biotechnology and energy sectors and environmental cleanup operations [128].
(c) ***Application in Electrochemical Systems:*** Electrochemical applications benefited from DESs, which demonstrated effectiveness as electrolytes in batteries, together with supercapacitors. These electrolytes present wide electrochemical windows as well as thermal stability factors, which enable their use in advanced energy storage devices. Continuous investigations focus on enhancing DESs' composition to enhance ionic conductivity, along with material electrode affinity [27].
(d) ***Computational Modeling and Machine Learning***: Computational methods received large attention to find the possible interaction mechanisms, optimal geometries, conformational energies, charge distributions, and electron cloud shapes. Furthermore, machine learning also helps to predict the possibilities of the hydrogen-bonding network, designing new DESs by varying different constituents [27].
(e) ***Environmental and Toxicological Assessments***: The validation of green credentials for DESs requires complete environmental impact assessments, together with toxicological tests. The research needs to systematically assess DES biodegradability as well as toxicity effects on different organisms and also conduct lifecycle evaluations to enable sustainable DES implementation [128].
(f) ***Industrial Scale-Up and Process Integration***: Operation of DES applications at industrial scales needs attention to solve issues with scalability, as well as process integration and cost-effectiveness requirements. For DESs to gain

broader industrial acceptance, it is essential to develop efficient recycling processes and integrate them within current production systems [129].

(g) ***Regulatory and Standardization Frameworks***: The acceptance of DES applications will increase within different industries through standardized production guidelines established by regulation. Research groups, alongside industry representatives and regulatory organizations, need to cooperate in designing the necessary frameworks to guarantee safe DES implementation [130].

3.7 Potential of DESs in Shaping Future Biotechnology

The biotechnology sector witnesses significant potential changes through DESs, which represent versatile, sustainable alternatives to traditional solvents. DESs are desirable for multiple applications because their characteristic properties include toxicity-free qualities, along with decomposition capability and customizable physicochemical aspects. The upcoming section presents foreseeable applications for DESs in biotechnology, which will guide their future development, according to recently published research findings.

(a) ***Enzyme Catalysis and Biotransformation***: The use of DESs has proven successful in stabilizing enzymes while improving catalysis and extending the operational lifespan of biocatalysts. The availability of DESs enables environmentally friendly enzymatic production methods for pharmaceuticals, food processing, and fuels. Researchers aim to create descriptions of enzymatic systems that replicate natural enzyme conditions to optimize reaction requirements and enable their industrial-scale adoption for enzymatic conversions.

(b) ***Drug Delivery and Pharmaceutical Applications***: Formulating drugs requires successful solubilization of medicines that show low water solubility. Hydrophobic drug bioavailability enhancement, together with controlled drug release operation, is a promising solution that DESs provide to the pharmaceutical industry. Future DES development will target drug delivery services with controlled pharmacokinetic properties that boost drug performance without adverse effects.

(c) ***Biomass Processing and Biofuel Production***: The emphasis on sustainable energy has focused investigators on studying DESs as efficient solutions for biomass pretreatment, together with biofuel generation. Degradation of specific biomass components into fermentable sugars by these solvents remains essential for producing second-generation biofuels. Enhancements in this field have the potential to create affordable, sustainable biofuel solutions [131].

(d) ***Biopolymer and Nanomaterial Synthesis***: The synthesis of biodegradable polymers, together with nanomaterials and functional biomaterials, benefits from using DESs as potential solvents. The latest research shows that DES developments create new opportunities within three crucial fields: tissue engineering, wound healing, and biosensor manufacturing. DESs require further development to exist as reaction media and structuring agents in polymerization processes because this will unlock new possibilities for advanced biomaterials with enhanced qualities [132].

(e) ***Sustainable Food Processing and Nutraceuticals***: The food industry operates a transformation to eco-friendly extraction methods in which DESs demonstrate potential as an effective system for extracting bioactive compounds, antioxidants, and essential oils from various sources. The market outlook indicates growing DES adoption for functional food and nutraceutical processing because it improves both product nutrition and safety.

Full exploitation of DESs for biotechnical applications needs team efforts from chemistry experts, together with microbiologists, engineers, and modelers. Key paths for DES application development need to focus on the following:

- *Computational Design of DESs:* Advanced molecular methods with artificial intelligence can improve the discovery of DESs that have ideal properties for particular task requirements through computational modeling.
- *Regulatory and Standardization Frameworks:* Safety and toxicity assessments, together with environmental impact standards, need to be developed as regulatory frameworks for industry adoption to proceed.
- *Scalability and Process Integration:* The widespread usage of DESs in biotechnology depends on developing cost-efficient, large-scale methods for production and recycling, combined with process integration strategies.
- *Hybrid Technologies:* Bio-based materials will experience new functionality through the combination of DESs with nanotechnology, biopolymers, and enzymatic systems.

3.8 Conclusion

The future of biotechnology depends heavily on DESs to create sustainable connections between green chemistry principles and industrial utility. The combination of sustainable solvent alternatives with superior reaction efficiency and bio-based process enhancement through DESs positions them as key elements for developing an environmentally friendly and innovative future. The future of the global economy depends on DESs as they advance through research and technology to being integrated into next-generation biotechnological solutions for sustainable ecological operations. This chapter analyzed DES classifications while presenting their synthetic processes and uses, encompassing enzymatic biocatalysis with drug developments and bio-separations, along with electrochemical operations. The evaluation of DES's advantages in comparison to traditional solvents, including ionic liquids, focused both on sustainability and their life cycle assessments. The research on DES faces ongoing issues and requires more investigations about their toxicity profiles as well as their recyclability mechanisms and industrial production operations. Commercial biotechnological processes need to overcome specific hurdles before they can successfully integrate DESs as part of their operations despite their vast potential. The successful implementation of DESs in biotechnology will depend on resolving their viscosity problems while enhancing recyclability and developing standard operating procedures. DES-based research, together with technological improvements, creates a path for sustainable biotechnological advancement in the future.

Conflict of Interest The authors declare no conflict of interest.

Declaration of Generative AI and AI-Assisted Technologies While preparing this work, the authors utilized the ChatGPT tool to enhance the language and readability of several sections of the manuscript. After using this tool, the authors reviewed and edited the content as needed, and they take full responsibility for the publication content.

References

1. Kumar JA, Patel D, More R et al (2019) New insight into experimental and computational studies of choline chloride-based 'green' ternary deep eutectic solvent (TDES). J Mol Struct 1181:295–299. https://doi.org/10.1016/j.molstruc.2018.12.106
2. Jangir AK, Mandviwala H, Patel P et al (2020) Acumen into the effect of alcohols on choline chloride: L-lactic acid-based natural deep eutectic solvent (NADES): a spectral investigation unified with theoretical and thermophysical characterization. J Mol Liq 317:113923. https://doi.org/10.1016/j.molliq.2020.113923
3. Jangir AK, Lad B, Dani U et al (2020) *In vitro* toxicity assessment and enhanced drug solubility profile of green deep eutectic solvent derivatives (DESDs) combined with theoretical validation. RSC Adv 10:24063–24072. https://doi.org/10.1039/C9RA10320A
4. Hansen BB, Spittle S, Chen B, Poe D, Zhang Y, Klein JM et al (2020) Deep eutectic solvents: a review of fundamentals and applications. Chem Rev 121:1232–1285. https://doi.org/10.1021/acs.chemrev.0c00385
5. Jangir AK, Nain AK, Kuperkar K (2021) Insight into structural properties and molecular interactions of maline (choline chloride + malonic acid) and 1, 4- butanediol based pseudo-binary mixture: a thermophysical, spectral, and simulation portrayal. J Mol Liq 334:116050. https://doi.org/10.1016/j.molliq.2021.116050
6. Omar KA, Sadeghi R (2020) Novel benzilic acid-based deep-eutectic-solvents: preparation and physicochemical properties determination. Fluid Phase Equilib 522:112752. https://doi.org/10.1016/j.fluid.2020.112752
7. Yeow AT, Hayyan A, Hayyan M, Junaidi MUM, Saleh J, Basirun WJ et al (2024) A comprehensive review on the physicochemical properties of deep eutectic solvents. Results Chem 7:101378. https://doi.org/10.1016/j.rechem.2024.101378
8. Hayyan A, Mjalli FS, AlNashef IM, Al-Wahaibi T, Al-Wahaibi YM, Hashim MA (2012) Fruit sugar-based deep eutectic solvents and their physical properties. Thermochim Acta 541:70–75. https://doi.org/10.1016/j.tca.2012.04.030
9. Hayyan A, Mjalli FS, AlNashef IM, Al-Wahaibi YM, Al-Wahaibi T, Hashim MA (2013) Glucose-based deep eutectic solvents: physical properties. J Mol Liq 178:137–141. https://doi.org/10.1016/j.molliq.2012.11.025
10. Ibrahim RK, Hayyan M, Alsaadi MA, Ibrahim S, Hayyan A, Hashim MA (2017) Diethylene glycol based deep eutectic solvents and their physical properties. Stud Univ Babes-Bolyai, Chem 62(4):433–450. https://doi.org/10.24193/subbchem.2017.4.37
11. Ibrahim RK, Hayyan M, AlSaadi MA, Ibrahim S, Hayyan A, Hashim MA (2019) Physical properties of ethylene glycol-based deep eutectic solvents. J Mol Liq 276:794–800. https://doi.org/10.1016/j.molliq.2018.12.032

12. Florindo C, Oliveira FS, Rebelo LPN, Fernandes AM, Marrucho IM (2014) Insights into the synthesis and properties of deep eutectic solvents based on cholinium chloride and carboxylic acids. ACS Sustain Chem Eng 2(10):2416–2425. https://doi.org/10.1021/sc500439w
13. AlOmar MK, Hayyan M, Alsaadi MA, Akib S, Hayyan A, Hashim MA (2016) Glycerol-based deep eutectic solvents: physical properties. J Mol Liq 215:98–103. https://doi.org/10.1016/j.molliq.2015.11.032
14. Zhu J, Yu K, Zhu Y, Zhu R, Ye F, Song N, Xu Y (2017) Physicochemical properties of deep eutectic solvents formed by choline chloride and phenolic compounds at T=(293.15 to 333.15) K: the influence of electronic effect of substitution group. J Mol Liq 232:182–187. https://doi.org/10.1016/j.molliq.2017.02.071
15. Omar KA, Sadeghi R (2021) Novel deep eutectic solvents based on pyrogallol: synthesis and characterizations. J Chem Eng Data 66(5):2088–2095. https://doi.org/10.1021/acs.jced.1c00023
16. Hayyan M, Aissaoui T, Hashim MA, AlSaadi MA, Hayyan A (2015) Triethylene glycol based deep eutectic solvents and their physical properties. J Taiwan Inst Chem Eng 50:24–30. https://doi.org/10.1016/j.jtice.2015.03.001
17. Cui Y, Li C, Yin J, Li S, Jia Y, Bao M (2017) Design, synthesis and properties of acidic deep eutectic solvents based on choline chloride. J Mol Liq 236:338–343. https://doi.org/10.1016/j.molliq.2017.04.052
18. Chemat F, Anjum H, Shariff AM, Kumar P, Murugesan T (2016) Thermal and physical properties of (choline chloride+ urea+ l-arginine) deep eutectic solvents. J Mol Liq 218:301–308. https://doi.org/10.1016/j.molliq.2016.02.062
19. Qin H, Hu X, Wang J, Cheng H, Chen L, Qi Z (2020) Overview of acidic deep eutectic solvents on synthesis, properties and applications. GEE 5:8–21. https://doi.org/10.1016/j.gee.2019.03.002
20. Abbott AP, Boothby D, Capper G, Davies DL, Rasheed RK (2004) Deep eutectic solvents formed between choline chloride and carboxylic acids: versatile alternatives to ionic liquids. J Am Chem Soc 126:9142–9147. https://doi.org/10.1021/ja048266j
21. Zhang Q, Vigier KDO, Royer S, Jérôme F (2012) Deep eutectic solvents: syntheses, properties and applications. Chem Soc Rev 41:7108–7146. https://doi.org/10.1039/C2CS35178A
22. Basaiahgari A, Panda S, Gardas RL (2018) Effect of ethylene, diethylene, and triethylene glycols and glycerol on the physicochemical properties and phase behavior of benzyltrimethyl and benzyltributylammonium chloride based deep eutectic solvents at 283.15–343.15 K. J Chem Eng Data 63:2613–2627. https://doi.org/10.1021/acs.jced.8b00213
23. Makoś P, Słupek E, Gębicki J (2020) Hydrophobic deep eutectic solvents in micro-extraction techniques–a review. Microchem J 152:104384. https://doi.org/10.1016/j.microc.2019.104384
24. Skulcova A, Russ A, Jablonsky M, Sima J (2018) The pH behavior of seventeen deep eutectic solvents. Bioresources 13:5042–5051. https://doi.org/10.15376/biores.13.3.5042-5051
25. Rao GLP, Mandal A, Pal N (2025) Choline chloride-urea based deep eutectic solvent: characterization, interfacial behavior and synergism in binary (surfactant) systems. Chem Phys 588:112496. https://doi.org/10.1016/j.chemphys.2024.112496
26. Luan J, Cheng Y, Xue F, Cui L, Wang D (2023) Refractive index of 48 neat deep eutectic solvents and of selected mixtures: effect of temperature, hydrogen-bonding donors, hydrogen-bonding acceptors, mole ratio, and water. ACS omega 8:25582–25591. https://doi.org/10.1021/acsomega.3c03502
27. Getie FA, Ayele DW, Habtu NG, Yemata TA, Yihun FA, Worku AK, Teshager MA (2024) Recent advances and various detection strategies of deep eutectic solvents for zinc air batteries. Heliyon 10. https://doi.org/10.1016/j.heliyon.2024.e40383
28. Sánchez PB, González B, Salgado J, Parajó JJ, Domínguez Á (2019) Physical properties of seven deep eutectic solvents based on L-proline or betaine. J Chem Thermodyn 131:517–523. https://doi.org/10.1016/j.jct.2018.12.017
29. Smith EL, Abbott AP, Ryder KS (2014) Deep eutectic solvents (DESs) and their applications. Chem Rev 114:11060–11082. https://doi.org/10.1021/cr300162p

30. Abbott AP, Barron JC, Ryder KS, Wilson D (2007) Eutectic-based ionic liquids with metal-containing anions and cations. Chem Eur J 13:6495–6501. https://doi.org/10.1002/chem.200601738
31. Zdanowicz M, Wilpiszewska K, Spychaj T (2018) Deep eutectic solvents for polysaccharides processing. A review. Carbohydr Polym 200:361–380. https://doi.org/10.1016/j.carbpol.2018.07.078
32. Hammond OS, Bowron DT, Edler KJ (2016) Liquid structure of the choline chloride-urea deep eutectic solvent (reline) from neutron diffraction and atomistic modelling. Green Chem 18:2736–2744. https://doi.org/10.1039/C5GC02914G
33. Dal Bosco C, Di Lisio V, D'Angelo P, Gentili A (2021) Hydrophobic eutectic solvent with antioxidant properties: application for the dispersive liquid–liquid microextraction of fat-soluble micronutrients from fruit juices. ACS Sustain Chem Eng 9:8170–8178. https://doi.org/10.1021/acssuschemeng.1c01473
34. Ferreira C, Sarraguça M (2024) A comprehensive review on deep eutectic solvents and its use to extract bioactive compounds of pharmaceutical interest. Pharmaceuticals 17:124. https://doi.org/10.3390/ph17010124
35. Murador DC, de Souza Mesquita LM, Vannuchi N, Braga ARC, de Rosso VV (2019) Bioavailability and biological effects of bioactive compounds extracted with natural deep eutectic solvents and ionic liquids: advantages over conventional organic solvents. Curr Opin Food Sci 26:25–34. https://doi.org/10.1016/j.cofs.2019.03.002
36. Choi YH, Verpoorte R (2019) Green solvents for the extraction of bioactive compounds from natural products using ionic liquids and deep eutectic solvents. Curr Opin Food Sci 26:87–93. https://doi.org/10.1016/j.cofs.2019.04.003
37. Jangir AK, Sethy P, Verma G et al (2021) An inclusive thermophysical and rheology portrayal of deep eutectic solvents (DES) for metal oxides dissolution enhancement. J Mol Liq 332:115909. https://doi.org/10.1016/j.molliq.2021.115909
38. Isaifan RJ, Amhamed A (2018) Review on carbon dioxide absorption by choline chloride/urea deep eutectic solvents. Adv Chem 2018(1):2675659. https://doi.org/10.1155/2018/2675659
39. Li P, Zhang Z, Zhang X, Li K, Jin Y, Wu W (2023) DES: their effect on lignin and recycling performance. RSC Adv 13(5):3241–3254. https://doi.org/10.1039/D2RA06033G
40. Gutiérrez MC, Ferrer ML, Mateo CR, del Monte F (2009) Freeze-drying of aqueous solutions of deep eutectic solvents: a suitable approach to deep eutectic suspensions of self-assembled structures. Langmuir 25(10):5509–5515. https://doi.org/10.1021/la900552b
41. Lončarić M, Jakobek L, Molnar M (2021) Deep eutectic solvents in the production of biopolymer-based materials. Croat Chem Acta 94(2):75–82. https://doi.org/10.5562/cca3836
42. Bhat MA, Naglah AM, Akber Ansari S, Al-Tuwajiria HM, Al-Dhfyan A (2021) ChCl: Gly (DESs) promote environmentally benign synthesis of xanthene derivatives and their antitubercular activity. Molecules 26(12):3667. https://doi.org/10.3390/molecules26123667
43. Ijardar SP, Singh V, Gardas RL (2022) Revisiting the physicochemical properties and applications of deep eutectic solvents. Molecules 27(4):1368. https://doi.org/10.3390/molecules27041368
44. González-Rivera J, Husanu E, Mero A, Ferrari C, Duce C, Tine MR et al (2020) Insights into microwave heating response and thermal decomposition behavior of deep eutectic solvents. J Mol Liq 300:112357. https://doi.org/10.1016/j.molliq.2019.112357
45. Sakti AS, Saputri FC, Mun'im A (2019) Optimization of choline chloride-glycerol based natural deep eutectic solvent for extraction bioactive substances from Cinnamomum burmannii barks and Caesalpinia sappan heartwoods. Heliyon 5(12):e02915. https://doi.org/10.1016/j.heliyon.2019.e02915
46. Batishcheva EV, Sokolova DN, Fedotova VS, Sokolova MP, Nikolaeva AL, Vakulyuk AY et al (2021) Strengthening cellulose nanopaper via deep eutectic solvent and ultrasound-induced surface disordering of nanofibers. Polymers 14(1):78. https://doi.org/10.3390/polym14010078

47. Al-Risheq DI, Nasser MS, Qiblawey H, Hussein IA, Benamor A (2021) Choline chloride based natural deep eutectic solvent for destabilization and separation of stable colloidal dispersions. Sep Purif Technol 255:117737. https://doi.org/10.1016/j.seppur.2020.117737
48. Wang DD, Lu ZH, Yang MNO, Guo HM, Yang ZH (2020) A choline chloride-ethylene glycol deep eutectic solvent based on magnetic polydopamine with preconcentration and determination for sulfonylurea herbicides in water samples. J Braz Chem Soc 31(7):1509–1517. https://doi.org/10.6084/m9.figshare.14303915.v1
49. Hong S, Yuan Y, Yang Q, Chen L, Deng J, Chen W et al (2019) Choline chloride-zinc chloride deep eutectic solvent mediated preparation of partial O-acetylation of chitin nanocrystal in one step reaction. Carbohydr Polym 220:211–218. https://doi.org/10.1016/j.carbpol.2019.05.075
50. Sarjuna K, Ilangeswaran D (2022) Preparation and physico-chemical studies of Ag2O nanoparticles using newly formed malonic acid and ZnCl2-based deep eutectic solvents. Mater Today Proc 49:2943–2948. https://doi.org/10.1016/j.matpr.2021.11.355
51. Elgharbawy AA, Syed Putra SS, Khan HW, Azmi NAN, Sani MSA, Ab Llah N et al (2023) Menthol and fatty acid-based hydrophobic deep eutectic solvents as media for enzyme activation. PRO 11(2):547. https://doi.org/10.3390/pr11020547
52. Hollenbach R, Delavault A, Gebhardt L, Soergel H, Muhle-Goll C, Ochsenreither K, Syldatk C (2022) Lipase-mediated mechanoenzymatic synthesis of sugar esters in dissolved unconventional and neat reaction systems. ACS Sustain Chem Eng 10(31):10192–10202. https://doi.org/10.1021/acssuschemeng.2c01727
53. Plastiras OE, Samanidou V (2022) Applications of deep eutectic solvents in sample preparation and extraction of organic molecules. Molecules 27(22):7699. https://doi.org/10.3390/molecules27227699
54. Weerasinghe UA, Wu T, Chee PL, Yew PYM, Lee HK, Loh XJ, Kai D (2024) Deep eutectic solvents towards green polymeric materials. Green Chem. https://doi.org/10.1039/d4gc00532e
55. Sari IP, Zulaicha AS, Triyanti SB (2024) Choline chloride-based deep eutectic solvents: physicochemical properties and spectroscopic insights. Analit Anal Environ Chem 9(02):23–34. https://doi.org/10.23960/analit.v9i02.200
56. Yadav A, Dong CD, Sharma D, Tsai ML, Sun PP, Nargotra P et al (2024) Integrated choline chloride/citric acid-microwave pretreatment for efficient nanolignin extraction and bioethanol production from cocoa pod husk waste. Energy Environ 0958305X241270269. https://doi.org/10.1177/0958305X241270269
57. Paiva Santiago P, Fernandes FAN, Santiago-Aguiar RSD (2025) Valorization of brewer's spent grain through ultrasound-assisted extraction of phenolic compounds using deep eutectic solvents. ACS Omega. https://doi.org/10.1021/acsomega.4c10526
58. Vuksanovic JM, Todorović N, Kijevčanin M, Serbanovic SP, Radović I (2017) Experimental investigation and modeling of thermophysical and extraction properties of choline chloride plus DL-malic acid based deep eutectic solvent. J Serb Chem Soc 82(11):1287–1302. https://doi.org/10.2298/JSC170316054V
59. Bedair HM, Samir TM, Mansour FR (2024) Antibacterial and antifungal activities of natural deep eutectic solvents. Appl Microbiol Biotechnol 108(1):198. https://doi.org/10.1007/s00253-024-13044-2
60. Ullah R, Atilhan M, Anaya B, Khraisheh M, García G, ElKhattat A et al (2015) A detailed study of cholinium chloride and levulinic acid deep eutectic solvent system for CO 2 capture via experimental and molecular simulation approaches. Phys Chem Chem Phys 17(32):20941–20960. https://doi.org/10.1039/c5cp03364k
61. Zhang M, Tian R, Han H, Wu K, Wang B, Liu Y et al (2022) Preparation strategy and stability of deep eutectic solvents: a case study based on choline chloride-carboxylic acid. J Clean Prod 345:131028. https://doi.org/10.1016/j.jclepro.2022.131028
62. Ghorpade VM, Hanna MA (1999). Method and apparatus for production of levulinic acid via reactive extrusion. U.S. Patent 5,859,263, issued January 12

63. Wang D, Tang RC (2018) Dissolution of wool in the choline chloride/oxalic acid deep eutectic solvent. Mater Lett 231:217–220. https://doi.org/10.1016/j.matlet.2018.08.056
64. Pham PT, Nguyen HT, Nguyen TT, Nguyen LHT, Dang MHD, Doan TLH et al (2022) Rapid and simple microwave-assisted synthesis of benzoxazoles catalyzed by [cholineCl][oxalic acid]. Catalysts 12(11):1394. https://doi.org/10.3390/catal12111394
65. Farooq MQ, Abbasi NM, Anderson JL (2020) Deep eutectic solvents in separations: methods of preparation, polarity, and applications in extractions and capillary electrochromatography. J Chromatogr A 1633:461613. https://doi.org/10.1016/j.chroma.2020.461613
66. Can A, Özlüsoylu İ, Antov P, Lee SH (2023) Choline chloride-based deep eutectic solvent-treated wood. Forests 14(3):569. https://doi.org/10.3390/f14030569
67. Jangir AK, Dani U, Chaudhary VK (2025) Exploring the influence of composition on physicochemical and biological properties of choline chloride-based deep eutectic solvents: an experimental and theoretical endorsement. J Mol Liq 427:127419. https://doi.org/10.1016/j.molliq.2025.127419
68. Hayyan M, Hashim MA, Hayyan A, Al-Saadi MA, AlNashef IM, Mirghani ME, Saheed OK (2013) Are deep eutectic solvents benign or toxic? Chemosphere 90:2193–2195. https://doi.org/10.1016/j.chemosphere.2012.11.004
69. Cheng QB, Zhang LW (2017) Highly efficient enzymatic preparation of daidzein in deep eutectic solvents. Molecules 22:186. https://doi.org/10.3390/molecules22010186
70. Opallo M, Lesniewski A (2011) A review on electrodes modified with ionic liquids. JEAC 656:2–16. https://doi.org/10.1016/j.jelechem.2011.01.008
71. Jangir AK, Bhawna VG et al (2022) Design and thermophysical characterization of betaine hydrochloride-based deep eutectic solvents as a new platform for CO_2 capture. New J Chem 46:5332–5345. https://doi.org/10.1039/D1NJ05373F
72. Gorke JT, Srienc F, Kazlauskas RJ (2008) Hydrolase-catalyzed biotransformations in deep eutectic solvents. Chem Commun 1235-1237. https://doi.org/10.1039/B716317G
73. Zhao H, Baker GA, Holmes S (2011) Protease activation in glycerol-based deep eutectic solvents. J Mol Catal B Enzym 72:163–167. https://doi.org/10.1016/j.molcatb.2011.05.015
74. Wachtmeister J, Rother D (2016) Recent advances in whole cell biocatalysis techniques bridging from investigative to industrial scale. Curr Opin Biotechnol 42:169–117. https://doi.org/10.1016/j.copbio.2016.05.005
75. Peters C, Rudroff F, Mihovilovic MD, Bornscheuer T, U. (2017) Fusion proteins of an enoate reductase and a Baeyer-Villiger monooxygenase facilitate the synthesis of chiral lactones. Biol Chem 398:31–37. https://doi.org/10.1515/hsz-2016-0150
76. Xu P, Zheng GW, Zong MH, Li N, Lou WY (2017) Recent progress on deep eutectic solvents in biocatalysis. Bioresour Bioprocess 4:1–18. https://doi.org/10.1186/s40643-017-0165-5
77. Weingärtner H, Cabrele C, Herrmann C (2012) How ionic liquids can help to stabilize native proteins. Phys Chem Chem Phys 14:415–426. https://doi.org/10.1039/C1CP21947B
78. Guo J, Zhao N, Zhao Y, Jin H, Sun G, Yu J et al (2023) Effect of deep eutectic solvents on the activity and stability of cellulases and pectinases. ACS Omega 8(48):45678–45686. https://doi.org/10.1021/acsomega.3c06088
79. Kim SH, Park S, Yu H, Kim JH, Kim HJ, Yang YH et al (2016) Effect of deep eutectic solvent mixtures on lipase activity and stability. J Mol Catal B Enzym 128:65–72. https://doi.org/10.1016/j.molcatb.2016.03.012
80. Yadav N, Bhakuni K, Bisht M, Bahadur I, Venkatesu P (2020) Expanding the potential role of deep eutectic solvents toward facilitating the structural and thermal stability of α-chymotrypsin. ACS Sustain Chem Eng 8(27):10151–10160. https://doi.org/10.1021/acssuschemeng.0c02213
81. Gajardo-Parra NF, Meneses L, Duarte ARC, Paiva A, Held C (2022) Assessing the influence of betaine-based natural deep eutectic systems on horseradish peroxidase. ACS Sustain Chem Eng 10(38):12873–12881. https://doi.org/10.1021/acssuschemeng.2c04045
82. Gajardo-Parra NF, Rodríguez G, Arroyo-Avirama AF, Veliju A, Happe T, Canales RI et al (2023) Impact of deep eutectic solvents on kinetics and folding stability of formate dehydrogenase. PRO 11(10):2815. https://doi.org/10.3390/pr11102815

83. Joana Gil-Chávez G, Villa JA, Fernando Ayala-Zavala J, Basilio Heredia J, Sepulveda D, Yahia EM, González-Aguilar GA (2013) Technologies for extraction and production of bioactive compounds to be used as nutraceuticals and food ingredients: an overview. CRFSFS 12:5–23. https://doi.org/10.1111/1541-4337.12005
84. Hu J, Lu W, Lv M, Wang Y, Ding R, Wang L (2019) Extraction and purification of astaxanthin from shrimp shells and the effects of different treatments on its content. Rev Bras 29:24–29. https://doi.org/10.1016/j.bjp.2018.11.004
85. Huang J, Guo X, Xu T, Fan L, Zhou X, Wu S (2019) Ionic deep eutectic solvents for the extraction and separation of natural products. J Chromatogr A 1598:1–19. https://doi.org/10.1016/j.chroma.2019.03.046
86. García-Roldán A, Piriou L, Jauregi P (2023) Natural deep eutectic solvents as a green extraction of polyphenols from spent coffee ground with enhanced bioactivities. Front Plant Sci 13:1072592. https://doi.org/10.3389/fpls.2022.1072592
87. Xu L, Liaqat F, Khazi MI, Sun J, Zhu D (2024) Natural deep eutectic solvents-based green extraction of vanillin: optimization, purification, and bioactivity assessment. Front Nutr 10:1279552. https://doi.org/10.3389/fnut.2023.1279552
88. Li D (2022) Natural deep eutectic solvents in phytonutrient extraction and other applications. Front Plant Sci 13:1004332. https://doi.org/10.1039/C5MD00551E
89. Fan Y, Li Q (2022) An efficient extraction method for essential oil from angelica sinensis radix by natural deep eutectic solvents-assisted microwave hydrodistillation. Sustain Chem Pharm 29:100792. https://doi.org/10.1016/j.scp.2022.100792
90. Cao J, Yang M, Cao F, Wang J, Su E (2017) Well-designed hydrophobic deep eutectic solvents as green and efficient media for the extraction of artemisinin from Artemisia annua leaves. ACS Sustain Chem Eng 5:3270–3278. https://doi.org/10.1021/acssuschemeng.6b03092
91. Su E, Yang M, Cao J, Lu C, Wang J, Cao F (2017) Deep eutectic solvents as green media for efficient extraction of terpene trilactones from Ginkgo biloba leaves. J Liq Chromatogr Relat Technol 40:385–391. https://doi.org/10.1080/10826076.2017.1309661
92. Zhao B-Y, Xu P, Yang FX, Wu H, Zong MH, Lou WY (2015) Biocompatible deep eutectic solvents based on choline chloride: characterization and application to the extraction of Rutin from Sophora japonica. ACS Sustain Chem Eng 3:2746–2755. https://doi.org/10.1021/acssuschemeng.5b00619
93. Duru KC, Slesarev GP, Aboushanab SA, Kovalev IS, Zeidler DM, Kovaleva EG et al (2022) An eco-friendly approach to enhance the extraction and recovery efficiency of isoflavones from kudzu roots and soy molasses wastes using ultrasound-assisted extraction with natural deep eutectic solvents (NADES). Ind Crop Prod 182:114886. https://doi.org/10.1016/j.indcrop.2022.114886
94. Popovic BM, Micic N, Potkonjak A, Blagojevic B, Pavlovic K, Milanov D et al (2022) Novel extraction of polyphenols from sour cherry pomace using natural deep eutectic solvents—ultrafast microwave-assisted NADES preparation and extraction. Food Chem 366:130562. https://doi.org/10.1016/j.foodchem.2021.130562
95. Xia B, Yan D, Bai Y, Xie J, Cao Y, Liao D et al (2015) Determination of phenolic acids in Prunella vulgaris L.: a safe and green extraction method using alcohol-based deep eutectic solvents. Anal Methods 7:9354–9364. https://doi.org/10.1039/C5AY02035B
96. Marchel M, Cieśliński H, Boczkaj G (2022) Deep eutectic solvents microbial toxicity: current state of art and critical evaluation of testing methods. J Hazard Mater 425:127963. https://doi.org/10.1016/j.jhazmat.2021.127963
97. Mouffok A, Bellouche D, Debbous I, Anane A, Khoualdia Y, Boublia A et al (2023) Synergy of garlic extract and deep eutectic solvents as promising natural antibiotics: experimental and COSMO-RS. J Mol Liq 375:121321. https://doi.org/10.1016/j.molliq.2023.121321
98. Radošević K, Čanak I, Panić M, Markov K, Bubalo MC, Frece J et al (2018) Antimicrobial, cytotoxic and antioxidative evaluation of natural deep eutectic solvents. ESPR 25:14188–14196. https://doi.org/10.1007/s11356-018-1669-z

99. Popović BM, Gligorijević N, Aranđelović S, Macedo AC, Jurić T, Uka D et al (2023) Cytotoxicity profiling of choline chloride-based natural deep eutectic solvents. RSC Adv 13:3520–3527. https://doi.org/10.1039/D2RA07488E
100. Sharma M, Mukesh C, Mondal D, Prasad K (2013) Dissolution of α-chitin in deep eutectic solvents. RSC Adv 3:18149–18155. https://doi.org/10.1039/C3RA43404D
101. McReynolds C, Adrien A, Castejon N, Fernandes SC (2022) Green in the deep blue: deep eutectic solvents as versatile systems for the processing of marine biomass. Green Chem Lett Rev 15:383–404. https://doi.org/10.1080/17518253.2022.2065890
102. Shirani M, Habibollahi S, Akbari A (2019) Centrifuge-less deep eutectic solvent based magnetic nanofluid-linked air-agitated liquid–liquid microextraction coupled with electrothermal atomic absorption spectrometry for simultaneous determination of cadmium, lead, copper, and arsenic in food samples and non-alcoholic beverages. Food Chem 281:304–311. https://doi.org/10.1016/j.foodchem.2018.12.110
103. Pradeepkumar P, Elgorban AM, Bahkali AH, Rajan M (2018) Natural solvent-assisted synthesis of amphiphilic co-polymeric nanomicelles for prolonged release of camptothecin delivery. New J Chem 42:10366–10375. https://doi.org/10.1039/C8NJ00901E
104. Lu W, Chen H (2022) Application of deep eutectic solvents (DESs) as trace level drug extractants and drug solubility enhancers: state-of-the-art, prospects and challenges. J Mol Liq 349:118105. https://doi.org/10.1016/j.molliq.2021.118105
105. Wang J, Li M, Duan L, Lin Y, Cui X, Yang Y, Wang C (2022) Deep eutectic systems as novel vehicles for assisting drug transdermal delivery. Pharmaceutics 14:2265. https://doi.org/10.3390/pharmaceutics14112265
106. Chen Q, Yang Z, Liu H, Man J, Oladejo AO, Ibrahim S et al (2024) Novel drug delivery systems: an important direction for drug innovation Research and Development. Pharmaceutics 16:674. https://doi.org/10.3390/pharmaceutics16050674
107. Mannu A, Blangetti M, Baldino S, Prandi C (2021) Promising technological and industrial applications of deep eutectic systems. Materials 14:2494. https://doi.org/10.3390/ma14102494
108. Lu C, Cao J, Wang N, Su E (2016) Significantly improving the solubility of non-steroidal anti-inflammatory drugs in deep eutectic solvents for potential non-aqueous liquid administration. MedChemComm 7(5):955–959. https://doi.org/10.1039/C5MD00551E
109. Shekaari H, Zafarani-Moattar MT, Shayanfar A, Mokhtarpour M (2018) Effect of choline chloride/ethylene glycol or glycerol as deep eutectic solvents on the solubility and thermodynamic properties of acetaminophen. J Mol Liq 249:1222–1235. https://doi.org/10.1016/j.molliq.2017.11.057
110. Phaechamud T, Tuntarawongsa S, Charoensuksai P (2016) Evaporation behavior and characterization of eutectic solvent and ibuprofen eutectic solution. AAPS PharmSciTech 17:1213–1220. https://doi.org/10.1208/s12249-015-0459-x
111. Li Z, Lee PI (2016) Investigation on drug solubility enhancement using deep eutectic solvents and their derivatives. Int J Pharm 505(1–2):283–288. https://doi.org/10.1016/j.ijpharm.2016.04.018
112. Pedro SN, Freire MG, Freire CS, Silvestre AJ (2019) Deep eutectic solvents comprising active pharmaceutical ingredients in the development of drug delivery systems. Expert Opin Drug Deliv 16(5):497–506. https://doi.org/10.1080/17425247.2019.1604680
113. Faggian M, Sut S, Perissutti B, Baldan V, Grabnar I, Dall'Acqua S (2016) Natural deep eutectic solvents (NADES) as a tool for bioavailability improvement: pharmacokinetics of rutin dissolved in proline/glycine after oral administration in rats: possible application in nutraceuticals. Molecules 21(11):1531. https://doi.org/10.3390/molecules21111531
114. Abbott AP, Ballantyne AD, Conde JP, Ryder KS, Wise WR (2012) Salt modified starch: sustainable, recyclable plastics. Green Chem 14:1302–1307. https://doi.org/10.1039/C2GC16568F
115. Qin H, Owyeung RE, Sonkusale SR, Panzer MJ (2019) Highly stretchable and nonvolatile gelatin-supported deep eutectic solvent gel electrolyte-based ionic skins for strain and pressure sensing. J Mater Chem C 7:601–608. https://doi.org/10.1039/C8TC05918G

116. Krainara S, Mistry AN, Malee C, Chavananikul C, Pinyakong O, Assavalapsakul W et al (2023) Development of a plastic waste treatment process by combining deep eutectic solvent (DES) pretreatment and bioaugmentation with a plastic-degrading bacterial consortium. J Hazard Mater 460:132507. https://doi.org/10.1016/j.jhazmat.2023.132507
117. Chyerochana N, Huynh QT, Jaitham U, Phitsuwan P, Aryusuk K, Hongsibsong S, Chang KL (2025) Sustainable production of high-performance bioplastics from agricultural and industrial biomass waste by integrating deep eutectic solvent (DES) pretreatment and acetylation processes. ACS Omega 10(11):10949–10961. https://doi.org/10.1021/acsomega.4c09059
118. Liu C, Liu H, Wang H, Yu Z, Yan M, Zhou X, Li R (2024) Deep eutectic solvent (DES) pretreatment and lignin regeneration for the development of a bamboo leaf-based bioplastic. Front Bioeng Biotechnol 12:1484585. https://doi.org/10.3389/fbioe.2024.1484585
119. Masdar ND, Yusof R, Ramzani NA (2022) Effect of deep eutectic solvent on tensile properties and biodegradation of pectin with eggshell bioplastic. Malay J Anal Sci 26(1):58–69
120. Leroy E, Decaen P, Jacquet P, Coativy G, Pontoire B, Reguerre AL, Lourdin D (2012) Deep eutectic solvents as functional additives for starch based plastics. Green Chem 14(11):3063–3066. https://doi.org/10.1021/acs.jced.1c00023
121. Zaib Q, Eckelman MJ, Yang Y, Kyung D (2022) Are deep eutectic solvents really green?: a life-cycle perspective. Green Chem 24:7924–7930. https://doi.org/10.1039/D2GC01752K
122. Cabezas R, Zurob E, Gomez B, Merlet G, Plaza A, Araya-Lopez C et al (2022) Challenges and possibilities of deep eutectic solvent-based membranes. Ind Eng Chem Res 61:17397–17422. https://doi.org/10.1021/acs.iecr.2c02747
123. Wang J, Zhang S, Ma Z, Yan L (2021) Deep eutectic solvents eutectogels: progress and challenges. Green Chem Eng 2:359–367. https://doi.org/10.1016/j.gce.2021.06.001
124. González de Castilla A, Bittner JP, Müller S, Jakobtorweihen S, Smirnova I (2019) Thermodynamic and transport properties modeling of deep eutectic solvents: a review on gE-models, equations of state, and molecular dynamics. J Chem Eng Data 65:943–967. https://doi.org/10.1021/acs.jced.9b00548
125. Mubashir M, D'Angelo FN, Gallucci F (2022) Recent advances and challenges of deep eutectic solvent based supported liquid membranes. Sep Purif Rev 51:226–244. https://doi.org/10.1080/15422119.2021.1901742
126. Tolmachev D, Lukasheva N, Ramazanov R, Nazarychev V, Borzdun N, Volgin I et al (2022) Computer simulations of deep eutectic solvents: challenges, solutions, and perspectives. Int J Mol Sci 23:645. https://doi.org/10.3390/ijms23020645
127. Shishov A, El-Deen AK, Godunov P, Bulatov A (2024) Atypical deep eutectic solvents: new opportunities for chemical analysis. TrAC 117752. https://doi.org/10.1016/j.trac.2024.117752
128. Wysokowski M, Luu RK, Arevalo S, Khare E, Stachowiak W, Niemczak M et al (2023) Untapped potential of deep eutectic solvents for the synthesis of bioinspired inorganic–organic materials. Chem Mater 35:7878–7903. https://doi.org/10.1021/acs.chemmater.3c00847
129. Paiva A, Matias AA, Duarte ARC (2018) How do we drive deep eutectic systems towards an industrial reality? Curr Opin Green Sustain Chem 11:81–85. https://doi.org/10.1016/j.cogsc.2018.05.010
130. Luu RK, Wysokowski M, Buehler MJ (2023) Generative discovery of de novo chemical designs using diffusion modeling and transformer deep neural networks with application to deep eutectic solvents. Appl Phys Lett 122. https://doi.org/10.1063/5.0155890
131. del Mar Contreras-Gámez M, Galán-Martín Á, Seixas N, da Costa Lopes AM, Silvestre A, Castro E (2023) Deep eutectic solvents for improved biomass pretreatment: current status and future prospective towards sustainable processes. Bioresour Technol 369:128396. https://doi.org/10.1016/j.biortech.2022.128396
132. Shen Y, Zhou H, He X, Shen F, Xu Z, Yang B et al (2024) Emerging applications of deep eutectic solvents in the preparation and functionalization of biomass-derived carbonaceous materials: challenges and prospects. Green Chem. https://doi.org/10.1039/D4GC01579G

Chapter 4
Micellization Behavior Altered by Deep Eutectic Solvents

Manoj Kumar Banjare, Kamalakanta Behera, Siddharth Pandey, and Kallol Kumar Ghosh

4.1 Introduction

Engineers and scientists are likely to design sustainable chemical development to produce less destructive materials and more environmentally friendly sources of energy production, as the chemical and energy industries have been the sources of environmental pollution for several decades. In several fields of science and technology, deep eutectic solvents (DESs), a novel class of green designer solvents, have recently gained a great deal of attention. In addition to being inexpensive, DESs are combustible, biodegradable, recyclable, and generated from biomass [1, 2]. DESs are typically ecologically benign liquids, and their primary feature is most likely their potential electron-donating or electron-accepting capabilities. They are great candidates for dissolving a wide range of substances, such as salts, proteins, polysaccharides, sugars, surfactants, amino acids, and medications, because of this specific characteristic. This occurrence happens at the critical micelle concentration (CMC), which is brought on by the surfactant's dual nature, which produces a range of extended properties in the solution, including "self-assembly," or micelles. Anti-corrosion, cosmetics, surface sciences, nanoscience, pharmaceuticals, medicine delivery, agrochemicals, and many other sectors employ surfactants.

M. K. Banjare (✉)
Chemistry Division, State Forensic Science Laboratory, Raipur, Chhattisgarh, India

K. Behera
Department of Chemistry, University of Allahabad, Prayagraj, Uttar Pradesh, India

S. Pandey
Department of Chemistry, Indian Institute of Technology Delhi, Hauz Khas, New Delhi, India

K. K. Ghosh
School of Studies in Chemistry, Pt. Ravishankar Shukla University,
Raipur, Chhattisgarh, India

K. Kuperkar et al. (eds.), *Deep Eutectic Solvents*, Environmental Footprints and
Eco-design of Products and Processes,
https://doi.org/10.1007/978-3-032-03801-2_4

4.2 Definition and Types of Deep Eutectic Solvents

A schematic representation of the formation of a deep eutectic solvent. DES is generally defined as a system made up of a mixture of at least two components, a hydrogen bond donor (HBD), illustrated in Fig. 4.1, and a quaternary ammonium salt as a hydrogen bond acceptor (HBA), which can self-associate to form a new eutectic phase with a melting point (below 1000 °C) lower than that of each component [3, 4]. These constituting components should also have low cost, low toxicity, renewability, and biodegradability [5, 6] (Fig. 4.2 and Table 4.1).

The formula Cat+X⁻zY, where Cat+ is the cation of any ammonium, sulfonium, or phosphonium salt and X^- is a Lewis base, often the halide anion of the salt, has been used frequently to characterize DESs was shown in Table 4.1. The number of

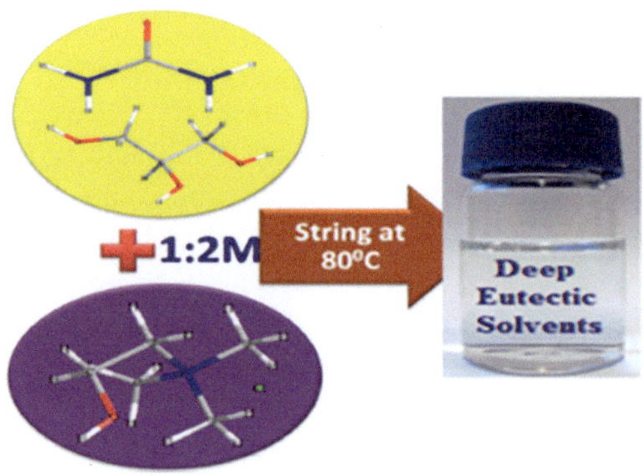

Fig. 4.1 Synthesis of deep eutectic solvents

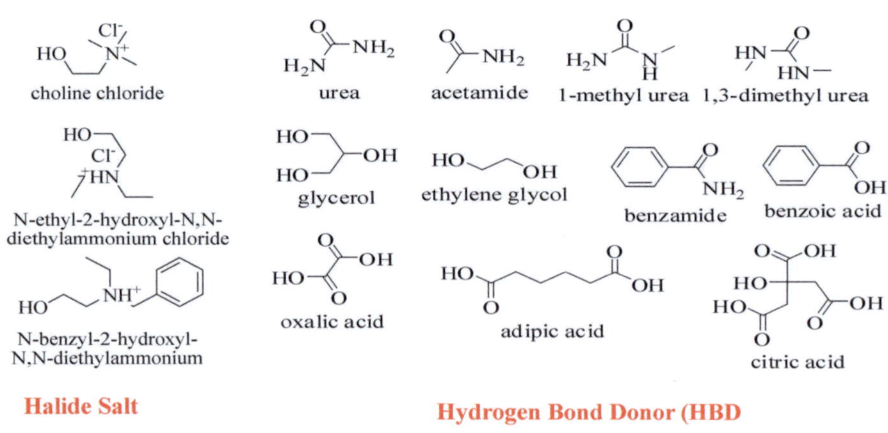

Fig. 4.2 Structure of some halide salts and hydrogen bond donor

Table 4.1 The main types of deep eutectic solvents

Type	Formula	Terms
I	$Cat^{+}X^{-}\ zMCl_{x}^{-}$	**M:** Zn, Sn, Fe, Al, Ga, In
II	$Cat^{+}X^{-}\ zMCl_{x}{\cdot}yH_{2}O$	**M:** Cr, Co, Cu, Ni, Fe
III	$Cat^{+}X^{-}\ zRZ$	**M:** $CONH_2$, COOH, OH
IV	$Cat_{x}+RZ = MCl_{x-1}^{+}{\cdot}RZ + MCl_{x+1}^{-}$	**M:** Al, Zn **Z:** $CONH_2$, OH

Y molecules is denoted by z, and Y can be either a Lewis or a Brønsted base. Between X and Y, a complex anionic class is created. Although sporadic electrostatic forces and van der Waals interactions are also probably conventional, hydrogen bonds constitute the primary interaction between the halide salt, or HBA, and HBD [7].

4.3 Deep Eutectic Solvent as a Subclass of Ionic Liquid

Ionic liquids (ILs) originally appeared around the turn of the twentieth century, but it was not until much later that their importance as environmentally friendly substitutes for typical organic solvents became apparent [8]. Some of their first presumed beneficial qualities began to be questioned [9] despite the early excitement demonstrated by a surge in publications in the decade 2000–2010 [10]. According to this viewpoint, DES was first considered a new generation of ILs but later proved to be a flexible substitute for traditional ILs. DESs differ from ILs in three key ways, although they have similar physicochemical characteristics: ILs are expensive and extremely poisonous, while DESs are less biocompatible, have low biodegradability, and are not as sustainable as ILs.

DESs are less expensive and have a significantly smaller environmental impact than ILs. Second, DESs and ILs are prepared in quite different ways [11–15]. Preparing ILs is much more challenging and costly than synthesizing DESs, which can be done by merely mixing the ingredients and does not require any purification procedures or involve any waste. Last, DES is the outcome of the complexation between an HBA and an HBD, whose interactions, as previously mentioned, mostly involve hydrogen bonds [16–20]. ILs are made up of ions that interact through ionic bonds (Fig. 4.3).

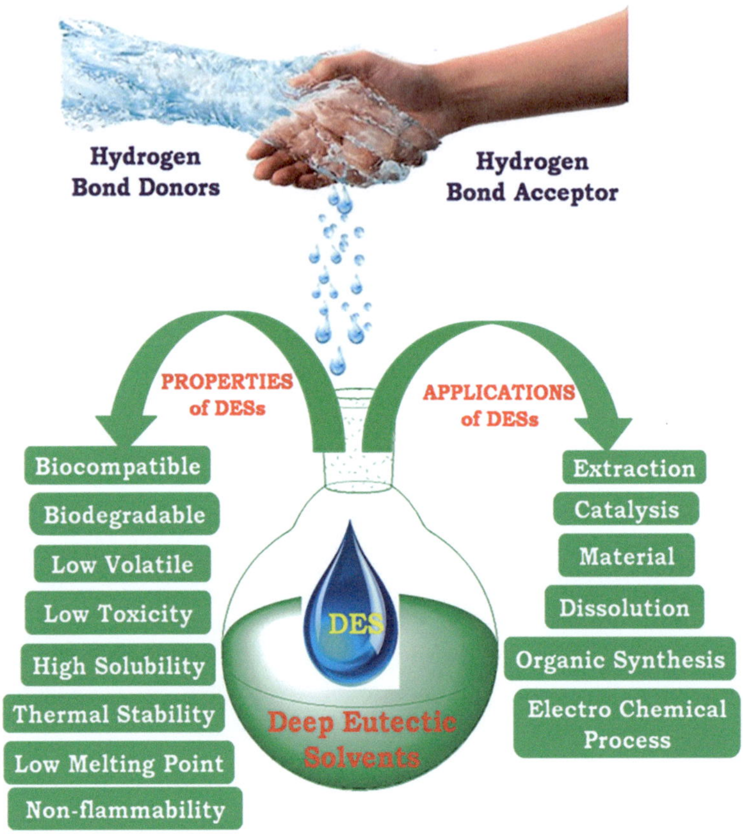

Fig. 4.3 Synthesis route and properties of deep eutectic solvents

4.4 Preparation and Properties of DESs

Lewis and Brønsted acid/base mixes that result in complex formation and low melting point systems are known as deep eutectic solvents (DESs). The hydrogen bond donor (HBD) and hydrogen bond acceptor (HBA) that make up a DES combine to generate a eutectic that has a melting point significantly lower than either of the constituent parts. Traditional DESs are made up of a binary mixture (1:1 and 1:2 ratio) of an HBD (such as alcohol, carboxylic acids, amines, and carbohydrates) and an HBA (such as a quaternary salt) [20, 21].

Recently, ternary DESs have attracted interest because they have lower melting points than conventional binary DESs. The characteristic properties of DESs include a melting point below that of the isolated components, low cost, less toxicity, high conductivity, relatively low viscosity, nonflammability, environmental friendliness, and biodegradability [22–26].

Numerous DESs have been made since Abbott's initial work in 2003 [27–29], in which the solid starting ingredients urea and choline chloride were heated at a 1:2 molar ratio to produce a combination that was liquid at ambient temperature. A thorough understanding of these solvents' physical characteristics is crucial for the right design of DESs and the logical expansion of their uses. DES is distinguished by its low melting points, low volatility, nonflammability, low vapor pressure, dipolar nature, high solubility, tuneability, low toxicity, and biodegradability.

Chemically adjustable DESs allow their qualities to be changed to create task-specific DESs with various physicochemical characteristics, based on the applications that are visualized. DESs have been utilized extensively in a variety of chemical reactions in recent years as catalysts, reagents, or solvents. A lot of work has gone into using ILs as biopolymer solvents.

4.5 Micelle Formation

Micellization is a well-known phenomenon in which micellar aggregates are associated with surface-active molecules [27]. The surfactant's dual nature results in a number of broad features in the solution, including "self-assembly," often referred to as micelles, and the concentration at which this occurrence takes place is known as CMC [28]. The manufacture of nanoparticles, drug encapsulation, and the creation of surfactant-ion-selective membranes are just a few of the uses for surfactants, which are known to form micelles with polar or ionic groups at the surface when in contact with water [29]. The reason behind surfactants' unusual behavior in the presence of water is their dual nature.

Adsorption is the tendency for a surfactant molecule to collect at an interface in an oriented fashion. The hydrophilic end remains absorbed in water, while the hydrophobic part is isolated. Very low concentrations of surfactants are soluble in water, and they form a true solution. CMC determination is an important physicochemical exercise for self-assembling surfactant solutions [30]. Various techniques can be used for the determination of the CMC, viz., surface tension, fluorescence, conductometry, viscosity, light scattering, sound velocity, calorimetry, spectrophotometry, small-angle neutron scattering (SANS), and dye solubilization. The most frequently used methods are tensiometry, conductometry, and fluorimetry (Fig. 4.4). The conductance method applies only to ionic surfactants.

4.5.1 Characterization Techniques of Micelle Formation

The CMC and micelle size in aqueous solutions may be measured using a variety of methods, comprising surface tension, conductometry, viscosity, light scattering, fluorescence, sound velocity, calorimetry, spectrophotometry, and SANS. The most common technique is dynamic light scattering (DLS), which uses laser light

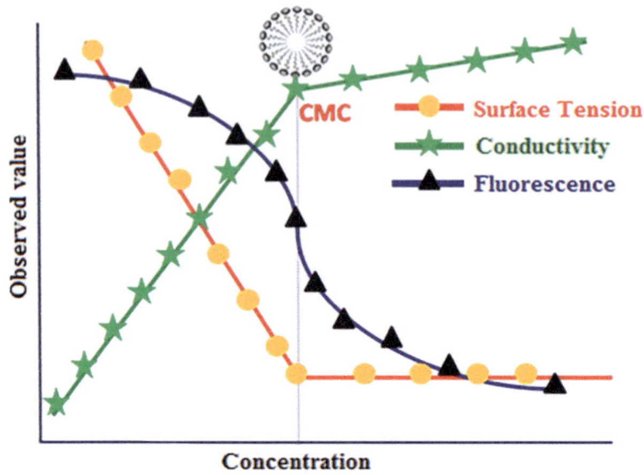

Fig. 4.4 Micelle formations using the surface tension, conductivity, and fluorescence methods

to measure the Brownian motion of the micelle particles in the solution [31] (Fig. 4.5).

The DLS method enables the polydispersity and average size of the micelles to be measured. Transmission electron microscopy (TEM), which enables direct micelle viewing, is an additional method that may be applied [32, 33]. The size, form, and surface charge of the micelles, may be ascertained using this method.

4.5.2 Physicochemical Properties of Surfactants

The unique properties of surfactant molecules have prompted extensive studies. Micelles are self-assembling molecules that are formed by surfactants. To aid in the production of surfactants, they may also adsorb and form a monolayer at the air–water interface [34, 35].

4.5.2.1 Interfacial Properties

Interfacial properties are characteristics of the interface between two different materials, such as an aqueous solution and an organic phase. Interfacial properties include surface tension, wettability, surface energy, wetting, spreading, adsorption, and adhesion [36]. The force required to move a line of water over a surface is known as surface tension, and the wettability of a liquid is how readily it spreads across a surface. Surface energy is the quantity of energy required to produce a unit area of the surface, whereas wetting and spreading describe how a liquid will spread

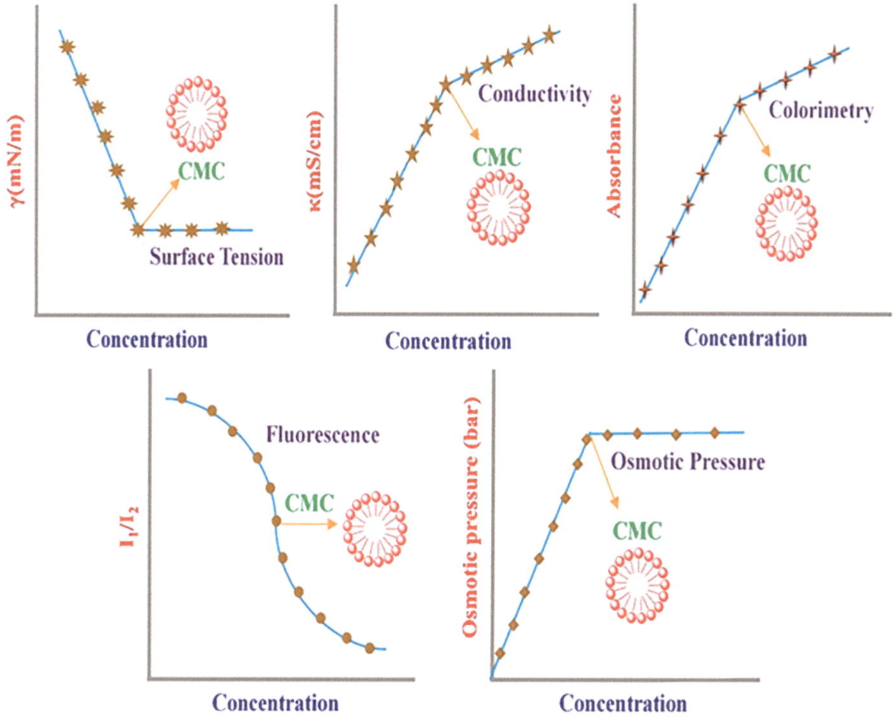

Fig. 4.5 Micelle formation of surfactants

out over a surface [37, 38]. Adsorption is the process of an adsorbate being attracted to a surface, while adhesion is the force between two surfaces. Understanding interfacial properties is important for many industries, as they determine the effectiveness of interactions between materials [39].

The Surface Tension at CMC (γ_{CMC})

Surface tension is a fundamental physical property that is used in surface, interface, and colloid research and is shown by amphiphilic molecules, such as ionic liquids and surfactants [40]. Surface tension has consistently decreased in several solvents when the molar ratio of amphiphilic compounds has grown. Equation (4.1) was utilized to compute the surface tension (γ) of amphiphilic systems [41].

$$\gamma_{CMC} = \frac{\text{Drop of Water X 72}}{\text{Drop of Solution}} \tag{4.1}$$

The Maximum Surface Excess Concentration (Γ_{max})

A chemical species' maximum concentration at the surface of a liquid or solid is known as its maximum surface excess concentration. In assessing a material's stability, solubility, and reactivity, it is a crucial characteristic. Measurements of molecule adsorption and desorption may also be made using it [42]. Temperature, material type, and surface area available for adsorption can be used to define it. Equation (4.2) was utilized to determine the (Γ_{max}) of amphiphilic systems based on the slope ($d\gamma/d\log_{10}C$).

$$\Gamma_{max} = \left(\frac{1}{2.303nRT}\right)\left(\frac{d\gamma}{d\log_{10} C}\right)Tp \quad (4.2)$$

where the surfactant concentration (C), the absolute temperature (T) in Kelvin, and the pre-factor (n) values for the nonionic surfactant (1), ionic liquids, and cationic and anionic surfactants (2) have been obtained.

The Minimum Surface Area Per Molecule (A_{min})

It speaks about the smallest space that a molecule might possibly occupy on a surface [43]. Using Eq. (4.3), the (A_{min}) of amphiphilic systems has been determined.

$$A_{min} = \left(\frac{1}{\Gamma_{max}}\right)N_A \quad (4.3)$$

Surface Pressure at CMC (π_{CMC})

The force a surface applies to a body per unit area is measured by the surface pressure at CMC. In addition, it determines the air density and air flow around an object [44]. Surface pressure can also be affected by temperature and humidity. Using Eq. (4.4), the (π_{CMC}) was determined,

$$\pi_{CMC} = \gamma_0 - \gamma_{CMC} \quad (4.4)$$

where γ_0 represents the surface tension of pure water and γ_{CMC} represents the surface tension at CMC.

The Efficiency of Adsorption (pC_{20})

The degree to which a material is adsorbed into a surface is determined by the effectiveness of adsorption (pC_{20}). A higher value of pC_{20} is shown to be more efficient in the adsorption process. Generally, adsorption is more efficient when the surface has a large area or when the material being adsorbed has a higher affinity to the surface [45]. The adsorption efficiency (pC_{20}) was determined by using Eq. (4.5).

$$\mathrm{pC}_{20} = -\log_{10} C_{20} \tag{4.5}$$

4.5.2.2 Thermodynamic Properties

The most important thermodynamic property of amphiphilic molecules is their tendency to self-assemble into ordered structures. The hydrophobic effect, which is caused by the molecules' hydrophobic groups interacting attractively, speeds up this self-assembly. Hydrogen bonds and other interactions between amphiphilic molecules can further stabilize their arrangement into ordered structures [46].

Degree of Micellization (α)

When the degree of micellization is low, the amphiphilic molecules remain separate, and the solubility and stability are lower [47]. In general, while maximizing a substance's solubility and stability, one must take into account the degree of micellization of an amphiphilic molecule. The slopes of the pre- and post-micellar curves of conductivity determine the degree of micellar ionization (α) using Eq. (4.6).

$$\alpha = \frac{S_2}{S_1} \tag{4.6}$$

where S_1 and S_2 represent the pre- and post-micellar slopes of the surfactant, respectively.

The Standard Gibbs Free Energy of Micellization ($\Delta G^\circ_{\mathrm{M}}$)

The ($\Delta G^\circ_{\mathrm{M}}$) of the amphiphilic system has been computed by using Eq. (4.7).

$$\Delta G^\circ_{\mathrm{M}} = (2-\alpha) RT \ln \frac{55.4}{\mathrm{CMC}} \tag{4.7}$$

The Standard Gibbs Free Energy of Adsorption (ΔG°_{ads})

The standard Gibbs free energy of adsorption (ΔG°_{ads}) of amphiphilic molecules is calculated by using Eq. (4.8)

$$\Delta G^{\circ}_{ads} = \Delta G^{\circ}_{M} - \frac{\pi_{CMC}}{\Gamma_{max}} \tag{4.8}$$

The Standard Gibbs Free Energy of Transfer (ΔG°_{trans})

ΔG°_{trans} is the energy released or absorbed when an amphiphilic molecule transfers from one medium to another. When studying how amphiphilic molecules behave in various settings and interact with their surroundings, ΔG°_{trans} is a useful tool [48–50]. ΔG°_{trans} is calculated by using Eq. (4.9).

$$\Delta G^{\circ}_{trans} = \Delta G^{\circ}_{M(\text{solvents mixed media})} - \Delta G^{\circ}_{M(\text{pure water})} \tag{4.9}$$

The Standard Gibbs Free Energy of the Given Air–Water Interface $\left(\Delta G^{(s)}_{min}\right)$

The standard Gibbs free energy of the specified air–water interface $\left(\Delta G^{(s)}_{min}\right)$ has been determined by using Eq. (4.10).

$$\Delta G^{(s)}_{min} = A_{min} \cdot \Upsilon_{CMC} \cdot N_A \tag{4.10}$$

The Standard Gibbs Free Energy of Micellization Per Alkyl Tail ($\Delta G^{\circ}_{m,\,tail}$)

$\Delta G^{\circ}_{m,\,tail}$ is used to calculate the energy required to form micelles from amphiphilic chemicals. It has to deal with the balance between the hydrophilic and hydrophobic parts of the molecule. This energy is dependent on the size, shape, stiffness, and solvent characteristics of the amphiphilic molecules [51, 52]. It is also affected by temperature and the ionic strength in the solution. Equation (4.11) has been utilized to compute the conventional ($\Delta G^{\circ}_{m,\,tail}$) of amphiphilic systems.

$$\Delta G^{\circ}_{m,tail} = \frac{\Delta G^{\circ}_{M}}{2} \tag{4.11}$$

4.5.2.3 Critical Packing Parameter (*P*)

The critical packing parameter (*P*) of amphiphilic molecules is calculated by using Eq. (4.12) [53, 54]. *P* and the minimum surface area of amphiphilic molecules were discovered by Israelachvili et al. [55].

$$P = \frac{V_0}{\mathrm{A}_{\min} l_c} \tag{4.12}$$

$$V_0 = \left(27.4 + 26.9C_n\right)\mathring{\mathrm{A}}^3 \text{ per hydrocarbon chain} \tag{4.13}$$

$$l_c = \left(1.5 + 1.265C_n\right)\mathring{\mathrm{A}} \text{ per hydrocarbon chain} \tag{4.14}$$

Tanford's equation [56] was used to determine the length of the long alkyl chain (hydrophobic tail) and the minimum area needed for each amphiphilic molecule to aggregate in the systems. V_0, l_c, and $A_{\min}$ represent the volume of long alkyl chain groups of the hydrophobic portion of an amphiphilic molecule in Eqs. (4.13) and (4.14).

4.6 Phase Behaviors

It is possible to alter the surface characteristics of materials by exposing surfactants to varying pressures, temperatures, and concentrations. The efficacy of surfactants' adsorption and their uses are significantly influenced by their phase behavior [57]. The degree of hydrophobicity and the molecular structure of surfactants play a major role in determining their phase behavior. Surfactants can create a lamellar phase, an isotropic liquid, and an isotropic gel at low temperatures. The higher-temperature surfactant molecules produce a reverse micellar phase, a micellar cubic phase, and eventually a micellar hexagonal phase. Increased oil recovery, detergents, and cosmetics are only a few of the applications that depend on these stages' effects on the solubilization capacity, interfacial tension reduction, and emulsion stability [58] (Table 4.2).

4.7 Viscosity

The degree of the hydrophobicity–hydrophilicity balance of the molecule, the kind of media, the temperature, and the concentration of the molecule all have an impact on the viscosity of amphiphilic molecules. Additionally, the viscosity of amphiphilic molecules can be impacted by the presence of other molecules in the medium.

Table 4.2 Different shapes of surfactant monomers and the corresponding shape of the micelle [59–63]

Packing parameter values	Critical packing shape	Shape	Characteristics	Shape of self-assembly entities
$P = <0.33$		Spherical micelle	Surfactants with a single chain and substantial head group	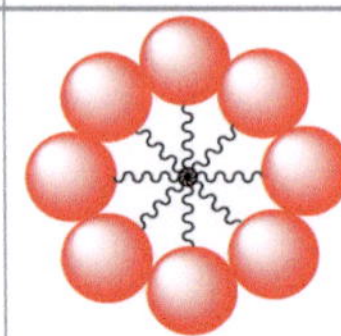
$P = 0.33–0.5$		Cylindrical micelle	Surfactants with a single chain and tiny head groups	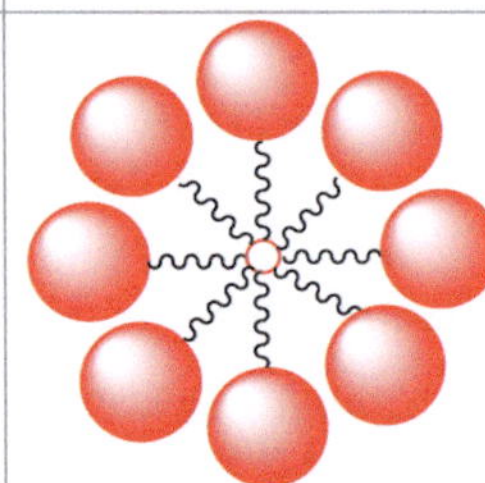
$P = 0.5–1.0$		Bilayer vesicle	Surfactants with two chains, each with a big head group and flexible chains	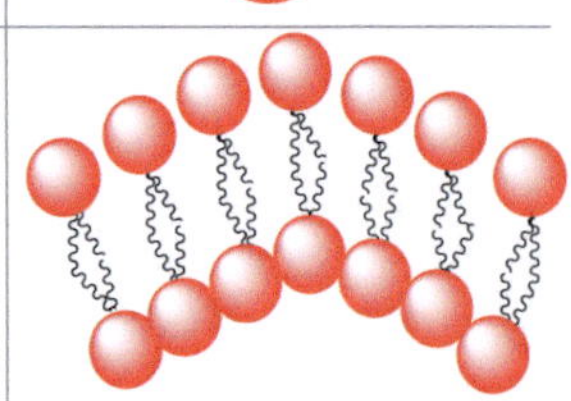
$P = 1.0$		Lamellar phase	Small head groups on double-chain surfactants or inflexible, immovable chains	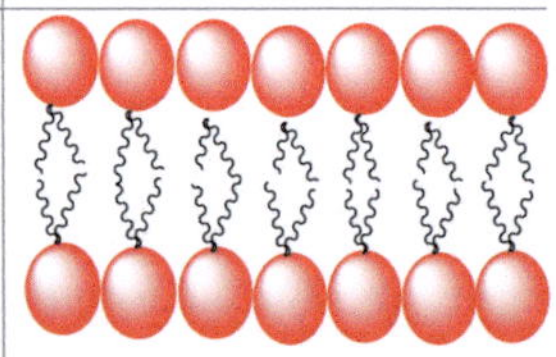
$P = >1.0$		Inverted/ reverse micelle	Surfactants having a double chain, big hydrophobic groups, and a tiny head group	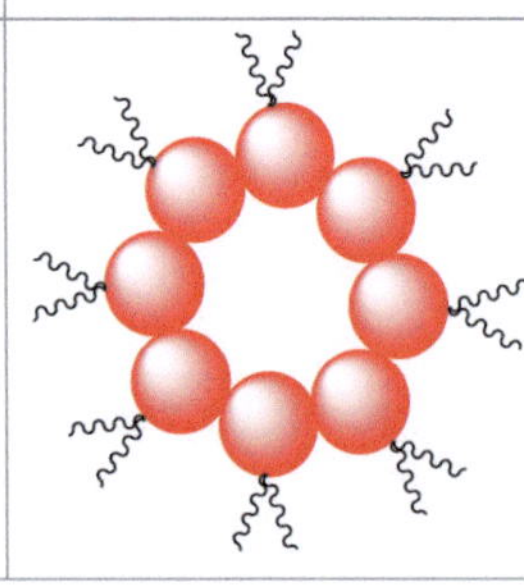

Varied amphiphilic systems have varied viscosity parameters, including relative, specific, decreased, and intrinsic viscosity [64].

4.7.1 *Relative Viscosity (η_r)*

A higher relative viscosity is exhibited by molecules whose hydrophilic and hydrophobic components are more affixed to one another. This relative viscosity is also influenced by the dimensions of the molecules and the geometry of their hydrophobic and hydrophilic constituents. The (η_r) of all systems computed from Eq. (4.15) is expressed as follows:

$$\eta_r = \frac{\eta_s}{\eta_0} \tag{4.15}$$

where η_r is the relative viscosity of the system, η_s is the viscosity of the solution, and η_0 is the viscosity of the pure solvent.

4.7.2 *Specific Viscosity (η_{sp})*

Specific viscosity is a measure of how much a material's viscosity increases when it is exposed to a certain concentration of an amphiphilic molecule. The increase in viscosity is due to the amphiphilic molecules' ability to interact with and increase the viscosity of the material [65]. The specific viscosity (η_{sp}) of all systems computed from Eq. (4.16) is expressed as follows:

$$\eta_{sp} = \eta_r - 1 \tag{4.16}$$

4.7.3 *Reduced Viscosity (η_{Red})*

The reduced viscosity of amphiphilic molecules occurs when the molecules are aggregates/clusters in an aqueous solution. When a molecule repels, its opposite end is drawn to water. This creates a structure in which the molecules arrange themselves in a way that reduces the overall viscosity of the solution [66]. Reduced viscosity can also be used to improve the performance of industrial processes. The reduced viscosity (η_{Red}) of all systems is calculated from Eqs. (4.17) and (4.18) and is expressed as follows:

$$\eta_{Red} = \frac{\eta_{sp}}{C} \tag{4.17}$$

$$\eta_{sp} = \eta_r - 1 \tag{4.18}$$

4.7.4 *Intrinsic Viscosity [η]*

The intrinsic viscosity values $[\eta]$ for the amphiphilic system were calculated from Eq. (4.19).

$$[\eta] = \lim_{C \to 0} \frac{\eta_{sp}}{C} \tag{4.19}$$

4.8 Mixed Micelle

Mixed micelles of two amphiphilic molecules (i.e., surfactants, ionic liquids, polymer, and copolymer) represent a unique type of self-assembled structure in which the hydrophilic and hydrophobic moieties are arranged distinctly. Various researchers have proposed several ideas to forecast the interaction in binary amphiphilic systems [67].

4.8.1 *Clint Model*

The Clint model is employed to assess the ideal micellar mole fraction of two combined surfactant systems, both of which clarify departures from ideality for researchers of mixed systems [68]. The Clint model of the predicted ideal mixing of CMC value is obtained from Eq. (4.20).

$$\frac{1}{\mathrm{CMC_{ideal}}} = \frac{\alpha_1}{\mathrm{CMC_1}} + \frac{\alpha_2}{\mathrm{CMC_2}} \tag{4.20}$$

where CMC_1 and CMC_2 are the critical micelle concentrations of the two amphiphilic components. A molar percentage of amphiphilic molecules, α_1 and α_2, make up CMC_{ideal}, which is a combination of CMC.

4.8.2 *Rubingh Model*

The Rubingh model is the phase separation model of mixed micelles. This theoretical model was developed by Dutch physicist Peter Rubingh in the early 2000s. It is used to explain the properties of mixed micelles, which are aggregates of amphiphilic molecules, such as lipids and surfactants. The model considers the interaction of hydrophobic and hydrophilic contacts, as well as the impact of electrostatic charge interactions between the amphiphilic molecule, to describe the behavior of a specific mixed micelle system [69]. The model also considers how these interactions modify the mixed micelles' size and structure. The Rubingh model may be used to measure these interactions and forecast the stability and solubility of mixed micelles in a given solution. The Rubingh theory may be used to determine a mixture of CMC for nonideal mixed micellization of amphiphilic components from Eq. (4.21).

$$\frac{1}{\mathrm{CMC}_{\mathrm{mix}}} = \frac{1}{f_1 \mathrm{CMC}_1} + \frac{1}{f_2 \mathrm{CMC}_2} \tag{4.21}$$

The activity coefficients of amphiphilic components, denoted as f_1 and f_2, may be calculated using the following Eqs. (4.22) and (4.23).

$$f_1 = \exp\left[\beta\left(1 - X_1\right)^2\right] \tag{4.22}$$

$$f_2 = \exp\left[\beta\left(X_1\right)^2\right] \tag{4.23}$$

The Rubingh model is used to calculate the micellar mole fraction of amphiphile 1:

$$\frac{\left(X_1\right)^2 \ln\left(\frac{\alpha_1 \mathrm{CMC}_{\mathrm{mix}}}{X_1 \mathrm{CMC}_1}\right)}{\left(1 - X_1\right)^2 \ln\left[\frac{\left(1 - \alpha_1\right)\mathrm{CMC}_{\mathrm{mix}}}{\left(1 - X_1\right)\mathrm{CMC}_2}\right]} = 1 \tag{4.24}$$

where X_1 is the mole percentage of component 1 of the mixed amphiphilic micelle; the X_1 equation may be solved by using Eqs. (4.24) and (4.25).

$$X_1^{\mathrm{ideal}} = \frac{\alpha_1 \mathrm{CMC}_2}{\alpha_1 \mathrm{CMC}_2 + \alpha_2 \mathrm{CMC}_1} \tag{4.25}$$

The attractive or repulsive contact between two amphiphilic components is confirmed by the Rubingh model using the regular solution theory by an interaction parameter computed by Eq. (4.26).

$$\beta = \frac{\ln\left(\frac{\alpha_1 C_{\mathrm{mix}}}{X_1 C_1}\right)}{\left(1 - X_1\right)^2} \tag{4.26}$$

β is a sign of how much the two surfactants interact.

4.8.3 Rosen Model

The micellar fraction of moles of surfactant 1 (X^{α}) at the mixed adsorption layer may be interactively determined using the Rosen model, which was estimated with the help of Eq. (4.27).

$$\frac{\left(X^{\alpha}\right)^{2}\ln\left(\frac{\alpha_1 \text{CMC}_{\text{mix}}}{X^{\alpha}\text{CMC}_1}\right)}{\left(1-X^{\alpha}\right)^{2}\ln\left[\frac{\left(1-\alpha_1\right)\text{CMC}_{\text{mix}}}{\left(1-X^{\alpha}\right)\text{CMC}_2}\right]}=1 \tag{4.27}$$

Pure surfactants, CMC_1 and CMC_2, respectively, are represented by C_{mix}, the mixed concentrations [70]. The proportion of pure surfactant values is 1 in solution with a certain γ value and α_1 stoichiometric mole fraction. Equation (4.28) was utilized to determine the value of X^{α}, which is the interaction parameter (β^{σ}) evaluated at the air–solution interface.

$$\beta^{\sigma}=\frac{\ln\left(\frac{\alpha_1 C_{\text{mix}}}{X^{\alpha}C_1}\right)}{\left(1-X^{\alpha}\right)^{2}} \tag{4.28}$$

In the mixed micelle system, the activity coefficients (f_1 and f_2) of the two surfactants are connected to the interaction parameter (β^{σ}). Equations (4.29) and (4.30) are used to calculate the activity coefficient.

$$f_1=\exp\left[\beta\left(1-X^{\alpha}\right)^{2}\right] \tag{4.29}$$

$$f_2=\exp\left[\beta\left(X^{\alpha}\right)^{2}\right] \tag{4.30}$$

4.8.4 Motomura Model

Motomura's method has been applied to determine the micellar mole fraction in the ideal condition by using Eq. (4.31) [71].

$$X_{\text{ideal}}=\frac{\alpha_1 \text{CMC}_2}{\alpha_1 \text{CMC}_2+\left(1-\alpha_2\right)\text{CMC}_1} \tag{4.31}$$

Eq. (4.32) to establish a connection between the activity coefficients and excess free energy of mixing.

$$\Delta G_{ex} = RT\left[X_1.\ln f_1 + \left(1 - X_1\right).\ln f_2\right] \tag{4.32}$$

4.9 Conclusions

Since DESs are new agents with significant promise in chemical engineering, green chemistry, and synthesis applications, their physicochemical properties are essential fundamental traits that demand further investigation and reporting. Given their physicochemical properties, a thorough assessment of DESs' sustainability as potential replacements for conventional organic solvents ought to be carried out. The synthesis, micelle generation, and other micellization properties are presented in this chapter. Dimeric surfactants have been shown in studies to have better qualities than traditional ones, which may result in significant cost reductions in some applications.

The potential for beneficial application is indicated by the decreased CMC, primarily in the sectors of oil recovery, cosmetics, pharmaceuticals, catalysis, and polymer synthesis (emulsion polymerization). The behavior of surface-active chemicals was changed by DESs, a new generation of green organic solvents, which is beginning to highlight the need for further advancements in areas where conventional surfactants are currently used. Emphasis should be placed on DESs' simplified synthetic routes to soon open markets and applications.

Conflict of Interest Statement The authors declare no conflict of interest.

Declaration of Generative AI and AI-Assisted Technologies in the Writing Process During the preparation of this work, we made use of the ChatGPT tool to improve the wording and readability of a number of the manuscript's statements. The writers take full responsibility for the publication's content after utilizing this tool/service and reviewing and editing it as necessary.

References

1. Paiva A, Craveiro R, Aroso I, Martins M, Reis RL, Duarte ARC (2014) Natural deep eutectic solvents—solvents for the 21st century. ACS Sustain Chem Eng 2:1063–1071. https://doi.org/10.1021/sc500096j
2. Verevkin SP, Sazonova AY, Frolkova AK, Zaitsau DH, Prikhodko IV, Held C (2015) Separation performance of biorenewable deep eutectic solvents. Ind Eng Chem Res 54:3498–3504. https://doi.org/10.1021/acs.iecr.5b00357
3. Ghorbanizamani F, Timur S (2018) Ionic liquids from biocompatibility and electrochemical aspects toward applying in biosensing devices. Anal Chem 90:640–648. https://doi.org/10.1021/acs.analchem.7b03596

4. Zhang H, Liu J, Guan M, Shang Z, Sun Y, Lu Z, Li H, An X, Liu H (2018) Nanofibrillated cellulose (NFC) as a pore size mediator in the preparation of thermally resistant separators for lithium ion batteries. ACS Sustain Chem Eng 6(4):4838–4844. https://doi.org/10.1021/acssuschemeng.7b04203
5. Vidal C, Merz L, Álvarez JG (2015) Deep eutectic solvents: biorenewable reaction media for Au(I)-catalysed cycloisomerisations and one-pot tandem cycloisomerisation/Diels–Alder reactions. Green Chem 17:3870–3878. https://doi.org/10.1039/C5GC00656B
6. Banjare MK, Barman B (2024) Effect of biologically active amino acids based deep eutectic solvents on sodium dodecyl sulfate: a comparative spectroscopic study. Spectrochim Acta Part A: Mol Biomol Spectrosc 308:123700. https://doi.org/10.1016/j.saa.2023.123700
7. Díaz A, Katsarava R, Puiggalí J (2014) Synthesis, properties and applications of biodegradable polymers derived from diols and dicarboxylic acids: from polyesters to poly(ester amide)s. Int J Mol Sci 15:7064–7123. https://doi.org/10.3390/ijms15057064
8. Tomé LIN, Baião V, Silva WD, Brett CMA (2018) Deep eutectic solvents for the production and application of new materials. Appl Mater Today 10:30–50. https://doi.org/10.1016/j.apmt.2017.11.005
9. Moity L, Durand M, Benazzouz A, Pierlot C, Molinier V, Aubry JM (2012) Panorama of sustainable solvents using the COSMO-RS approach. Green Chem 14:1132–1145. https://doi.org/10.1039/C2GC16515E
10. Kang L, Chen J, Teng T, Chen XL, Yu R, Lu CZ (2015) Experimental and theoretical studies of highly emissive dinuclear Cu(I) halide complexes with delayed fluorescence. Dalton Trans 44:11649–11659. https://doi.org/10.1039/C5DT01292A
11. Lazzeretti P (2004) Assessment of aromaticity via molecular response properties. Phys Chem Chem Phys 6:217–223. https://doi.org/10.1039/B311178D
12. Abbott P, Boothby D, Capper G, Davies DL, Rasheed RK (2004) Deep eutectic solvents formed between choline chloride and carboxylic acids: versatile alternatives to ionic liquids. J Am Chem Soc 126:9142–9147. https://doi.org/10.1021/ja048266j
13. Abbott AP, Capper G, Davies DL, Rasheed RK, Tambyrajah V (2004) Novel solvent properties of choline chloride/urea mixtures. Chem Commun:70–71. https://doi.org/10.1039/B210714G
14. Abbott AP, Capper G, McKenzie KJ, Ryder KS (2006) Electropolishing of stainless steels in a choline chloride based ionic liquid: an electrochemical study with surface characterisation using SEM and atomic force microscopy. Phys Chem Chem Phys 8:4214–4221. https://doi.org/10.1039/B607763N
15. Gutierrez MC, Ferrer ML, Mateo R, Monte FD (2009) Freeze-drying of aqueous solutions of deep eutectic solvents: a suitable approach to deep eutectic suspensions of self-assembled structures. Langmuir 25:5509–5515. https://doi.org/10.1021/la900552b
16. Hayyan A, Mjalli FS, AlNashef IM, Al-Wahaibi T, Al-Wahaibi YM, Ali Hashim MA (2012) Fruit sugar-based deep eutectic solvents and their physical properties. Thermochim Acta 541:70–75. https://doi.org/10.1016/j.tca.2012.04.030
17. Hayyan A, Mjalli FS, AlNashef IM, Al-Wahaibi YM, Al-Wahaibi T, Hashim MA (2013) Glucose-based deep eutectic solvents: physical properties. J Mol Liq 178:137–141. https://doi.org/10.1016/j.molliq.2012.11.025
18. Guo W, Hou Y, Ren S, Tian S, Wu W (2013) Formation of deep eutectic solvents by phenols and choline chloride and their physical properties. J Chem Eng Data 58:866–872. https://doi.org/10.1021/je300997v
19. Jibril B, Mjalli F, Naser J, Gano Z (2014) New tetrapropylammonium bromide-based deep eutectic solvents: synthesis and characterizations. J Mol Liq 199:462–469. https://doi.org/10.1016/j.molliq.2014.08.004
20. Liu YT, Chen YA, Xing YJ (2014) Synthesis and characterization of novel ternary deep eutectic solvents. Chin Chem Lett 25:104–106. https://doi.org/10.1016/j.cclet.2013.09.004
21. Parnica J, Antalika M (2014) Urea and guanidine salts as novel components for deep eutectic solvents. J Mol Liq 197:23–26. https://doi.org/10.1016/j.molliq.2014.04.016

22. Seyedi N, Khabazzadeh H, Saeednia S (2015) ZnCl2/urea as a deep eutectic solvent for the preparation of bis(indolyl)methanes under ultrasonic conditions. Synth React Inorg Metal 45:1501–1505. https://doi.org/10.1080/15533174.2013.862828
23. Bagh FSG, Shahbaz K, Mjalli FS, Hashim MA, AlNashef IM (2015) Zinc (II) chloride-based deep eutectic solvents for application as electrolytes: preparation and characterization. J Mol Liq 204:76–83. https://doi.org/10.1016/j.molliq.2015.01.025
24. Gano ZS, Mjallia FS, Al-Wahaibia T, Al-Wahaibia Y, Nashef IMA (2015) Extractive desulfurization of liquid fuel with $FeCl_3$-based deep eutectic solvents: experimental design and optimization by central-composite design. Chem Eng Process 93:10–20. https://doi.org/10.1016/j.cep.2015.04.001
25. Kow KK, Sirat K (2015) Novel manganese(II)-based deep eutectic solvents: synthesis and physical properties analysis. Chin Chem Lett 26:1311–1314. https://doi.org/10.1016/j.cclet.2015.05.049
26. Hayyan M, Aissaoui T, Hashim MA, AlSaadi MA, Hayyan A (2015) Triethylene glycol based deep eutectic solvents and their physical properties. J Taiwan Inst Chem Eng 50:24–30. https://doi.org/10.1016/j.jtice.2015.03.001
27. Naqvi AZ, Noori S, Din KU (2016) Mesoporous NiCo2O4 nanospheres with a high specific surface area as electrode materials for high-performance supercapacitors. RSC Adv 6:20324–20336. https://doi.org/10.1039/C6RA14099H
28. Benrraou M, Bales BL, Zana R (2003) Effect of the nature of the counterion on the properties of anionic surfactants. 1. Cmc, ionization degree at the Cmc and aggregation number of micelles of sodium, cesium, tetramethylammonium, tetraethylammonium, tetrapropylammonium, and tetrabutylammonium dodecyl sulfates. J Phys Chem B 107:13432–13440. https://doi.org/10.1021/jp021714u
29. Tanford C (1974) Theory of micelle formation in aqueous solutions. J Phys Chem 78:2469–2479. https://doi.org/10.1021/j100617a012
30. Dominguez A, Fernandez A, Gonzalez N, Iglesias E, Montenegro L (1997) Determination of critical micelle concentration of some surfactants by three techniques. J Chem Educ 74:1227–1231. https://doi.org/10.1021/ed074p1227
31. Tanwar LKS, Banjare MK, Sharma S, Ghosh KK (2021) Physicochemical studies on the micellization of anionic surfactants in the presence of long alkyl chain ionic liquid. Chem Phys Lett 769:138399. https://doi.org/10.1016/j.cplett.2021.138399
32. Sun T, Gao S, Chen Q, Shen X (2014) Investigation on the interactions between hydrophobic anions of ionic liquids and Triton X-114 micelles in aqueous solutions. Colloids Surf A Physicochem Eng Asp 456:18–25. https://doi.org/10.1016/j.colsurfa.2014.05.002
33. Shafi M, Bhat PA, Dar AA (2009) Solubilization capabilities of mixtures of cationic gemini surfactant with conventional cationic, nonionic and anionic surfactants towards polycyclic aromatic hydrocarbons. J Hazard Mater 167:575–581. https://doi.org/10.1016/j.jhazmat.2009.01.022
34. Li Q, Wang X, Yue X, Chen X (2013) Wormlike micelles formed using gemini surfactants with quaternary hydroxyethyl methylammonium headgroups. Soft Matter 9:9667–9674. https://doi.org/10.1039/C3SM51722E
35. Banjare MK, Behera K, Kurrey R, Banjare RK, Satnami ML, Pandey S, Ghosh KK (2018) Self-aggregation of bio-surfactants within ionic liquid 1-ethyl-3-methylimidazolium bromide: a comparative study and potential application in antidepressants drug aggregation. Spectrochim Acta A Mol Biomol Spectrosc 199:376–386. https://doi.org/10.1016/j.saa.2018.03.079
36. Geng Y, Romsted LS, Menger F (2006) Specific ion pairing and interfacial hydration as controlling factors in gemini micelle morphology chemical trapping studies. J Am Chem Soc 128:492–501. https://doi.org/10.1021/ja056807e
37. Pandya SJ, Kapitanov IV, Banjare MK, Behera K, Borovkov V, Ghosh KK, Karpichev Y (2022) Mixed oxime-functionalized IL/16-s-16 Gemini surfactants system: physicochemical study and structural transitions in the presence of promethazine as a potential chiral pollutant. Chemosensors 10:46. https://doi.org/10.3390/chemosensors10020046

38. Kralova I, Sjoblom J (2009) Surfactant used in food industry: a review. J Dispers Sci Technol 30:1363–1383. https://doi.org/10.1080/01932690902735561
39. E X, Pei ZI, Schmidi KA (2010) Ice cream foam formation and stabilization—a review. Food Rev Int 26:122–137. https://doi.org/10.1080/8755912090356447
40. Schramm LL, Stasiak EN, Marangoni DG (2003) 2 surfactant and their applications. Annu Rep Prog Chem Sect C 99:3–48. https://doi.org/10.1039/B208499F
41. Zelena V, Veverk K (2007) Effect of surfactant and liquid utilization on transcuticular penetration of fungicides. Plant Protect Sci 43:151–156. https://doi.org/10.17221/2236-PPS
42. Banjare MK, Behera K, Rub MA, Ghosh KK (2025) Micellization behavior of imidazolium surface active ionic liquid within aqueous solution of deep eutectic solvents: a comparative spectroscopic study. RSC Adv 15:15879–15892. https://doi.org/10.1039/d5ra01940k
43. Banjare MK, Barman B, Behera K, Banjare RK, Pandey S, Ghosh KK (2024) Molecular interaction between three novel amino acid based deep eutectic solvents with surface active ionic liquid: a comparative study. Heliyon 10(15):e35598. https://doi.org/10.1016/j.heliyon.2024.e35598
44. Singh R (2012) Solubilization of organic dyes in surfactants micelles. Div Appl Surf Chem:1–45. https://doi.org/10.3390/ma6020580
45. Manojlovic JZ (2012) The kraft temperature of surfactant solution. Therm Sci 16:31–40. https://doi.org/10.2298/TSCI120427197M
46. Duarte JCS, Oliveira LL, Mayor MS, Okada DY, Varesche MBA (2010) Degradation of detergent (linear alkyl benzene sulfonate) in an anaerobic stirred sequencing-batch reactor containing granular biomass. Int Biodeterior Biodegradation 64:129–134. https://doi.org/10.1016/j.ibiod.2009.12.003
47. Mishra S, Tyagi VK (2007) Ester quats: the novel class of cationics fabric softners. J Oleo Sci 56:269–276. https://doi.org/10.5650/jos.56.269
48. Levitt D, Jackson A, Heinson C, Britton LN, Malik T, Dwarakanath V, Pope GA (2006) Identification and evaluation of high-performance EOR surfactant. SPE Reserv Eval Eng 12:243–253. https://doi.org/10.2118/100089-PA
49. Ko KM, Chon BH, Jang SB, Jang HY (2014) Surfactant flooding characteristics of dodecyl alkyl sulfate for enhanced oil recovery. J Ind Eng Chem 20:228–233. https://doi.org/10.1016/j.jiec.2013.03.043
50. Lu W, Bazin B, Ma D, Liu Q, Han D, Wu K (2011) Static and dynamic adsorption of anionic and amphoteric surfactants with and without the presence of alkali. J Pet Sci Eng 77:209–218. https://doi.org/10.1016/j.petrol.2011.03.006
51. Ginzburg VV (2022) Mesoscale modeling of micellization and adsorption of surfactants and surfactant-like polymers in solution: challenges and opportunities. Ind Eng Chem Res 61:15473–15487. https://doi.org/10.1021/acs.iecr.2c01512
52. Banipal TS, Kaur H, Kaur A, Banipal PK (2016) Effect of tartarate and citrate based food additives on the micellar properties of sodium dodecylsulfate for prospective use as food emulsifier. Food Chem 1:599–606. https://doi.org/10.1016/j.foodchem.2015.05.130
53. Israelachvili JN, Mitchell DJ, Ninham BW (1976) Theory of selfassembly of hydrocarbon amphiphiles into micelles and bilayers. J Chem Soc Faraday Trans 2:1525–1568. https://doi.org/10.1039/F29767201525
54. Tanford C (1980) The hydrophobic effect: formation of micelles and biological membranes. Wiley, New York. https://doi.org/10.1002/pol.1980.130181008
55. Bhattacharya S, Haldar J (2004) Thermodynamics of micellization of multiheaded single-chain cationic surfactants. Langmuir 20:7940–7947. https://doi.org/10.1021/la0495433
56. Zhu YP, Masuyama A, Kobata Y, Nakatsuji Y, Okahara M, Rose MJ (1993) Double-chain surfactants with two carboxylate groups and their relation to similar double-chain compounds. J Colloid Interface Sci 158:40–45. https://doi.org/10.1006/jcis.1993.1226
57. Al Jahdaly BA (2023) Synthesis and characterization of a new nonionic surfactant with anti-corrosive activity of aluminum in 1.0 M hydrochloric acid solution. Chem Pap 77:33–45. https://doi.org/10.1007/s11696-022-02429-0

58. Koya PA, Khan ZA (2010) Conductometric studies of micellization of gemini surfactant pentamethylene-1,5-bis (tetradecyldimethylammonium bromide) in water and water–organic solvent mixed media. J Colloid Interface Sci 342:340–347. https://doi.org/10.1016/j.jcis.2009.10.056
59. Kumar G, Chauhan MS (2018) Conductometric investigations of surfactant behaviour in aqueous polar aprotic organic additives. J Mol Liq 1:710–715. https://doi.org/10.1016/j.molliq.2017.11.090
60. Wang J, Zhang L, Wang H, Wu C (2011) Aggregation behaviour modulation of 1-dodecyl-3-methylimidazolium bromide by organic solvents in aqueous solution. J Phys Chem B 115(17):4955–4962. https://doi.org/10.1021/jp201604u
61. Sharma V, Cantero-López P, Yañez-Osses O, Rojas-Fuentes C, Kumar A (2018) Influence of BSA on micelle formation of SDBS and CPC: an experimental–theoretical approach of its binding properties. J Mol Liq 271:443–451. https://doi.org/10.1016/j.molliq.2018.09.003
62. Sharma V, Yañez O, Alegría-Arcos M, Kumar A, Thakur RC, Cantero-López P (2020) A physicochemical and conformational study of co-solvent effect on the molecular interactions between similarly charged protein surfactant (BSA-SDBS) system. J Chem Thermodyn 1:106022. https://doi.org/10.1016/j.jct.2019.106022
63. Shah SK, Chakraborty G, Bhattarai A, De R (2022) Synergistic and antagonistic effects in micellization of mixed surfactants. J Mol Liq 29:120678. https://doi.org/10.1016/j.molliq.2022.120678
64. Wang C, Ma S, Hu Y, Wang R (2017) Hierarchical colloidal polymeric structure from surfactant-like amphiphiles in selective solvents. Langmuir 33:3427–3433. https://doi.org/10.1021/acs.langmuir.6b04509
65. Niraula TP, Shah SK, Chatterjee SK, Bhattarai A (2018) Effect of methanol on the surface tension and viscosity of sodiumdodecyl sulfate (SDS) in aqueous medium at 298.15–323.15 K. Karbala Int J Mod Sci 4:26–34. https://doi.org/10.1016/j.kijoms.2017.10.004
66. Ren H, Huang J, Luo Y, Zheng YC, Mei P, Yu WC, Lai L, Chang YL, Li FX (2016) Effect of isopropanol on the micellization of binary mixture containing amino sulfonate amphoteric surfactant in aqueous solution: mixing with sodium dodecylbenzene sulfonate. J Taiwan Inst Chem Eng 1:482–487. https://doi.org/10.1016/j.jtice.2016.05.018
67. Ren ZH, Huang J, Zheng YC, Lai L, Hu LL (2017) Effect of isopropanol on the micellization of binary mixture containing amino sulfonate amphoteric surfactant in aqueous solution: mixing with octylphenol polyoxyethylene ether (7). J Mol Liq 1:101–106. https://doi.org/10.1016/j.molliq.2017.03.083
68. Huang J, Ren ZH (2019) Micellization and interactions for ternary mixtures of amino sulfonate surfactant and nonionic octylphenol polyoxyethylene ethers in aqueous solution: 1 blending with nonionic surfactants with smaller numbers of hydrophilic unit. J Mol Liq 15:53–60. https://doi.org/10.1016/j.molliq.2019.01.035
69. Tiwari K, Saha SK (2014) Aggregation properties and thermodynamics of micellization of gemini surfactants with diethyl ether spacer in water and water–organic solvent mixed media. J Chem Thermodyn 70:24–32. https://doi.org/10.1016/j.jct.2013.10.010
70. Sheng R, Ding QY, Ren ZH, Li DN, Fan SC, Quan XF, Wang Y, Yi MT, Zhang YX, Cao YX, Wang H (2021) Interfacial and micellization behavior of binary mixture of amino sulfonate amphoteric surfactant and octadecyltrimethyl ammonium bromide: effect of short chain alcohol and its chain length. J Mol Liq 15:116064. https://doi.org/10.1016/j.molliq.2021.116064
71. Li N, Zhang YX, Ren ZH, Huang J, Li BB, Zhang QH, Yi MT, Quan XF, Wang YX, Wang BR, Qian ZB (2021) Molecular interaction for quasi-binary mixture of N-acyl amino sulfonate amphoteric surfactant from castor oil and stearyltrimethyl ammonium bromide. J Mol Liq 1:116813. https://doi.org/10.1016/j.molliq.2021.116813

Chapter 5
Toxicological Landscape of Deep Eutectic Solvents: From Cellular Impacts to Ecological Considerations

Anshu Sharma and Bong-Seop Lee

5.1 Introduction

Deep eutectic solvents (DESs) have received a lot of interest in recent years as potential green alternatives to traditional solvents [1], addressing a variety of difficulties in sustainable chemistry [2, 3], materials science [4], and biotechnology [5] in a promising way. They are useful solvents in a variety of applications, such as pharmaceuticals, electrochemistry, catalysis, and biomass processing, due to their versatility, affordability, and ease of manufacture [6]. When compared to ionic liquids (ILs), DESs have comparatively low toxicity and lesser volatility, which has led to their widespread promotion as being environmentally safe [7], and their possible breakdown and their widespread usage necessitate an in-depth knowledge of their toxicological and ecological implications [8]. The concept that all DES formulations are essentially harmless and biodegradable is excessively simplistic, as their toxicity varies greatly depending on their chemical components' molar proportions and the environmental conditions in which they decompose [9]. It is uncertain how DESs will affect the environment in the long run because some of their constituents, particularly those made from the quaternary salts of ammonium, polyols, and specific hydrogen bond donors (HBDs), may be extremely detrimental to ecosystems and species [10]. Therefore, a thorough, methodical investigation of DES toxicity profiles across a variety of biological structures and ecological contexts is necessary [11, 12].

The swift growth of DES usage has prompted research into their toxicity, leaving a vacuum in comprehending their whole environmental imprint [13]. Unlike

A. Sharma · B.-S. Lee (✉)
Department of Chemical Engineering, Kangwon National University, Chuncheon, Kangwon, Republic of Korea
e-mail: bslee0425@kangwon.ac.kr

K. Kuperkar et al. (eds.), *Deep Eutectic Solvents*, Environmental Footprints and Eco-design of Products and Processes,
https://doi.org/10.1007/978-3-032-03801-2_5

traditional solvents, DESs are created by simply mixing hydrogen bond acceptors (HBAs) and hydrogen bond donors (HBDs), producing a eutectic mixture with a substantially lower melting point than its components [14]. This technique frequently uses nontoxic and biodegradable starting ingredients, such as choline chloride (ChCl), urea (Ur), and organic acids; nevertheless, the final DES combination may have emergent hazardous attributes that are not present in the initial compounds [15]. The existence of free ions, hydrogen-bonding interactions, and the DES's physicochemical properties [16, 17] can all influence its bioavailability, cellular uptake, and toxicity to a variety of life forms [18], including bacteria, algae, plants, and higher mammals. The detrimental effect of DESs on microbial communities [19, 20] is especially significant since these organisms play an indispensable part in their biodegradation and cycling of nutrients in natural ecosystems. According to studies, certain DESs enhance the growth of bacteria and metabolic activities [21, 22], while others impede biochemical functions and cellular equilibrium [5]. Therefore, creating a comprehensive toxicological profile that takes into consideration the structural variation of DESs and how they interact with biological processes is essential.

A holistic approach that takes into account cellular, molecular, and ecological components is necessary to comprehend the toxicity of DESs. Due to their effects on membrane integrity, protein function, and oxidative stress, DESs can even cause cytotoxicity at the cellular level [23]. Certain DES formulations have been shown to damage the DNA and cause mitochondrial dysfunction, which lowers cell viability and, ultimately, leads to death. The molecular makeup of DESs frequently dictates their mode of action; DESs derived from quaternary ammonium are more cytotoxic than those made from natural ingredients like sugars and amino acids. Furthermore, DESs' cellular absorption is influenced by their hydrophilicity or hydrophobicity; hydrophobic DESs are more likely to penetrate lipid membranes and build up inside cells. For industrial applications, these mechanistic insights are essential for creating DESs with reduced biological risks while maintaining their functional qualities. To determine how environmentally friendly DESs are, several analytical tools, including the Green Star approach, the Eco-Scale tool, and life cycle assessment (LCA), examine their toxicity, biodegradability, and environmental impact [24]. Computational techniques like quantitative structure–activity relationship (QSAR) modeling and molecular docking investigations enable the logical development of safer formulations by forecasting DES toxicity based on molecular descriptors. It is still challenging to standardize toxicity assessment processes, nevertheless, because disparate test organisms, exposure situations, and methodologies make it challenging to compare results [25].

DESs can be made more sustainable by choosing less toxic and more biocompatible HBDs and HBAs. Natural deep eutectic solvents (NADESs), which are safer alternatives, have good solvent characteristics. They are made up of organic acids, carbohydrates, and amino acids, among other bioderived substances. Structural modifications, including the inclusion of hydrophilic functional groups, can improve water solubility and microbial biodegradability, which further reduces environmental risks. Similar to this, hybrid DES systems aim to balance ecological security and performance by combining biodegradable components with functional additions. In order to support the sustainable growth and use of DESs in scholastic and industrial sectors, this chapter offers a thorough overview of the toxicology landscape of DESs, emphasizing their cellular effects, ecological consequences, and methods for improving their greenness. By employing DESs as solvents in the processes of enzyme synthesis and bio-fermentation, the toxicity evaluations of DESs on these various strains can also pave the way for novel commercial uses in the future.

5.2 Mapping the Toxic Trajectory: From the Microcosm of Cells to the Macrocosm of Ecosystems

Assuming that deep eutectic solvents (DESs) are nontoxic to various organisms may be incorrect, as the formation of hydrogen bonds can generate a new supramolecular structure [26]. This structural change necessitates further investigation into DESs' potential toxicity. Additionally, the accuracy and reliability of toxicity assessments depend heavily on the testing methods employed. Currently, most studies on DES toxicity focus on cytotoxicity, cell viability assays (using human cell lines), or antimicrobial activity tests [27]. Researchers need to understand that not all toxicity assessment techniques are appropriate for DESs. For example, some of the employed procedures are not appropriate due to pronounced viscoelastic properties and phase instability of water-based solutions, among other factors. Put another way, using protocols does not always serve the intended objective. Therefore, the results drawn from these kinds of studies are just incorrect.

5.2.1 Cell Trajectory

DESs are emerging green designer solvents with very few cytotoxicity profiles. The MTT assay (3-(4,5-dimethyl-2-thiazolyl)-2,5-diphenyl-2*H*-tetrazolium bromide) is a colorimetric method that assesses cellular metabolic activity by measuring mitochondrial function, which serves as an indicator of viable cell count [28]. This technique is widely used to evaluate in vitro cytotoxicity in cell cultures. In a study examining choline chloride (ChCl)-based DES—synthesized with sugars, alcohols, and organic acids—low cytotoxicity was observed in the channel catfish ovary (CCO) cell line, particularly when compared to ionic liquids (ILs) [29]. A separate investigation evaluated the cytotoxicity of three deep eutectic solvents (DESs) using both channel catfish ovary (CCO) cells and human breast adenocarcinoma (MCF-7) cells. The results revealed that oxalic acid (OA)-based DESs, when used as a hydrogen bond donor (HBD), exhibited substantially greater toxicity (Half Maximal Effective Concentration EC_{50} <5 mM) compared to the other tested formulations (EC_{50} >10 mM). Considering cosmetic industry, a study was carried out to test the toxicity on human cell lines HaCaT (keratinocyte cells) of a non-tumoral model and MNT-1 (melanoma cells) at high doses, with [Chol]Cl- and [N_{1111}]Cl-based formulations showing potential for skin regeneration via increased viability, whereas [N_{4444}]Cl-based DESs exhibited harmful cytotoxicity [23]. Temperature fluctuations and the addition of water as a third component both disrupt the hydrogen-bonding network in deep eutectic solvents (DESs). Research indicates that aqueous DES formulations typically demonstrate lower toxicity than their anhydrous counterparts. Studies have revealed a direct relationship between cytotoxicity and DES viscosity, where higher viscosity correlates with increased toxicity. Conductor-like Screening Model for Real Solvents (COSMO-RS) computational modeling of interactions between natural deep eutectic solvents (NADESs) and membrane phospholipids revealed strong solvent–cell surface interactions. This suggests that cytotoxicity may arise primarily from cellular aggregation mechanisms [27]. In a comprehensive study applying MTT assays to 28 choline chloride (ChCl)-based DESs with various hydrogen bond donors (HBDs), including alcohols and sugars, researchers evaluated cytotoxicity in HEK-293 human kidney cells. Quantitative structure–activity relationship (QSAR) analysis identified key structural determinants of toxicity: rotatable bond number (RBN), influencing molecular flexibility and membrane permeability, and mean atomic van der Waals volume of HBD components. These factors, along with the HBA:HBD molar ratio, were shown to significantly impact cytotoxic effects [30]. Table 5.1 summarizes recent cytotoxicity data from various in vitro and in vivo assessment methods.

Table 5.1 Toxicity profile of various DESs on different living cell lines

S. no.	HBA	HBD	Add-on	Ratio	Tested cell lines	Results	References
1.	[N_{4444}] Cl	Butanoic acid		1:1	HaCaT, MNT-1	Toxic to both cell lines	[23]
2.	ChCl	Fructose Glucose (Glc) Sucrose Glycerol (Gly) Malonic acid (MA)	Water water Water Water	5:2:5 5:2:5 4:1:4 1:2:1 1:1	HelaS3, CaOV3, B16F10, MCF-7	ChCl:MA (1:1) was found most toxic on all cells lines	[31]
3.	ChCl	Ascorbic acid		2.:1	HT-29, Caco-2, MCF-7, MRC-5	Very high toxicity	[32]
4.	Tetrabutylammonium bromide (TBAB) TBAB TBAB	Gly Ethylene glycol (EG) Diethylene glycol (DEG)		1:2 1:2 1:2	CRL-1658	TBAB:Gly was found to be nontoxic	[33]

Table 5.1 (continued)

S. no.	HBA	HBD	Add-on	Ratio	Tested cell lines	Results	References
5.	ChCl ChCl ChCl ChCl	Gly Ur Glycerin (Gl) Triethylene glycol (TEG)		1:3 1:3 1:3 1:3	MCF-7, A375, OKF6, HT29, H413, PC3, HepG2	Higher IC50 values were found for ChCl:TEG, ChCl:Gly, ChCl:EG, and ChCl:Ur with OKF6, MCF-7, A375, HT29, and H413, respectively	[34]
6.	ChCl ChCl ChCl	Glc OA Gly		2:1 1:1 1:2	CCO, MCF-7	ChCl:OA shows the highest toxicity	[35]
7.	Citric acid monohydrate	Trehalose anhydrous	Water	2:1:3	*Danio rerio*	Did not show toxicity	[36]
8.	*N,N*-diethylammonium chloride (DAC) DAC DAC DAC DAC ChCl ChCl ChCl	EG Gly Ur Zinc chloride MA EG Gly Ur		1:2 1:2 1:2 1:2 1:1 1:2 1:2 1:2	HelaS3, A375, AGS, WRL-68, MCF-7, PC3	DAC:MA shows the maximum toxicity and ChCl:Gly shows the least	[37]
9.	ChCl ChCl ChCl	Malic acid (Mal) Citric acid (CA) Lactic acid Fructose Xylose Mannose		(1:1, 1.5:1) 2:1 1:1 3:2 2:1 5:2	CCO	All the DESs show relative toxicity, but ChCl:malic acid shows the highest value	[29]
10.	ChCl ChCl ChCl ChCl Betaine (Bt) Bt Bt CA CA CA	OA Ur Xylitol Sorbitol Glc Malic acid (Mal) Mal Pr Glc Fructose	Proline (Pr) Glc Gly Gly	1:1 1:2 5:2 2:3 5:2 1:1:1 1:1:1 1:1 1:1:1 1:1:1	HeLa, MCF-7, HEK293T	ChCl:OA, ChCl:Ur shows significant toxicity	[38]

5.2.2 Microorganism Trajectory

The initial toxicity assessment of choline chloride (ChCl)-based deep eutectic solvents (DESs) employed four model microorganisms—two gram-positive (*Bacillus subtilis* and *Staphylococcus aureus*) and two gram-negative (*Escherichia coli* and *Pseudomonas aeruginosa*) bacteria—along with Artemia salina lethality tests. Studies revealed distinct toxicity patterns based on the DES composition: acid-containing DESs demonstrated significant antimicrobial activity against all tested bacterial strains (*E. coli*, *S. aureus*, *Proteus mirabilis*, *P. aeruginosa*, and *Salmonella typhimurium*) and human cell lines (MCF-1, HEK293T, and HeLa), while other formulations remained nontoxic [39]. Contrastingly, in fungal models (*Lentinus tigrinus, Aspergillus niger, Phanerochaete chrysosporium,* and *Candida albicans*), DESs with basic hydrogen bond donors (HBDs) exhibited greater toxicity than acidic variants [40]. This differential sensitivity likely stems from DES interactions with cellular surfaces through ionic and van der Waals forces, potentially disrupting critical processes like respiration, growth, and spore germination—particularly relevant for the observed antifungal effects [41]. Further investigations identified key structural determinants: Increasing HBD alkyl chain length enhanced toxicity in *E. coli* but not *Listeria monocytogenes*, while molar ratios showed no consistent correlation. Notably, the HBD type exerted greater influence on toxicity than the hydrogen bond acceptor selection, with glycerol-containing ChCl-DES consistently showing minimal toxicity [42]. These findings, along with additional microbial toxicity data summarized in Table 5.2, highlight the complex relationship between DES composition and biological effects.

Table 5.2 Toxicity profile of various DESs on different microorganisms

S. no.	HBA	HBD	Add-on	Ratio	Tested microorganism	Results	References
1.	Tetrabutylammonium bromide (TBAB)	Gly EG DEG		1:2 1:2 1:2	*Escherichia coli, Pseudomonas aeruginosa, Staphylococcus aureus, Listeria monocytogenes*	TBAB:Gly has the lowest toxicity, and TBAB:EG has the highest	[33]
2.	Methyltriphenylphosphonium bromide (MTPB)	Gly EG Triethylene glycol (TEG)		1:3 1:3 1:3	*Escherichia coli, Pseudomonas aeruginosa, Bacillus subtilis, Staphylococcus aureus*	Only MTPB:ethylene glycol DESs demonstrated effective toxicity toward *Bacillus subtilis* and *Staphylococcus aureus*, whereas all DESs showed a relatively harmful effect on *Escherichia coli* and *Pseudomonas aeruginosa*	[43]
3.	Benzalkonium chloride (BC) BC	Acrylic acid Methacrylic acid		1:2.5 1:2	*Escherichia coli* ATCC 25922, *Candida albicans* ATCC 18804, *Staphylococcus aureus* NRS234	Every DES exhibited a comparatively harmful impact on the examined bacterial and fungal genera	[44]
4.	Menthol	Lactic acid		1:2	*Escherichia coli, Pseudomonas aeruginosa, Staphylococcus epidermis*	Every DES had a harmful effect on the examined bacterial genus	[45]
5.	Bt	Ur		1:1	*Escherichia coli*	DES had a comparatively harmful impact on the studied bacterial genus	[46]
6.	ChCl ChCl ChCl Bt Bt Bt	Oxalic acid (OA) Ur Xylitol Sorbitol Glucose (Glc) Malic acid	Proline	1:1 1:1 5:2 2:3 5:2 1:1:1	*Escherichia coli 3014, Proteus mirabilis 3008, Salmonella typhimurium 3064, Pseudomonas aeruginosa 3024, Staphylococcus aureus 3048, Candida albicans 86*	On the examined genus of bacteria, all DESs containing acid demonstrated a relative harmful effect. Only ChCl:oxalic acid DESs prevented *Candida albicans* from growing	[38]

(continued)

Table 5.2 (continued)

S. no.	HBA	HBD	Add-on	Ratio	Tested microorganism	Results	References
7.	ChCl	1,2-propanediol		1:2	*Staphylococcus aureus* ATCC 25923, *Clostridium perfringens* ATCC 13124, *Listeria monocytogenes* ATCC 7644, *Escherichia coli* ATCC 25922, *Salmonella* spp. ATCC 13076	For the studied bacterial genus, this DES demonstrated a relatively harmful impact	[47]
8.	ChCl ChCl ChCl ChCl ChCl ChCl ChCl ChCl ChCl ChCl	$ZnCl_2$ Ur Gly EG DEG TEG Fructose (Fru) Glc *p*-toluenesulfonic acid MA		1:2 1:2 1:3 1:3 1:2 1:3 2:1 2:1 1:3 1:1	*Phanerochaete chrysosporium*, *Aspergillus niger*, *Lentinus tigrinus*, *Candida cylindracea*	All studied fungal genera were toxically affected by zinc salts and acid-containing DESs. *A. niger*, *L. tigrinus*, and *P. chrysosporium* were not toxically affected by the other DESs. Chlorine, calcium chloride, EG, DEG, and TEG DESs exhibited a comparatively harmful impact on *C. cylindracea*. The growth of *C. cylindracea* was suppressed by EG, DEG, TEG, and fructose, whereas $ZnCl_2$, *p*-toluenesulfonic acid, and malonic acid individually showed a relatively harmful effect on all investigated fungal genera	
9.	ChCl ChCl ChCl ChCl ChCl ChCl ChCl ChCl	Gly EG OA Malic acid Glc Fru Xylose Ur		1:2 1:2 1:1 1:1 2:1 3:2 2:1 1:2	*Saccharomyces cerevisiae*	DES containing urea and acid significantly reduced yeast cell viability, resulting in a hazardous effect on the tested yeast genus	[48]

S. no.	HBA	HBD	Add-on	Ratio	Tested microorganism	Results	References
10.	ChCl ChCl ChCl ChCl ChCl	Ur Gly Ethanediol TEG 1,4-butanediol		1:2 1:2 1:2 1:4 1:4	*Kurthia gibsonii*	At a 2% concentration, the number of damaged cells was significantly increased by ChCl:Ur, ChCl:TEG, and ChCl:1,4-butanediol DESs. At a 2% concentration, ChCl:Ethanediol and, particularly, ChCl:Gly significantly reduced the viability of bacterial cells	[49]
11.	ChCl ChCl	EG MA		1:2 1:2	*Bacillus cereus* EMB20	ChCl:EG exhibited a 54% growth inhibition and a relatively harmful effect on the tested bacterial genus. All cells died as a result of the extremely poisonous ChCl:MA	[50]
12.	ChCl ChCl ChCl	Ur Gly EG		1:2 1:2 1:2	*Arthrobacter simplex* TCCC 11037	For the examined bacterial genus, all of the DESs showed a relatively toxic effect, and membrane integrity dropped to 70%, 51%, and 39% for ChCl:Gly, ChCl:EG, and ChCl:Ur, respectively	[51]

5.3 Biodegradability Analysis

The term "biodegradability" describes the process by which a chemical loses its biological activity due to breakdown or conversion by microbes and enzymes. Natural decomposition is essential to environmental sustainability because it prevents chemicals from building up in ecosystems, which lowers the risk of pollution and toxicity. Biodegradation is crucial in several industries, such as industrial chemistry, pharmaceuticals, and agriculture, where persistent chemicals can present serious environmental hazards. The ecotoxicity, accumulating potential, and overall environmental impact of a molecule are determined by how well it decomposes. A substance is considered to be "readily biodegradable" by the Organisation for Economic Co-operation and Development (OECD) rules if it passes particular screening procedures for ultimate biodegradability [52].

A popular technique for determining the easy biodegradability of organic compounds/DESs in an aerobic aquatic environment is the closed bottle test (CBT), which adheres to the OECD 301D standard [53]. Upon introducing the DES into a mineral medium that contains a microbial inoculum, usually derived from wastewater treatment plant effluent, the test determines the gradual decrease in dissolved oxygen (DO) due to microbial respiration. To prevent extra oxygen from entering the system, the test system is made up of sealed bottles with a predetermined initial oxygen concentration. While a reference component (such as sodium acetate, NaOH, or KOH) is used to validate the test, a control blank, which only includes the inoculum and no test material, is provided to account for the oxygen consumption of endogenous microbial activity. To stop any photodegradation, the test bottles are kept at a steady 20 °C in the dark for 28 days. A dissolved oxygen meter is used to measure oxygen depletion regularly, usually every 7 days. According to its theoretical oxygen demand (ThOD), the test is deemed legitimate if the reference compound degrades by at least 60% in 14 days. If a chemical undergoes ≥60% biodegradation in 28 days, it is considered rapidly biodegradable [54]. Because of its limited oxygen availability and closed system setting, the CBT may not be appropriate for volatile or poorly soluble compounds, but it is very helpful for determining the biodegradability of water-soluble compounds like DESs. It is straightforward, yet by assessing the microbial breakdown of compounds in aerobic environments, it offers important information on the possible environmental purpose of chemicals. The procedure to calculate DO concentration is described elsewhere [55].

The environmental sustainability of deep eutectic solvents (DESs) must be rigorously evaluated before they can be considered truly "green" for large-scale industrial applications. Current research reveals that biodegradability varies significantly depending on multiple factors, including DES composition, concentration, molar ratios, and environmental conditions. Among various formulations, choline chloride (ChCl)-based DESs demonstrate superior biodegradability, attributed to the inherently biodegradable nature of their components [35, 56]. Structural features play a crucial role, with hydroxyl groups and specific hydrogen bond donor (HBD) configurations particularly enhancing biodegradation. Standardized tests classify certain DESs as "readily biodegradable," with degradation exceeding 69.3% within 28 days, following a clear pattern: sugar-based > alcohol-based > acid-based DESs. Notably, the ChCl:urea (Ur) system shows exceptional performance, achieving 97.1% biodegradation [39]. However, aqueous formulations exhibit more complex behavior—while dilution reduces biodegradability below 60% for ChCl:glycerol (Gly) and ChCl:ethylene glycol (EG) systems, the ChCl:Ur mixture maintains its readily biodegradable status with even improved degradation rates [57].

A comprehensive assessment framework must account for several critical factors. First, different DES categories (natural, polymeric, hydrophobic, responsive, and pharmaceutical/therapeutic) require tailored evaluation approaches based on their intended applications. Second, environmental conditions significantly influence outcomes, as the degradative capacity varies among microbial communities (gram-positive vs. gram-negative bacteria) and between aerobic/anaerobic environments. Third, physicochemical properties, including molar ratios, alkyl chain length, molecular weight, and carbon content, all impact biodegradation kinetics [58]. These considerations highlight the need for application-specific testing protocols to accurately predict DES environmental behavior across different use scenarios.

5.4 Analytical Metric Tools to Assess Greenness

Green chemistry (GC) has received a lot of attention in recent decades since it focuses on sustainable development in laboratories and enterprises. Green analytical chemistry (GAC) is an important component of this trend since it strives to reduce the environmental and health risks associated with analytical techniques. The key difficulty in GAC is achieving a balance between decreasing the environmental effect of analytical techniques and preserving or improving the quality of analytical results.

Employing greener analytical methods is not just ecologically friendly but also economically practical, as eliminating pollution is significantly less expensive than cleaning up a contaminated environment. As a result, green analytical methodologies offer an appealing answer from both an economic and practical standpoint. Traditional GAC measures, on the other hand, frequently fail to assess the overall greenness of analytical procedures, restricting their usefulness.

To overcome these shortcomings, various greenness evaluation metrics have been developed to ensure a holistic and accurate assessment of the environmental impact of analytical processes. These newer metrics facilitate a more quantitative and structured evaluation, thereby enhancing the effectiveness of green chemistry in analytical sciences. Through continued innovation and adoption of greener analytical techniques, laboratories and industries can achieve both sustainability and high analytical performance, ultimately contributing to a safer and cleaner environment. The greenness evaluation metric tools (Fig. 5.1) that are addressed in the literature

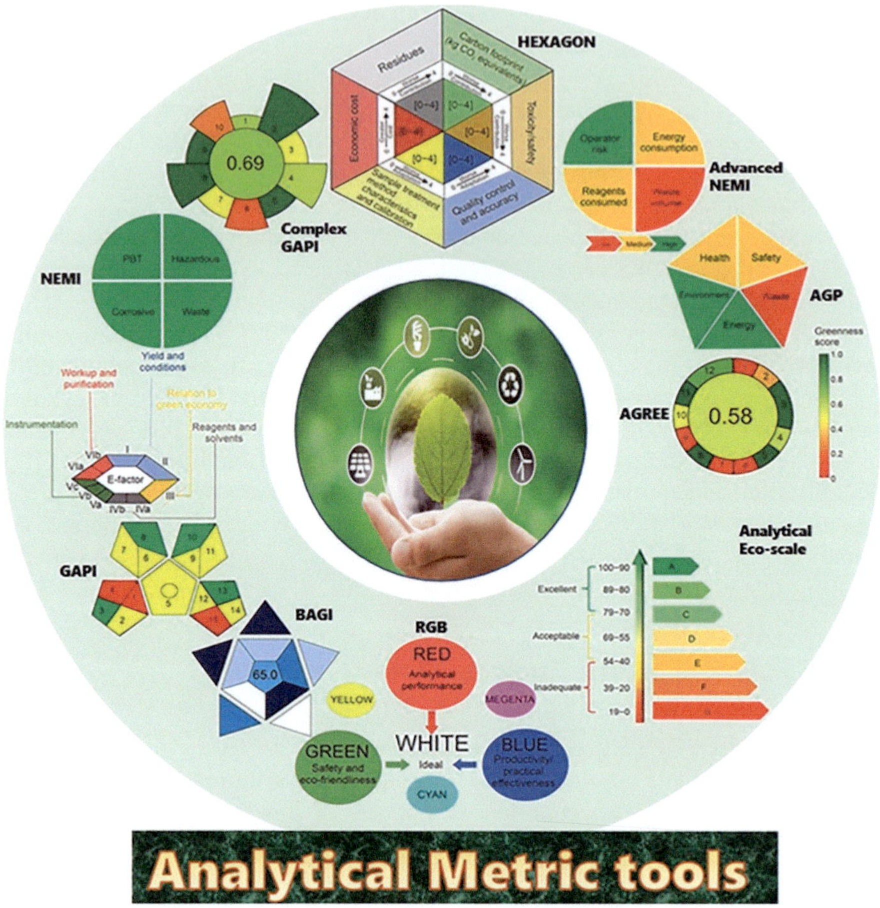

Fig. 5.1 Analytical metric tools to assess greenness

are tabulated in Table 5.3. It presents and elucidates the principles, merits, and limitations of using GAC metrics to assess the greenness of different analytical assays.

Table 5.3 presents and elucidates the principles, merits, limitations, and practical examples of using GAC metrics to assess the greenness of different analytical assays. Additionally, it highlights the current needs and future perspectives of GAC tools and provides valuable insights into the current GAC metrics.

Table 5.3 Existing analytical metric tools, with their merits and demerits for greenness evaluation

S. no.	Method	Criteria and scope	Merits	Demerits	References
1.	Chemical hazard evaluation for management strategies (CHEMS-1)	Provide a relative assessment of chemical hazards to human health and the environment, operator risks, toxicity, and environmental risks	Simple, effective, and widely used metric	CHEMS-1 only considers the hazard of reagents and has limitations in comprehensively assessing the various factors that contribute to the greenness of analytical methods	[59]
2.	National Environmental Methods Index (NEMI)	A pictogram-based tool evaluating methods based on toxicity, hazardous waste, corrosivity (pH), and persistence (PBT)	Simple pictogram-based evaluation, qualitative assessment, fast to interpret	Lacks quantitative scoring, does not provide details on solvent hazards, limited to a few parameters	[60]
3.	Advanced NEMI	Expands on NEMI with green-yellow-red scaling for semiquantitative analysis by adding energy consumption and environmental impact scoring	Provides better visualization with semiquantitative assessment, including energy factor, which is more comprehensive than NEMI	Still lacks full quantitative scoring, limited information on sources of threats, requires manual evaluation	[61]
4.	Assessment of Green Profile (AGP)	An extension of NEMI that assesses safety, health, energy, waste, and environmental impact using a five-section pictogram	Provides a broader environmental impact assessment, considers waste and energy use	Subjective color coding, lacks a detailed numerical assessment	[62]
5.	Analytical Eco-Scale	Uses a penalty point system for nongreen parameters (solvents, energy, and waste) with a maximum score of 100	Quantitative, easy to use, suitable for most analytical methods	Does not provide hazard source information, lacks a full environmental impact breakdown	[63]
6.	High-Performance Liquid Chromatography (HPLC)–Environmental Assessment Tool (EAT)	Focuses on HPLC solvent toxicity, energy use, and health risks	Free software-based tool, specific to liquid chromatography (LC)	Limited to LC methods, does not assess sample preparation, instrumentation	[64]

(continued)

Table 5.3 (continued)

S. no.	Method	Criteria and scope	Merits	Demerits	References
7.	Analytical Method Volume Intensity (AMVI)	Calculates solvent consumption per chromatographic peak, a higher value means a worse impact	Quantifies waste, simple to calculate, effective for liquid chromatography	Ignores solvent toxicity, does not assess method efficiency	[65]
8.	Analytical Method Greenness Score (AMGS)	Score-based tool evaluating solvents, energy consumption, safety, and waste	Comprehensive quantitative approach, software-based	Complex calculations, requires detailed input data	[66]
9.	Green Analytical Procedure Index (GAPI)	Comprehensive pictogram-based tool, evaluates sample preparation, solvents, instrumentation, waste, and energy consumption	Covers all process steps, provides a holistic approach	Complex interpretation, limited detail on specific hazards, some subjectivity in scoring	[67]
10.	Complex GAPI	Enhances GAPI by adding pre-analytical steps, economic aspects, and instrumentation details	More detailed than GAPI, accounts for E-factor and sustainability	Requires extensive data collection, difficult to implement	[68]
11.	Red-Green-Blue (RGB) Additive Color Model	Color-based tool that assesses methods using red (performance), green (safety), and blue (practicality)	Simple visual representation, effective for quick assessments	Limited quantitative output, may overlook some method details	[69]
12.	RGB 12 Algorithm	An extended version of the RGB model that incorporates 12 criteria for greenness evaluation	More comprehensive than the standard RGB model	Requires Excel-based processing, not widely adopted	[70]
13.	Analytical Greenness (AGREE)	Uses 12 GAC principles, assigns a 0–1 scale per criterion, produces a circular pictogram	Comprehensive, visually intuitive, balances different green chemistry factors	The weighting system is not fully transparent, requires subjective input	[71]
14.	Analytical greenness metric for sample preparation (AGREEprep)	Adaptation of AGREE for sample preparation with ten weighted factors (solvents, waste, and energy)	Best quantitative tool for assessing pre-analytical stages, customizable weight distribution	Lacks method performance metrics, does not fully balance greenness vs. accuracy	[72]
15.	Blue Applicability Grade Index (BAGI)	Uses scores, pictograms, and colors to assess method practicality and greenness	Rapid evaluation, easy color-coded output	Does not include safety, health, and environment (SHE) assessment	[73]

Table 5.3 (continued)

S. no.	Method	Criteria and scope	Merits	Demerits	References
16.	ChlorTox Scale	Evaluates the hazards of chemicals using chloroform as a reference	Provides direct risk values for chemical safety	Does not evaluate waste treatment or method efficiency	[74]
17.	HEXAGON	Multi-criteria metric evaluating carbon footprint, cost, safety, and figures of merit in a hexagonal pictogram	Standardized assessment, considers economic and safety impacts	Complex evaluation process, requires more data collection	[75]
18.	Other green assessment approaches	Include TOPSIS, PROMETHEE, Spider Diagram, Hasse Diagram, etc.	Multi-criteria decision analysis (MCDA) techniques for ranking green methods	Not widely adopted, require statistical expertise	[76]

5.5 Conclusion

DESs are often assumed to be nontoxic due to their composition, which typically consists of natural, biodegradable, and low-toxicity compounds, such as sugars, natural carboxylic acids, ChCl, and amino acids. However, despite their perceived low toxicity and biodegradability, the long-term environmental impact of DESs remains largely unexamined, and regulatory frameworks primarily focus on traditional solvents and ILs, leaving DESs largely unregulated. Furthermore, their interactions with environmental pollutants, such as heavy metals and microplastics, are not well-understood, raising concerns about their broader ecological consequences.

Addressing these gaps requires a multidisciplinary approach involving chemists, toxicologists, environmental scientists, and policymakers to establish standardized guidelines for DES toxicity, biodegradability, and environmental safety. While DESs have been widely promoted as a green alternative to conventional solvents, they cannot be universally classified as readily biodegradable without comprehensive biodegradation studies. Research must be expanded to include novel DES formulations, ensuring that they meet stringent biodegradability standards before widespread industrial application. In addition, their toxicity and biodegradable testing procedures must be standardized, to establish physicochemical parameters of DES, such as density and viscosity. This standardization will aid in determining whether a DES can actually be termed a green solvent, unlike relying entirely on misconceptions about its constituent components.

A key limitation in current DES biodegradability and toxicity assessments is the overreliance on Material Safety Data Sheets (MSDS) to determine the degradation properties of DES components. The techniques employed for MSDS may differ greatly from those used in specialized DES biodegradation studies, resulting in potentially misleading outcomes. Furthermore, cross-reactivity among DESs and salts or minerals in media for cultivation has not been taken into account in toxicity studies, making it difficult to determine their real environmental impact. Another overlooked factor is the preadaptation of microbial cells to DES exposure, which has not been systematically tested in biodegradability studies.

Moreover, most microbial strains used in previous DES biodegradability assessments are either microaerobic, anaerobic, or capable of switching their metabolism between oxic and anoxic conditions. Such adaptations to metabolism are rarely explored in existing biodegradability research, potentially leading to incorrect assessments of DES persistence in various environmental situations. To fully realize the potential benefits of DESs and expand their use across diverse industries, an extensive study is required to deepen our knowledge of their biodegradation processes while creating DES formulations with higher biodegradability.

Finally, extensive toxic and biological degradation studies should become a fundamental element of DES studies, ensuring that these solvents keep their green chemistry credentials while reducing their environmental impact. Creating clear regulatory requirements is critical to ensuring that DESs can be broadly deployed in a manner that is secure and environmentally friendly. All of the advantages of DESs may only be realized while limiting potential environmental hazards through formulation optimization, comprehensive toxicity testing, and an understanding of biodegradability mechanisms.

Acknowledgments This research was supported by the Basic Science Research Program of the National Research Foundation (NRF) of Korea, funded by the Ministry of Education (Grant No. NRF-2020R1I1A2069224).

Competing Interests The authors declare no competing interests.

References

1. Prabhune A, Dey R (2023) Green and sustainable solvents of the future: deep eutectic solvents. J Mol Liq 379:121676. https://doi.org/10.1016/j.molliq.2023.121676
2. Liu J, Li X, Row KH (2022) Development of deep eutectic solvents for sustainable chemistry. J Mol Liq 362:119654. https://doi.org/10.1016/j.molliq.2022.119654
3. Sharma A, Lee B-S, Shin HY (2025) Assessing suitability of glycerol-derived green solvent for the separation of n-hexane+ ethanol azeotropic mixture, accompanied by VLE studies using machine learning. J Mol Liq 423:127003. https://doi.org/10.1016/j.molliq.2025.127003
4. Ge X, Gu C, Wang X, Tu J (2017) Deep eutectic solvents (DESs)-derived advanced functional materials for energy and environmental applications: challenges, opportunities, and future vision. J Mater Chem A 5:8209. https://doi.org/10.1039/C7TA01659J
5. Mbous YP, Hayyan M, Hayyan A, Wong WF, Hashim MA, Looi CY (2017) Applications of deep eutectic solvents in biotechnology and bioengineering—promises and challenges. Biotechnol Adv 35:105. https://doi.org/10.1016/j.biotechadv.2016.11.006
6. Farooq MQ, Abbasi NM, Anderson JL (2020) Deep eutectic solvents in separations: methods of preparation, polarity, and applications in extractions and capillary electrochromatography. J Chromatogr A 1633:461613. https://doi.org/10.1016/j.chroma.2020.461613

7. Płotka-Wasylka J, De la Guardia M, Andruch V, Vilková M (2020) Deep eutectic solvents vs ionic liquids: similarities and differences. Microchem J 159:105539. https://doi.org/10.1016/j.microc.2020.105539
8. Hayyan M, Hashim MA, Hayyan A, Al-Saadi MA, AlNashef IM, Mirghani ME, Saheed OK (2013) Are deep eutectic solvents benign or toxic? Chemosphere 90:2193. https://doi.org/10.1016/j.chemosphere.2012.11.004
9. Sharma A, Lee B-S (2024) Toxicity test profile for deep eutectic solvents: a detailed review and future prospects. Chemosphere 350:141097. https://doi.org/10.1016/j.chemosphere.2023.141097
10. Mbous YPV (2017) University of Malaya (Malaysia)
11. Ullah N, Haseeb A, Tuzen M (2024) Application of recently used green solvents in sample preparation techniques: a comprehensive review of existing trends, challenges, and future opportunities. Crit Rev Anal Chem 54:2714. https://doi.org/10.1080/10408347.2023.2197495
12. Sharma A, Sharma MMM, Sharma P, Husen A (2024) Plant response to silver nanoparticles: plant growth, development, production, and protection. Springer, Berlin, pp 123–136
13. Abdelrahman MM (2022) Green chemical analysis and sample preparations: procedures, instrumentation, data metrics, and sustainability. Springer, Berlin, pp 29–99
14. Martins MA, Pinho SP, Coutinho JA (2019) Insights into the nature of eutectic and deep eutectic mixtures. J Solut Chem 48:962. https://doi.org/10.1007/s10953-018-0793-1
15. Pohanish RP (2019) Sittig's handbook of toxic and hazardous chemicals and carcinogens. William Andrew, Norwich
16. Sharma A, Li L, Garg A, Lee BS (2024) Probe into the volumetric properties of binary mixtures: essence of regression-based machine learning algorithms. J Mol Liq 399:124498. https://doi.org/10.1016/j.molliq.2024.124498
17. Verma S, Sharma A, Maken S (2023) Volumetric, transport and acoustic properties of binary mixtures containing alkanol at 298.15–318.15 K. J Mol Liq 390:123029. https://doi.org/10.1016/j.molliq.2023.123029
18. Torregrosa-Crespo J, Marset X, Guillena G, Ramón DJ, Martínez-Espinosa RM (2020) New guidelines for testing "deep eutectic solvents" toxicity and their effects on the environment and living beings. Sci Total Environ 704:135382. https://doi.org/10.1016/j.scitotenv.2019.135382
19. Sha H, Cao S, Zhao B, Dong Z, Wang G, Duan J (2024) Effect of alkaline deep eutectic solvents pretreatment on CH4 yield from anaerobic digestion of corn stover. Energy 302:131683. https://doi.org/10.1016/j.energy.2024.131683
20. Qiao Y, Cai H-L, Yang X, Zang Y-Y, Chen Z-G (2018) Effects of natural deep eutectic solvents on lactic acid bacteria viability during cryopreservation. Appl Microbiol Biotechnol 102:5695. https://doi.org/10.1007/s00253-018-8996-3
21. Xie Y, Liu X, Liu L, Zhou Y, Wang Z, Huang C, He H, Zhai Y (2024) Deep eutectic solvents pretreatment enhances methane production from anaerobic digestion of waste activated sludge: effectiveness evaluation and mechanism elucidation. J Environ Manag 356:120615. https://doi.org/10.1016/j.jenvman.2024.120615
22. Cao W, Fang S, Wu Q, Wu Y, Feng L, Xie Z, Feng Q, Fang F, Xu Z, Luo J (2023) Choline chloride pretreatment on volatile fatty acids promotion from sludge anaerobic fermentation: in-situ deep eutectic solvents-like formation for EPS disintegration and associated microbial functional profiles upregulation. Chem Eng J 467:143556. https://doi.org/10.1016/j.cej.2023.143556
23. Macário I, Oliveira H, Menezes A, Ventura S, Pereira J, Gonçalves A, Coutinho J, Gonçalves F (2019) Cytotoxicity profiling of deep eutectic solvents to human skin cells. Sci Rep 9:3932. https://doi.org/10.1038/s41598-019-39910-y
24. Ferreira IJ, Oliveira F, Jesus AR, Paiva A, Duarte ARC (2022) Current methodologies for the assessment of deep eutectic systems toxicology: challenges and perspectives. J Mol Liq 362:119675. https://doi.org/10.1016/j.molliq.2022.119675

25. García CB, Concha J, Culleré L, Lomba L, Sangüesa E, Ribate MP (2023) Has the toxicity of therapeutic deep eutectic systems been assessed? Appl Sci 13:5980. https://doi.org/10.3390/app13105980
26. Gutiérrez MC, Ferrer ML, Yuste L, Rojo F, del Monte F (2010) Bacteria incorporation in deep-eutectic solvents through freeze-drying. Angew Chem 49:2158. https://doi.org/10.1002/ange.200905212
27. Martínez GM, Townley GG, Martínez-Espinosa RM (2022) Controversy on the toxic nature of deep eutectic solvents and their potential contribution to environmental pollution. Heliyon 8:e12567. https://doi.org/10.1016/j.heliyon.2022.e12567
28. Stockert JC, Horobin RW, Colombo LL, Blázquez-Castro A (2018) Tetrazolium salts and formazan products in cell biology: viability assessment, fluorescence imaging, and labeling perspectives. Acta Histochem 120:159. https://doi.org/10.1016/j.acthis.2018.02.005
29. Radošević K, Železnjak J, Bubalo MC, Redovniković IR, Slivac I, Srček VG (2016) Comparative in vitro study of cholinium-based ionic liquids and deep eutectic solvents toward fish cell line. Ecotoxicol Environ Saf 131:30. https://doi.org/10.1016/j.ecoenv.2016.05.005
30. Ahmadi R, Hemmateenejad B, Safavi A, Shojaeifard Z, Mohabbati M, Firuzi O (2018) Assessment of cytotoxicity of choline chloride-based natural deep eutectic solvents against human HEK-293 cells: a QSAR analysis. Chemosphere 209:831. https://doi.org/10.1016/j.chemosphere.2018.06.103
31. Hayyan M, Mbous YP, Looi CY, Wong WF, Hayyan A, Salleh Z, Mohd-Ali O (2016) Natural deep eutectic solvents: cytotoxic profile. Springerplus 5(1):913. https://doi.org/10.1186/s40064-016-2575-9
32. Popović BM, Gligorijević N, Aranđelović S, Macedo AC, Jurić T, Uka D, Mocko-Blažek K, Serra AT (2023) Cytotoxicity profiling of choline chloride-based natural deep eutectic solvents. RSC Adv 13:3520. https://doi.org/10.1039/D2RA07488E
33. Inayat S, Ahmad SR, Awan SJ, Nawshad M, Ali Q (2023) In vivo and in vitro toxicity profile of tetrabutylammonium bromide and alcohol-based deep eutectic solvents. Sci Rep 13:1777. https://doi.org/10.1038/s41598-023-28928-y
34. Hayyan M, Looi CY, Hayyan A, Wong WF, Hashim MA (2015) In vitro and in vivo toxicity profiling of ammonium-based deep eutectic solvents. PLoS One 10:e0117934. https://doi.org/10.1371/journal.pone.0117934
35. Radošević K, Bubalo MC, Srček VG, Grgas D, Dragičević TL, Redovniković IR (2015) Evaluation of toxicity and biodegradability of choline chloride based deep eutectic solvents. Ecotoxicol Environ Saf 112:46. https://doi.org/10.1016/j.ecoenv.2014.09.034
36. Ferreira IJ, Meneses L, Paiva A, Diniz M, Duarte ARC (2022) Assessment of deep eutectic solvents toxicity in zebrafish (Danio rerio). Chemosphere 299:134415. https://doi.org/10.1016/j.chemosphere.2022.134415
37. Mbous YP, Hayyan M, Wong WF, Hayyan A, Looi CY, Hashim MA (2020) Simulation of deep eutectic solvents' interaction with membranes of cancer cells using COSMO-RS. J Phys Chem B 124:9086. https://doi.org/10.1021/acs.jpcb.0c04801
38. Radošević K, Čanak I, Panić M, Markov K, Bubalo MC, Frece J, Srček VG, Redovniković IR (2018) Antimicrobial, cytotoxic and antioxidative evaluation of natural deep eutectic solvents. Environ Sci Pollut Res 25:14188. https://doi.org/10.1007/s11356-018-1669-z
39. Zhao B-Y, Xu P, Yang F-X, Wu H, Zong M-H, Lou W-Y (2015) Biocompatible deep eutectic solvents based on choline chloride: characterization and application to the extraction of rutin from Sophora japonica. ACS Sustain Chem Eng 3:2746. https://doi.org/10.1021/acssuschemeng.5b00619
40. Juneidi I, Hayyan M, Mohd Ali O (2016) Toxicity profile of choline chloride-based deep eutectic solvents for fungi and Cyprinus carpio fish. Environ Sci Pollut Res 23:7648. https://doi.org/10.1007/s11356-015-6003-4
41. Torri C, Samorì C, Adamiano A, Fabbri D, Faraloni C, Torzillo G (2011) Preliminary investigation on the production of fuels and bio-char from Chlamydomonas reinhardtii biomass residue after bio-hydrogen production. Bioresour Technol 102:8707. https://doi.org/10.1016/j.biortech.2011.01.064

42. Ghaedi H, Ayoub M, Sufian S, Hailegiorgis S, Krishnan S (2017) Toxicity of several potassium carbonate and phosphonium-based deep eutectic solvents towards Escherichia coli and Listeria monocytogenes bacteria. J Environ Anal Toxicol 7:0525. https://doi.org/10.4172/2161-0525
43. Hayyan M, Hashim MA, Al-Saadi MA, Hayyan A, AlNashef IM, Mirghani ME (2013) Assessment of cytotoxicity and toxicity for phosphonium-based deep eutectic solvents. Chemosphere 93:455. https://doi.org/10.1016/j.chemosphere.2013.05.013
44. Wang J, Xue J, Dong X, Yu Q, Baker SN, Wang M, Huang H (2020) Antimicrobial properties of benzalkonium chloride derived polymerizable deep eutectic solvent. Int J Pharm 575:119005. https://doi.org/10.1016/j.ijpharm.2019.119005
45. Alsaud N, Shahbaz K, Farid M (2021) Antioxidant and antibacterial evaluation of Manuka leaves (Leptospermum scoparium) extracted by hydrophobic deep eutectic solvent. Chem Eng Res Des 174:96. https://doi.org/10.1016/j.cherd.2021.08.004
46. Liang Y, Pan Z, Chen Z, Fei Y, Zhang J, Yuan J, Zhang L, Zhang J (2020) Ultrasound-assisted natural deep eutectic solvents as separation-free extraction media for hydroxytyrosol from olives. ChemistrySelect 5:10939. https://doi.org/10.1002/slct.202002026
47. Wojeicchowski JP, Marques C, Igarashi-Mafra L, Coutinho JA, Mafra MR (2021) Extraction of phenolic compounds from rosemary using choline chloride–based deep eutectic solvents. Sep Purif Technol 258:117975. https://doi.org/10.1016/j.seppur.2020.117975
48. Bubalo MC, Mazur M, Radošević K, Redovniković IR (2015) Baker's yeast-mediated asymmetric reduction of ethyl 3-oxobutanoate in deep eutectic solvents. Process Biochem 50(11):1788. https://doi.org/10.1016/j.procbio.2015.07.015
49. Peng F, Chen Q-S, Li F-Z, Ou X-Y, Zong M-H, Lou W-Y (2020) Using deep eutectic solvents to improve the biocatalytic reduction of 2-hydroxyacetophenone to (R)-1-phenyl-1, 2-ethanediol by Kurthia gibsonii SC0312. Mol Catal 484:110773. https://doi.org/10.1016/j.mcat.2020.110773
50. Sadaf A, Kumari A, Khare S (2018) Potential of ionic liquids for inhibiting the growth and β-lactamase production by Bacillus cereus EMB20. Int J Biol Macromol 107:1915. https://doi.org/10.1016/j.ijbiomac.2017.10.053
51. Mao S, Yu L, Ji S, Liu X, Lu F (2016) Evaluation of deep eutectic solvents as co-solvent for steroids 1-en-dehydrogenation biotransformation by Arthrobacter simplex. J Chem Technol Biotechnol 91:1099. https://doi.org/10.1002/jctb.4691
52. Khamidulina KK, Tarasova EV, Lastovetsky ML (2024) Prediction of the biodegradation of chemicals using OECD QSAR Toolbox software. Toxicol Rev 32:20. https://doi.org/10.47470/0869-7922-2024-32-1-20-30
53. Pagga U (1997) Testing biodegradability with standardized methods. Chemosphere 35:2953. https://doi.org/10.1016/S0045-6535(97)00262-2
54. Coleman D, Gathergood N (2010) Biodegradation studies of ionic liquids. Chem Soc Rev 39:600. https://doi.org/10.1039/B817717C
55. Ferro Orozco AM, Contreras EM, Zaritzky NE (2016) Monitoring the biodegradability of bisphenol A and its metabolic intermediates by manometric respirometry tests. Biodegradation 27:209. https://doi.org/10.1007/s10532-016-9767-4
56. Juneidi I, Hayyan M, Hashim MA (2015) Evaluation of toxicity and biodegradability for cholinium-based deep eutectic solvents. RSC Adv 5:83636. https://doi.org/10.1039/C5RA12425E
57. Candia-Lomelí M, Covarrubias-Garcia I, Aizpuru A, Arriaga S (2023) Preparation and physicochemical characterization of deep eutectic solvents and ionic liquids for the potential absorption and biodegradation of styrene vapors. J Hazard Mater 441:129835. https://doi.org/10.1016/j.jhazmat.2022.129835
58. Azzouz A, Hayyan M (2023) Are deep eutectic solvents biodegradable? Process Saf Environ Prot 176:1021. https://doi.org/10.1016/j.psep.2023.06.079
59. Swanson MB, Davis GA, Kincaid LE, Schultz TW, Bartmess JE, Jones SL, George EL (1997) A screening method for ranking and scoring chemicals by potential human health and environmental impacts. Environ Toxicol Chem 16:372. https://doi.org/10.1002/etc.5620160237

60. Yin L, Yu L, Guo Y, Wang C, Ge Y, Zheng X, Zhang N, You J, Zhang Y, Shi M (2024) Green analytical chemistry metrics for evaluating the greenness of analytical procedures. J Pharm Anal 14:101013. https://doi.org/10.1016/j.jpha.2024.101013
61. de la Guardia M, Garrigues S (2012) Handbook of green analytical chemistry. Wiley Online Library, New York
62. Kammoun AK, Khayat MT, Almalki AJ, Youssef RM (2023) Development of validated methods for the simultaneous quantification of Finasteride and Tadalafil in newly launched FDA-approved therapeutic combination: greenness assessment using AGP, analytical eco-scale, and GAPI tools. RSC Adv 13:11817. https://doi.org/10.1039/D3RA01437A
63. Gałuszka A, Migaszewski ZM, Konieczka P, Namieśnik J (2012) Analytical eco-scale for assessing the greenness of analytical procedures. TrAC Trends Anal Chem 37:61. https://doi.org/10.1016/j.trac.2012.03.013
64. Gaber Y, Törnvall U, Kumar M, Amin MA, Hatti-Kaul R (2011) HPLC-EAT (Environmental Assessment Tool): a tool for profiling safety, health and environmental impacts of liquid chromatography methods. Green Chem 13:2021. https://doi.org/10.1039/C0GC00667J
65. Hartman R, Helmy R, Al-Sayah M, Welch CJ (2011) Analytical method volume intensity (AMVI): a green chemistry metric for HPLC methodology in the pharmaceutical industry. Green Chem 13:934. https://doi.org/10.1039/C0GC00524J
66. Hicks MB, Farrell W, Aurigemma C, Lehmann L, Weisel L, Nadeau K, Lee H, Moraff C, Wong M, Huang Y (2019) Making the move towards modernized greener separations: introduction of the analytical method greenness score (AMGS) calculator. Green Chem 21:1816. https://doi.org/10.1039/C8GC03875A
67. Płotka-Wasylka J (2018) A new tool for the evaluation of the analytical procedure: Green Analytical Procedure Index. Talanta 181:204. https://doi.org/10.1016/j.talanta.2018.01.013
68. Mansour FR, Omer KM, Płotka-Wasylka J (2024) A total scoring system and software for complex modified GAPI (ComplexMoGAPI) application in the assessment of method greenness. Green Anal Chem 10:100126. https://doi.org/10.1016/j.greeac.2024.100126
69. Nowak PM, Arduini F (2024) RGBfast—a user-friendly version of the Red-Green-Blue model for assessing greenness and whiteness of analytical methods. Green Anal Chem 10:100120. https://doi.org/10.1016/j.greeac.2024.100120
70. Nowak PM, Wietecha-Posłuszny R, Pawliszyn J (2021) White analytical chemistry: an approach to reconcile the principles of green analytical chemistry and functionality. TrAC Trends Anal Chem 138:116223. https://doi.org/10.1016/j.trac.2021.116223
71. Pena-Pereira F, Wojnowski W, Tobiszewski M (2020) AGREE—Analytical GREEnness metric approach and software. Anal Chem 92:10076. https://doi.org/10.1021/acs.analchem.0c01887
72. Wojnowski W, Tobiszewski M, Pena-Pereira F, Psillakis E (2022) AGREEprep—analytical greenness metric for sample preparation. TrAC Trends Anal Chem 149:116553. https://doi.org/10.1016/j.trac.2022.116553
73. Manousi N, Wojnowski W, Płotka-Wasylka J, Samanidou V (2023) Blue applicability grade index (BAGI) and software: a new tool for the evaluation of method practicality. Green Chem 25:7598. https://doi.org/10.1039/D3GC02347H
74. Nowak PM, Wietecha-Posłuszny R, Płotka-Wasylka J, Tobiszewski M (2023) How to evaluate methods used in chemical laboratories in terms of the total chemical risk?—a ChlorTox Scale. Green Anal Chem 5:100056. https://doi.org/10.1016/j.greeac.2023.100056
75. Ballester-Caudet A, Campíns-Falcó P, Pérez B, Sancho R, Lorente M, Sastre G, González C (2019) A new tool for evaluating and/or selecting analytical methods: summarizing the information in a hexagon. TrAC Trends Anal Chem 118:538. https://doi.org/10.1016/j.trac.2019.06.015
76. Shen Y, Lo C, Nagaraj D, Farinato R, Essenfeld A, Somasundaran P (2016) Development of Greenness Index as an evaluation tool to assess reagents: evaluation based on SDS (Safety Data Sheet) information. Miner Eng 94:1. https://doi.org/10.1016/j.mineng.2016.04.015

Chapter 6
Extraction and Separation Processes Using Deep Eutectic Solvents

Darshna Hirpara, Vishwajit Chavda, Sanjeev Kumar, and Kushan Parikh

6.1 Introduction

Various extraction and separation methods may be divided broadly into traditional and modern approaches [1–3]. Traditional techniques include percolation, solvent extraction, maceration, and Soxhlet extraction [4]. These methods have been considered complicated with long extraction processes; they require high volumes of solvents, which increases the cost, and yields are usually low due to thermal degradation. Besides that, these methods require harmful and flammable solvents, and there are also problems with the removal and reprocessing of the solvents. The new methods of extraction include subcritical fluid extraction, ultrasound-assisted extraction (UAE), high-pressure-assisted extraction (HPAE), enzyme-assisted extraction, supercritical fluid extraction (SFE), and microwave-assisted extraction (MAE) [4, 5]. These newer processes typically tend to exhibit higher efficiency, less time of extraction, lesser costs, and excessive purity of the extracted analytes. Despite such benefits, both traditional and modern methods have their setbacks concerning the toxicity of solvents, polarity, thermal instability, selectivity, and solubility issues. Other variables that influence the yield of extraction are the type

D. Hirpara · S. Kumar
Department of Applied Chemistry, Faculty of Technology and Engineering, The Maharaja Sayajirao University of Baroda, Vadodara, Gujarat, India

V. Chavda
Department of Chemistry, Institute of Science, Humanities and Liberal Studies, Indus University, Shilaj, Ahmedabad, Gujarat, India

K. Parikh (✉)
Department of Chemical Science, Faculty of Applied Sciences, Parul University, Vadodara, India

K. Kuperkar et al. (eds.), *Deep Eutectic Solvents*, Environmental Footprints and Eco-design of Products and Processes,
https://doi.org/10.1007/978-3-032-03801-2_6

and amount of the solvent, water content, recovery of bioactive components, and structural form variation of a compound induced by ionization, oxidation, or hydrolysis [6].

Water is considered the primary solvent employed by the agricultural, pharmaceutical, food, and cosmetic industries due to its physicochemical characteristics, which fulfill the criteria of the FDA and European Union standards [5]. However, this is mainly suitable for hydrophilic and polar substances, thus turning out to be inefficient in extracting nonpolar and aquaphobic compounds. Moreover, impurities from water-soluble substances could pose additional problems with purification [2]. Over recent years, research on alternative chemical methodologies with a low environmental impact has attracted much attention as practical alternatives to conventional chemical processes.

One of the strategies is using eco-friendly solvents to replace harmful traditional solvents in industrial applications. Among them, ionic liquids (ILs) have generated much interest due to their interesting physicochemical characteristics [7, 8]. ILs are typically salts with melting temperatures below 100 °C, which consist of asymmetric, large organic cations and an organic or inorganic anion [9, 10]. These solvents possess beneficial properties, such as nonflammability, good chemical and thermal stability, and less volatility [7, 11]. However, there are several disadvantages of ILs, like their high cost of production, poor biodegradability, and toxicity [11]. Deep eutectic solvents (DESs) have become a viable substitute for ILs owing to these drawbacks.

The physicochemical characteristics of DESs exhibit a prominent similarity to those of ILs [11]. Nevertheless, it has been observed that DESs possess superior biodegradability, reduced toxicity, and enhanced cost-effectiveness compared to ILs [11–13]. Abbott and team [14] defined DESs as solvents that resulted from combining two or more Lewis and/or Brønsted–Lowry bases and acids, with a resultant depression in the freezing point, considerably lower than each component. DESs are eutectic liquids formed by the blending of a quaternary salt (hydrogen bond acceptor (HBA); e.g., choline chloride (ChCl)) with a hydrogen bond donor (HBD), a recently developed category of green solvents [7]. DESs are widely exploited for extracting and separating specific compounds from complex mixtures. Most DESs are cost-effective, easily synthesized, and biodegradable, leading to many innovative DES formulations [7].

Different novel DESs were recently invented based on primary metabolites and bio-renewable feedstock compounds, like sugars, organic acids, amino acids, and sugar alcohols [15, 16]. This novel class of solvents, termed "natural deep eutectic solvents" (NADESs), has explained several puzzling phenomena, which were up to now poorly understood, like the biosynthesis of water- and lipid-insoluble intermediate-polarity compounds [15–17]. DESs and NADESs have found important uses in biotechnology and the chemical industry, mainly for their promise as "green" solvents in extraction and separation processes [18–20]. With their applications covering more ground, DESs increasingly become promising alternatives to volatile and hazardous organic solvents [7, 12, 21]. Until now, several DESs have been elaborately designed for extraction and separation and can be effectively utilized in place of conventional organic solvents under certain conditions. Such techniques based on DESs have been validated for various compounds, such as metals, bioactive compounds, biodiesel, and aromatic hydrocarbons, for extraction and separation [22, 23]. Recently, DESs, containing NADESs and their derivatives, have been applied as wash agents for lead-contaminated soil remediation [24, 25].

In addition, DESs function as effective media for modifying materials such as polymers and silica, which, when treated with DESs, exhibit improved extraction and separation efficiency due to distinctive molecular interactions, including π–π interactions, anion exchange, and hydrogen bonding [26, 27]. Given their adaptability and environmentally friendly characteristics, DESs have found applications across diverse domains of chemistry, including electrochemistry [7, 28], organic synthesis [29, 30], wastewater treatment [31–33], pharmaceuticals [34, 35], fluorescent sensors [36], micellization [37–39], gas separation [40], solubilization [41], and enzymatic reactions [42, 43]. Other research studies have also explored their application in extracting and separating aromatic hydrocarbons, biodiesel, gases, and bioactive substances. This chapter offers a detailed discussion of the significance of DESs in extraction and separation processes (Fig. 6.1).

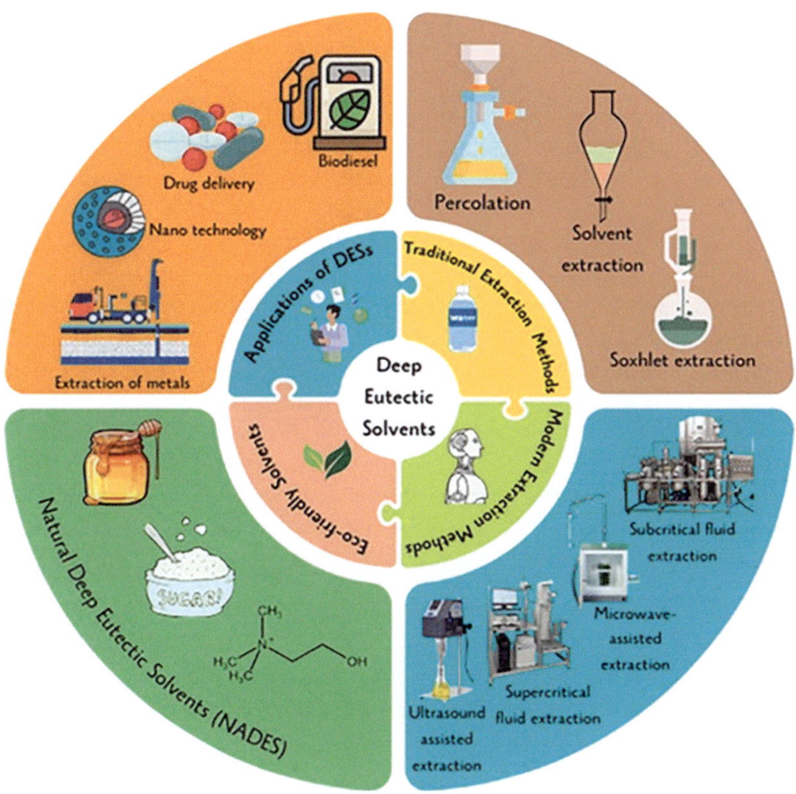

Fig. 6.1 Illustrative representation of deep eutectic solvents (DESs)

6.2 Key Characteristics of DESs as Extractive Agents

DESs have been recognized as greener solvents with tunable physicochemical characteristics. Two crucial factors affect the extraction efficiency: viscosity and polarity [44, 45]. Both factors significantly impact the direct interactions between DESs and target analytes. High viscosity has been considered one of the main drawbacks since it may hamper the diffusion of target analytes from matrices into the extractant [46]. However, the DES's polarity has been very influential in determining the solubility of bioactive compounds [44]. In general, DES formulations consider the type of salt and HBD utilized, with a specific molar ratio, which mainly affects their viscosity and polarity [47, 48].

DESs have higher viscosities than ILs and other molecular solvents [7, 8]. This is believed to result from high hydrogen-bonding forces between the constituents that restrict molecular mobility in the solvent system. Possible electrostatic interactions and van der Waals interactions among molecules can also support this. For developing DESs of low viscosity, the "hole theory" has been proposed as an approach [7]. According to this concept, molecular instability causes the formation of random free volumes or "holes" in the structure of the solvent, which is responsible for determining the overall viscosity. Incorporating fluorinated HBDs or small cations has proven to be an attractive strategy for viscosity reduction in DES formulations [49]. DESs generally show Arrhenius behavior, where viscosity decreases exponentially with the increase in temperature [7, 49]. Viscosity is also very sensitive to the composition of the solvent. The lower the viscosity, the better the diffusivity of compounds into the extraction medium, which may improve the extraction efficiency [50].

Apart from viscosity, polarity is the other important property that depends not only on the DES composition but also on the HBD type during the preparation [7, 49]. Temperature also governs the polarity; with an increase in temperature, the polarity decreases owing to increased thermal reorientation of dipoles [47]. Moreover, at higher temperatures, the acidity provided by the donation of hydrogen ions reduces, while its accepting basicity does not show a variation. The polar nature of DESs developed by the mixing of ChCl with different HBDs like urea, acids, ethylene glycol (EG), and glycerol (Gly) has been reported [51]. DES polarity is very diverse; thus, from values similar to methanol, several can reach values equal to or even higher than water. Further research and new combinations of DESs are expected to keep increasing the polarity range, hence expanding their possible use as extractive agents.

6.2.1 Influence of Water Addition on Extraction Efficiency

One major drawback of DESs is their elevated viscosity, which significantly decreases the transfer rate of bioactive composites from plant matrices into the extraction medium [2]. Further, this increased viscosity has brought more

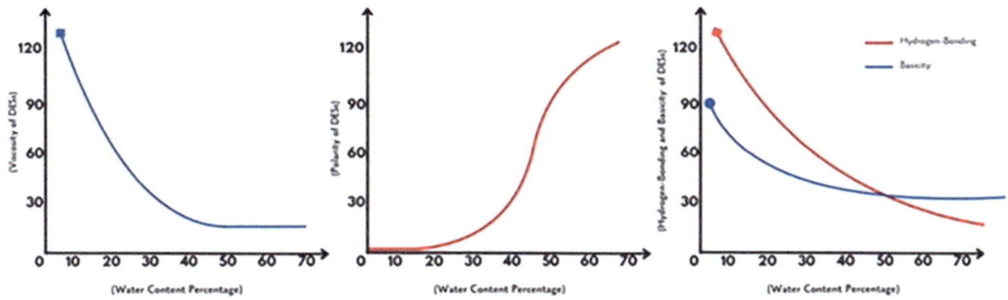

Fig. 6.2 Representative viscosity, polarity, hydrogen bonding (H-bonding), and basicity with varying water content. (Data illustrated as per the data published in [52, 54])

difficulties in many technical steps, such as decantation, dissolution, and filtration [52]. While modifications to the DES composition and variation of the salt-to-HBD molar ratio enable optimization of the viscosity, incorporating water can also be a feasible approach [53]. It is indeed very important that the water content be controlled stringently, as excessive dilution can break the intermolecular interactions within DESs [46, 53].

Dai et al. [46] investigated structural and physicochemical changes in DESs upon adding water and demonstrated that at about 50% (v/v) dilution (Fig. 6.2), the original interactions among DES components are mostly lost. Adding water can also reduce hydrogen-bond-accepting basicity while increasing polarizability and overall polarity [54]. Incorporating increasing amounts of water into the system enhances the polarity of DESs, and the increase in their polarity could result in dramatic variations in their dissolution capacity for some target compounds [52].

The polarity of DESs covers a wide range, from values above those of water to values comparable with both water and methanol [17].

Notably, water is either added to the system during system preparation or introduced later into preprepared DESs, which triggers varying impacts on the structural and physicochemical characteristics of DESs. The resulting differences, therefore, affect their performance and applications in extracting bioactive target compounds of interest.

6.3 DESs as Extraction Media for Different Compounds

The development of separation methodologies is critically important in dealing with complicated liquid systems, which can hardly be processed efficiently by traditional methods. Recently, DESs have become very popular in separating a broad range of components [5]. Due to their distinctive characteristics, DESs have shown outstanding performance in many applications, such as extraction and separation processes [45, 55]. Their very low vapor pressure makes them a sustainable replacement to conventional organic solvents by reducing the emission of volatile organic compounds, thus providing a cleaner environment and reducing possible health risks.

Table 6.1 Extraction methods for bioactive compounds using DESs and ILs

Methods of extractions		
ILs	DESs	References
HPAE	MAE	[5, 58, 59]
Ultrasound-assisted extraction	UAE	[60–62]
Micelle-mediated system	SFE	[63, 64]
Precipitation-based method	Ball mill-assisted	[53, 65]

Analogous to ILs, DESs can be modified by adjusting the types and molar ratios of HBAs and HBDs to satisfy the specific demands of a particular separation system [56, 57]. When working with liquid systems, the DES should be prepared beforehand to use as an extractant to separate a target compound into the DES phase. Liquid–liquid extraction has been considered the best method of separation whenever DESs are used because various mixtures of HBAs and HBDs provide a wide range of options for DESs. Various types of DESs and ILs were applied to target compounds with multiple methods of extraction, as outlined in Table 6.1.

When the target solute possesses either HBA or HBD nature, it interacts with the corresponding counterpart in the DESs and makes up a new DES phase. Therefore, through selective transportation of the desired solute, the solute is transferred from the combined product into the freshly synthesized DES phase. The mode of transferring the solute may be dissimilar due to the composition and decomposition of the DES. According to the nature of the constituents, DESs can act as extractants. The hydrophilic constituent promotes the transfer of the solute from the water phase into the organic phase. However, since water and hydrophilic DES constituents interact strongly, high water content can disrupt the structure of DESs and, eventually, cause their decomposition. This method is widely applied in the microextraction techniques of analytical chemistry, which require tiny amounts of DESs. The separation processes involving DESs are conventionally divided into three main groups: direct liquid–liquid extraction using a previously synthesized DES as a solvent, associative extraction when DES forms in situ with the analyte participated, and DES decomposition-based extraction [66].

6.3.1 *Extraction of Valuable Compounds from Vegetable and Fruit Waste Using DESs*

The Food and Agriculture Organization (FAO) valued 1.3 billion tonnes of yearly global food-based waste, comprising around 0.5 billion tonnes of fruits and vegetables. Peels, fibers, shells, seeds, skin, cores, vines, pomace, pods, and stones are the elements that constitute this type of waste. These by-products are rich in bioactive substances like alkaloids, carotenoids, flavonoids, phenolic acids, and glucosinolates [67]. These secondary metabolites play an important role in metabolic regulation and exhibit antioxidant, anticancer, anti-inflammatory, and anti-allergic properties [68]. Traditionally, these bioactive compounds have been extracted with the help of conventional organic solvents like methanol, benzene, hexane, petroleum ether, acetone, or chloroform. Nevertheless, their application has been increasingly questioned due to their hazardous properties, which are described as flammable, volatile, poorly biodegradable, and possibly toxic to human health. As a safer substitute, safer solvents have been recommended. Unfortunately, only a few greener solvents are available, and their highly polar nature is unsuitable for extracting some bioactive substances.

Considering the increased environmental problems linked to climate change, urgent attention should be directed to carbon emission minimization and minimum energy, water, and solvent consumption. As a result of this awareness, an increased concentration on green, environmentally friendly extraction techniques has emerged. For green technologies to be truly practical, they must be eco-efficient and aim to enhance selectivity and extraction throughput while maintaining low energy consumption. One of the promising ways to achieve this involves using ILs, including NADESs and DESs, often termed "green solvents" [69]. These can offer another avenue for extracting bioactive substances from vegetable- and fruit-based waste. Among the different DESs, most of the research has been conducted in various extraction applications to recover valuable substances from the by-products of agriculture [70].

6.3.2 Application of DESs in the Extraction of Bioactive Compounds

Fruits and vegetables serve as a source of healthy micronutrients, including magnesium, calcium, and potassium, in addition to bioactive compounds, including non-nutritive phytochemicals like carotenoids, vitamins, dietary fiber, and polyphenols [71]. In the past, acetone, hexane, and ethanol have been used in extraction processes for such compounds. Nevertheless, in the past decade, DESs have proven efficient alternatives for the extraction of useful compounds in residues in the agro-industry [72]. The extraction of useful bioactive phenols from residues generated during the processing of olive oils was examined with DESs by Bonacci et al. [73]. According to them, the environmentally friendly use of DESs with MAE is an efficient extraction technique for phenolic compound extraction from the residues of olive oils.

In a separate work, three NADESs have been used to extract anthocyanins from sour cherry pomace, and the ChCl:malic acid system has been proven to have a 62.33% yield for extraction [74]. Soluble phenols and flavonoids, such as caffeic acid, 1-caffeoylquinic acid, and chlorogenic acid, have been detected the most in phenolic compounds. These phenolic compounds have been effectively extracted with DESs made up of lactic acid and ChCl, in conjunction with UAE from apricot pomace [75].

Coffee and cocoa processing produce a high level of residues, like coffee pulp, cocoa pod husks, and cocoa husks, with huge contents of bioactive substances like fibers, phenolics, and alkaloids [72]. Betaine-based DESs, synthesized through a mixture of betaine as an HBA and a mixture of diols (propane-1,3-diol, propane-1,2-diol, and butane-1,2-diol) and acids (levulinic acid and lactic acid) as HBDs, have been used for phenolic compound extraction in coffee grounds from its residues (used coffee grounds). In agreement with experimental observations, a mixture of betaine with betaine-1,3-diol (1:5), butane-1,2-diol (1:7), and betaine-lactic acid (1:7) exhibited high antioxidant performance with IC50 (half maximal inhibitory concentration) values of 1.53, 1.59, and 1.31 ppm, respectively [76].

Vázquez-González et al. [77] conducted a research to isolate bioactive components from raspberries and strawberries using DESs. Seven DESs have been studied, and ChCl, oxalic acid, and glycolic acid DESs performed the best in phenolic (36.7%), anthocyanin (1.28 mg/100 mg of extract), and sugar extraction yields from extrudates. In another case, polyphenols in mango peel were extracted with an MAE protocol, with a mixture of DESs, comprising lactic acid, sodium acetate, and water (in a molar ratio of 3:1:4). In such a case, efficiency in the extraction of polyphenols was maximized. An extraction yield of 56.17 mg gallic acid equivalents (GAE)/g dry weight was achieved. Besides, when unprocessed mango peel extracts were mixed with sunflower oil, oxidative stability was enhanced by two times [78].

Other studies have analyzed the viability of NADESs in UAE of isoflavones, including daidzein, puerarin, and genistein, in kudzu root and molasses residues of soybeans. In such studies, high concentrations of water in NADESs (10%, 20%, and 30%) facilitated the extraction of puerarin, with yields of 9.7 ± 0.2, 12.6 ± 0.2, and 14.8 ± 0.3 mg/g, respectively. Most importantly, even with a low level of isoflavones compared to methanolic extracts, extracts with NADESs exhibited a high antioxidant activity [60].

The activity of DESs in the MAE of phenolic substances from pineapple peels was investigated by Vargas-Serna et al. [58]. Water, methanol, methanol–water (50%), ethanol–water (50%), and two DESs, ChCl:malic acid and ChCl:glycerol, were employed and compared with traditional solvents. With a phenolic yield of 7.98 mg GAE/g of dry weight, ChCl:Gly outperformed the other solvents.

The extraction of pigments, polyphenols, and anthocyanins in red grape pomace has been studied, too. Two glycerol-DESs with HBAs with variable polarity have been compared in terms of effectiveness in the extraction of polyphenols. There was increased efficiency in the extraction with an ultrasonic pretreatment during extraction. In solvents, a blend of glycerol and sodium benzoate exhibited better efficacy when compared with a mixture of sodium acetate and glycerol. The refinement of extraction temperature and times maximized rich pigments and polyphenol extract yields of red grape pomace [79].

Alibade et al. [79] performed betalain extraction, and stabilization was performed with DESs in beetroot residues. In one such process, magnesium chloride hexahydrate and urea have been used for preparing DESs. Betalains recovered in such a process showed high stability, with 75% of their red-violet color for 340 days when stored in yellowish-brown light and 150 days when stored in daylight. On the other hand, betalains recovered with water showed degradation in 5–7 days [80].

A work conducted by Vasyliev and colleagues [81] investigated the use of DESs based on ChCl for their antioxidant activity and phenolic component extraction from tomato pomace. Tomato pomace extracts were added to cosmetic compositions (colloidal solutions) in the form of antioxidants. Antioxidant activity in cosmetic emulsions was measured with the use of cyclic voltammetry, and it exhibited high phenol and flavanol contents in tomato pomace extracts, representing 26–37% of the contents in general. Chlorogenic acid represented a prevalent phenolic compound, with contents in the range of 37.23–52.33 μg/g. Out of the cosmetic emulsions prepared, emulsions with the addition of tomato pomace extract extracted with the use of ChCl:lactic acid DES exhibited the best antioxidant activity, followed by emulsions with extracts extracted with the use of ChCl:1,2-propanediol DES.

Another investigation was conducted by Tzani and team [82], who optimized the extraction of the bioactive compounds of Greek propolis (Peloponnese region) through UAE with the use of DESs. L-proline/D,L-lactic acid/water (Pro/LA/W) system, out of the solvents considered, performed best in terms of antioxidant performance and total phenolic content. An optimized extract of propolis-DES was mixed directly with a cosmetic cream, and organoleptic analysis through a sensory evaluation was performed in an attempt to assess its organoleptic characteristics, including odor, color, and texture. A range of in vitro techniques was utilized to determine the antioxidant performance of the DES-propolis extract, both in its pure form and after it was added as a cream. All such observations conclude that the antioxidant activity in cosmetic preparation with an added extract of DES-propolis is increased over a period and continues to exhibit efficacy even at 120 min. There are different kinds of DESs that are relevant in the extraction of bioactive compounds (Table 6.2).

6.3.3 Pectin Extraction Using DESs

Microwave-ultrasound-assisted extraction (MUAE) was used to extract pectin from dragon fruit (*Hylocereus polyrhizus*) peels in DESs. The top pectin yield was utilized in the MUAE method with a solvent system composition of 5:2:5 ChCl:glucose:water. Under ideal conditions, the pectin yield was 19.39% when the liquid-to-solid ratio was 35.25 mL/g, the water-to-DES ratio was 3.37 mL/mL, microwave irradiation was applied for 14.26 min at an energy value of 240 W, and ultrasonication was used for an additional 46.07 min [106].

Another study investigated pectin extraction in pomelo citrus peels. While comparing various formulations of DESs at different mole ratios, lactic acid:glucose:water (6:1:6 molar ratio), which is an acid-based DES, yielded an exceptionally high amount of pectin to the tune of 23.04% [107]. Major solvents found for the process of extraction in follow-up experiments are mostly ChCl:malonic acid and ChCl:glucose:water. At the optimized condition of pH 3, 80 °C, time 60 min, and 1:4.5 DES–water ratio, a similar pectin yield (94%) with a 52% degree of esterification was attained using both cosolvents [108]. Additionally, Chen and Lahaye [109] employed a variety of DESs, like ChCl:urea, ChCl:oxalic acid, and ChCl:lactic acid

Table 6.2 DESs used for extracting bioactive compounds

DES components				
HBAs	HBDs	Eutectic molar ratio	Amount of water added (%)	References
ChCl	Urea	1:2	–	[83, 84]
ChCl	1,2-Propanediol	1:1	–	[85, 86]
ChCl	1,2-Propanediol	1:3	–	[67, 87, 88]
ChCl	1,2-Butanediol	1:5	30	[53]
ChCl	1,3-Butanediol	1:6	10	[89]
ChCl	1,4-Butanediol	1:4	20	[89]
ChCl	1,4-Butanediol	1:4	30	[90]
ChCl	1,6-Hexanediol	7:1	30	[91]
ChCl	Citric acid	1:1	20	[92, 93]
ChCl	Acetamide	1:2	20	[92]
ChCl	D-sorbitol	1:1	20	[92]
ChCl	Ethylene glycol	1:4	36	[94]
ChCl	Ethylene glycol	1:4	30	[94]
ChCl	Ethylene glycol	1:2	–	[92]
ChCl	Glycerol	1:1	–	[95, 96]
ChCl	Glycerol	1:2	20	[92]
ChCl	Lactic acid	1:2	–	[97–99]
ChCl	Lactic acid	1:3	20	[100]
ChCl	Lactic acid	1:2	20	[17, 98]
ChCl	Levulinic acid	1:2	20	[92]
ChCl	Malic acid	1:1	20	[92]
ChCl	Oxalic acid	1:1	20	[92]
ChCl	Malonic acid	1:1	–	[101]
ChCl	Malonic acid	1:3	–	[102]
ChCl	Malonic acid	1:1	20	[92]
ChCl	Maltose	1:2	20	[103]
ChCl	Sucrose	5:5	25	[104]
Lactic acid	Glucose	3:1	25	[104]
Lactic acid	Sodium acetate	3:1	20	[100]
L-proline	Glycerol	2:5	10	[105]

(1:2 ratio) during the pretreatment phase before the extraction with hot water of pectin from apple pomace. This process showed that pectin is functionalized after treatment with apple pomace from ChCl:urea compared to ChCl:lactic acid, and thus, the pectin obtained after hot water extraction was modified by the DES pretreatment and did not resemble the typical pectin extracted using mineral acids.

Another research by Yusof and colleagues [81] prepared two different DESs to extract pectin from guava peels. In their research, DESs were prepared as 10% ChCl:EG (molar ratio of 1:2) and 10% ChCl:urea (molar ratio of 1:2). The maximum amount of pectin was obtained from the fourth formulation (10% citric acid, 29.2%; 10% ChCl:urea, 16.4%; and 10% ChCl:EG, 37.1%). The ability of the extracted

Table 6.3 Various DESs in extraction applications

Category	Active compounds	Source material	DES	Ratio	Water	Extraction rate	References
Terpene lactones	Artemisinin	Artemisia	Methyltrioctylammonium chloride–1-butane alcohol	1:4	–	7.9936 ± 0.0364 mg/g	[110]
Flavonoids	Rutin	*Sophora japonica*	ChCl–triethylene glycol	1:4	20%	194.17 ± 2.31 mg/g	[92]
Volatile oil	Essential oil	Cumin	ChCl and L-lactic acid	1:3	40%	40 g/g	[111]
Alkaloids	Boldine	*Peumus boldus*	L-proline–oxalic acid	1:1	20%	2.36 mg/g	[112]
Polysaccharides	Polysaccharide	Yam	ChCl-butane-1,4-diol	1:4	32.89%	15.98 ± 0.15 mg/g	[61]
Phenolics	Caffeic acid, quercetin, rosmarinic acid, gallic acid, luteolin, and epicatechin	*Thymus serpyllum* L	L-proline glycerin	1:2	25%	20.61, 25.83, 21.06, 524.18, 28.27, and 42.27 mg/g	[113]
Isoflavone	Daidzein, genistein, and puerarin	Kudzu roots and soy molasses	ChCl–citric acid	1:1	10%, 20%, 30%	9.7 ± 0.2, 12.6 ± 0.2, and 14.8 ± 0.3 mg/g	[60]
Anthocyanin	Cyanidin 3-rutinoside, quercetin 3-glucoside, quercetin 3-rutinoside, cyanidin 3-sophoroside, cyanidin 3-glucosylrutinoside	Sour cherry pomace	ChCl–malic acid	1:1	20%	812.76 ± 13.36, 53.80 ± 3.17, 172.42 ± 9.01, 146.98 ± 4.69, and 1483.20 ± 4.04 μg/g	[74]
Protein	Fava bean protein	Fava bean seeds	ChCl–glycerol	1:2	–	103.38 ± 2.31 mg/g	[114]

pectin to gel rapidly makes it a useful gelling agent for the food-processing sector. Table 6.3 lists the procedures that use DESs in extraction and sorption methods.

6.3.4 DESs in the Liquid-Phase Extraction of Volatile Organic Acids

These volatile fatty acids (FAs) can be prepared by the fermentation of organic waste materials and play a crucial role in acting as intermediates for synthesizing value-added products. To create a successful strategy for their extraction from diluted aqueous solutions, van Osch and colleagues [13] synthesized hydrophobic DESs by combining different HBAs (quaternary ammonium salts) with decanoic acid (HBD) at a molar ratio of 1:2. It was observed from the research output that the chain length of alkyls increased by the quaternary ammonium salts, and the amount of water and salt leaching was thus reduced with simultaneous efficiency enhancement. The extraction capabilities were considerably superior when compared to the conventional solvent trioctylamine.

Van den Bruinhorst et al. [115] conducted additional research on the effectiveness of hydrophobic DESs in extracting volatile FAs from diluted liquid media using liquid–liquid-based extraction techniques. Their findings indicated that DESs containing trioctylphosphine oxide exhibited stable extraction efficiency across a wide pH range, making them suitable for commercial organic acid recovery.

6.3.5 DESs in the Extraction of Metal Ions and Metal Oxides

In the majority of molecular solvents, metal oxides are insoluble. However, being amphoteric, they readily dissolve in an aqueous acid or alkaline solution. Again, dissolution plays an important role in various methods regarding catalyst preparation and metal recovery; it is involved in mitigating corrosion. Additionally, Abbott and colleagues [116] proved that ChCl and urea-based DES were able to dissolve a variety of metal oxides, such as ZnO, PbO_2, and CuO. Metals in a combination of metal oxide matrices were also made possible via electrodeposition. It was noted that in DESs, the reduction potential of metals would deviate from their standard reduction potentials in the aqueous form due to their coordination with an oxide/hydroxide–chloride and urea–ligand mixed complex.

Van Osch et al. [117], in another research, synthesized hydrophobic DESs based on capric acid and lidocaine and applied them to the extraction of metal ions from non-buffered water. The extraction occurred through an ion exchange mechanism, and the transition metal ions were efficiently removed even at a low mass and high

Co^{2+} concentration ratios of the DES to water. In 5 s, maximum extraction occurred, and a very stable and regenerable DES is possible.

6.3.6 DESs in the Extraction of Glycerin/Glycerol from Biodiesel

Biodiesel has received considerable attention as a future alternative fuel, but its complex character and purification challenges remain key limitations. Glycerol, a general by-product, negatively impacts purity in biodiesel. A ChCl:Gly-based DES was utilized to extract glycerol from palm oil-based biodiesel to address this problem [62]. In this study, the role of extraction efficiency in terms of variation in the molar ratio of DES to biodiesel and DES composition was analyzed and discussed. Optimum working conditions at a 1:1 molar ChCl:Gly DES/biodiesel effectively conformed to traditional standards for a biodiesel fuel through a considerable reduction in glycerol contents. Figure 6.3 represents the procedure suggested by DESs for the purification of biodiesel.

Subsequent studies conducted by the same group involved alternative DESs, namely ChCl:2,2,2-trifluoroacetamide and ChCl:EG, which effectively removed all free glycerol in palm oil-based biodiesel [118]. In a follow-up, three new DESs were synthesized with a salt base of methyl triphenyl phosphonium bromide and various HBDs, such as Gly, EG, and triethylene glycol. Out of synthesized DESs,

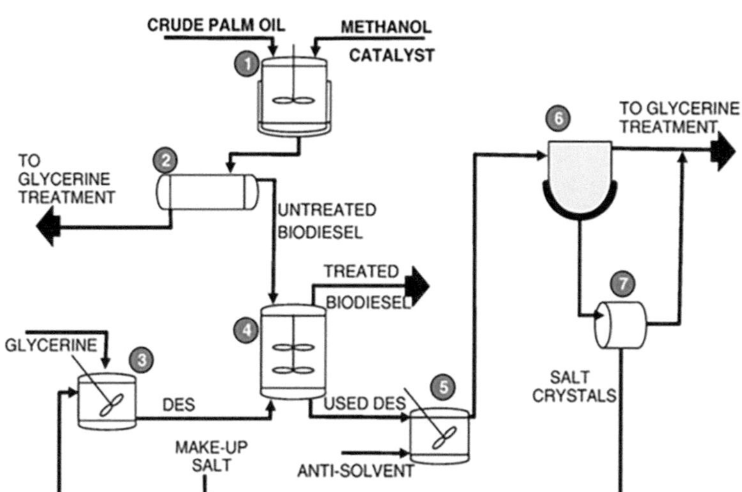

Fig. 6.3 Illustration of the proposed approach for refining biodiesel. 1: Reactor for transesterification, 2: decanter, 3: mixing tank for DES preparation, 4: extractor, 5: combining tank, 6: crystallization chamber, and 7: rotating filter. (Reproduced with the license granted by Elsevier from [62], Copyright 2009)

triphenyl phosphonium-containing ones with ethylene and triethylene glycols showed high efficiency in free glycerol extraction in terms of biodiesel [119].

In biodiesel production via transesterification, a catalyst is required, and its removal is essential for final purification. Shahbaz and team [120] investigated the utilization of nine ChCl-based DESs and nine methyltriphenylphosphonium bromide-based DESs to remove residual KOH from crude biodiesel. Their findings indicated that increasing the DES/biodiesel molar ratio and the mole fraction of the HBD in DESs improved KOH removal efficiency, except for DESs containing 2,2,2-trifluoroacetamide. The mean KOH elimination efficiencies for ChCl- and phosphonium-based DESs were reported to be 98.59% and 97.57%, respectively.

Additionally, Shahbaz et al. [121] have utilized artificial intelligence methodologies for the prediction of glycerol extraction with phosphonium- and ammonium-containing DESs. On an experimental basis, an artificial neural network (ANN) model was developed with predictive factors such as DES composition, biodiesel, and DES's mole fractions. With an optimized feed-forward neural network and the use of the Levenberg–Marquardt algorithm, an absolute average deviation of 6.46%, with high agreement with experimental values, was produced. In general, it was seen that phosphonium DESs performed better in efficiency in glycerol extraction compared to ammonium DESs.

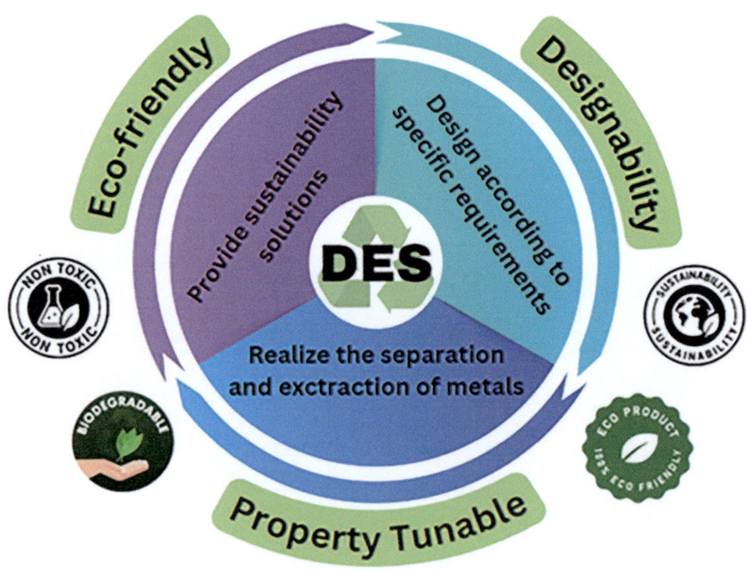

Fig. 6.4 Physicochemical properties of DESs relevant to metal extraction

6.3.7 Recovering Metals from E-Waste Using DESs

DESs have a great benefit over high-temperature molten salts and traditional aqueous solutions by overcoming their intrinsic limitations. While being used as electrolytes for the electrochemical extraction of alloys and metals, DESs enable the efficient recovery of active metals and their compounds with minimal or no water interference [122]. The distinctive solvent characteristics of DESs contribute to their activity in metal recovery, as depicted in Fig. 6.4**.**

The utilization of DESs for extracting valuable metals from electronic waste (e-waste) has gained attention and has become an increasingly studied approach [123]. As depicted in Fig. 6.5, this portion discusses the utilization of DESs in the recovery of vital resources from e-waste, like end-of-life permanent magnets, printed circuit boards, used batteries, and lamp phosphor waste.

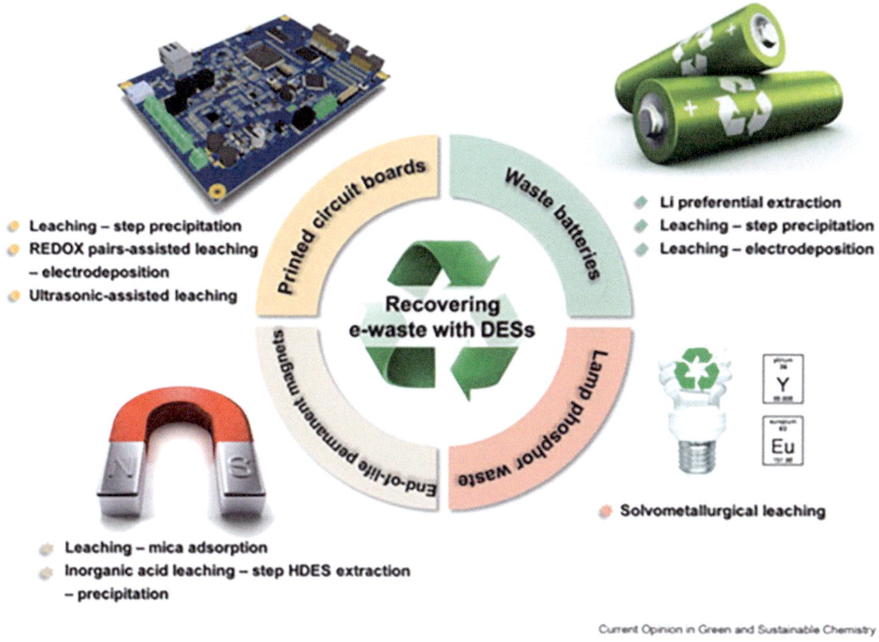

Fig. 6.5 Metal extraction from conventional e-waste using DESs. (Reproduced with the license granted by Elsevier from [124], Copyright 2024)

6.3.7.1 Valuable Metal Recovery from Waste Batteries Using DESs

DESs have been widely developed for effectively extracting valuable metals from waste batteries because of their unique physicochemical characteristics [125, 126]. A well-established DES system, composed of ChCl and EG, has demonstrated high efficiency in leaching metals from $LiCoO_2$ (LCO) batteries [127]. The β-hydrogen radical in ChCl facilitates the reduction of Co^{3+} to its soluble Co^{2+} state, promoting effective metal dissolution. Additionally, the oxidative Cl_3^- species formed in this DES system contribute to a leaching efficiency of 80% for lithium (Li) and over 90% for cobalt (Co), manganese (Mn), and nickel (Ni) from ternary nickel–cobalt–manganese (NCM) batteries [128]. However, challenges such as a low solid-to-liquid (S/L) ratio (~20 g/L) and the requirement for elevated temperatures (>180 °C) have limited the scalability of this process.

Acidic DES formulations have been explored to enhance metal leaching efficiency [125, 129, 130]. By releasing H^+ ions, these DESs lessen the reliance on traditional organic and inorganic acids by substituting Li^+ in cathode materials. Furthermore, the incorporation of super-concentrated hydrochloric acid (HCl) into DESs and the application of ozone have significantly improved dissolution capacities [131, 132]. Notably, an HCl-assisted DES system has achieved a high atomic economy of 94.9 ± 5.7% in lithium-ion battery (LIB) recycling, aligning with green chemistry principles [132].

The high viscosity of DESs frequently reduces their leaching efficacy, although this can be reduced by carefully adjusting certain components to accelerate the reaction kinetics [133, 134]. Recent advancements, such as microwave-assisted DES leaching, have improved mass transfer rates and enhanced the dissolution process [59]. After leaching, oxalic acid (OA) and water, respectively, can be used to precipitate precious metals like Li^+ and Co^{2+} selectively [135]. DES composed of caprolactam:urea:formic acid (3:1:1) has demonstrated exceptional recyclability, effectively recovering lithium over 20 cycles [136]. DESs made from oxalic acid and urea have much potential for large-scale commercial applications because these substances are readily available commodity chemicals. Moreover, certain hydrophobic DESs have been prepared for efficient metal extraction from waste battery-leaching solutions [137].

6.3.7.2 Metal Recovery from Waste Printed Circuit Boards Using DESs

Waste printed circuit boards (WPCBs) are a main source of e-waste, containing abundant metals like aluminum (Al), copper (Cu), iron (Fe), zinc (Zn), tin (Sn), lead (Pb), and various valuable metals [138]. The recovery of Pb, Zn, Sn, and Cu from WPCBs can be competently achieved by applying a two-stage leaching process with DESs based on ChCl:oxalic acid and ChCl:glycolic acid [139].

Abbott's research group demonstrated that low-power sonication of WPCBs in ChCl:EG DES generated bubbles at the interface, leading to cavitation-induced bubble collapse. This effect facilitated the rapid stripping of Cu from the substrate [140]. However, the stripping efficiency was highly dependent on precise sonication parameters, such as liquid height, tip diameter, and immersion depth, which limited its applied scalability.

To enhance metal dissolution from WPCBs, oxidants such as ferric chloride ($FeCl_3$), cupric chloride ($CuCl_2$), and iodine have been introduced into DESs, enabling catalytic metal leaching [141, 142]. The water content significantly influences the solvent properties and leaching efficiency of DESs. A nickel oxide/hydroxide passivation layer that prevents metal breakdown may form on WPCBs as a result of the high water content. Nevertheless, this passivation layer is successfully removed by ultrasonic treatment, which greatly increases leaching rates. The rate at which Cu dissolves from WPCBs in $CaCl_2{\cdot}6H_2O$:EG-based DES with $CuCl_2$ as an oxidant is almost 10,000 times faster in ultrasonic circumstances than in non-ultrasonic ones [143].

Additionally, selective oxidation and electrodeposition of Cu from WPCBs can be obtained using ChCl:EG-based DES with 5 wt.% hydrogen peroxide (H_2O_2) [144]. This procedure is set with sustainable and negative-carbon economy strategies, demonstrating a significant reduction in carbon emissions. In comparison to conventional recovery methods, CO_2 emissions were reduced by 93.6 kg per 1000 kg of processed WPCBs, highlighting the environmental benefits of DES-based metal recovery.

6.3.7.3 Rare Earth Element Recovery from Lamp Phosphor Wastes Using DESs

Lamp phosphor wastes serve as a significant secondary source of rare earth elements (REEs), such as lanthanum (La), europium (Eu), yttrium (Y), cerium (Ce), and terbium (Tb) [145]. It has been estimated that 1.0 kg of spent phosphors contains an REE content equivalent to that found in 11.9–22.5 kg of primary ore [146]. DESs have become viable nonaqueous substitutes for REE extraction using solvometallurgical methods because of their exceptional complexation characteristics [147].

Despite their potential, most existing studies primarily focus on the leaching efficiency of artificial phosphor resources while neglecting the impact of impurities present in real waste samples [148]. Comparative analyses reveal notable discrepancies in the leaching performance and selectivity between synthetic and actual lamp phosphor waste. To ensure commercial viability, future research should emphasize the development of advanced DES-based extraction technologies that are specifically tailored for real waste materials with more multifaceted compositions.

6.3.7.4 Recovery of Rare Earth Elements from End-of-Life Permanent Magnets Using DESs

End-of-life permanent magnets represent an additional significant secondary source abundant in rare earth elements (REEs). ChCl- and salicylic acid-based DESs (1:2 molar ratio) have demonstrated a strong leaching capacity for neodymium (Nd) from NdFeB magnets [149]. This is attributed to the ion–dipole interactions between the Ch^+ cation of ChCl and the surface Nd^{3+} and Fe^{3+} ions in the magnet, which disrupt the ionic lattice structure, thereby promoting Nd leaching. The specific composition of DESs and the type of substituents they include determine the extent to which they leach permanent magnets. For example, EG- and maleic acid (MA)-based DES in a 2:1 molar ratio shows greater leaching performance, likely due to the strong coordination abilities of EG and MA, enhancing selectivity for Nd extraction [150].

Furthermore, it has been shown that hydrophobic DESs are effective lixiviants for the selective extraction of REEs from leachates of magnetic materials. To selectively extract Nd and Sm from nitric acid leachates of Nd/Fe and Sm/Co magnets, for example, the aliquot 336:glycerol (1:2) hydrophobic DES works well with hydrometallurgical procedures [151]. With a recovery efficiency of 99% for Nd, praseodymium (Pr), and dysprosium (Dy), the trioctylphosphine oxide (TOPO) in decan-1-ol/myristyl alcohol/TOPO (molar ratio of 5:5:4) hydrophobic DES has also demonstrated considerable promise for REE extraction [152]. While encouraging, additional work is required to create more efficient hydrophobic DES systems to ensure high recovery efficiency for REEs and avoid losses during the extraction process.

6.4 Criteria for Evaluating DESs for Separation

6.4.1 Estimate the Melting and Freezing Points of DESs

For DESs to apply effectively in liquid–liquid extraction, the prepared DESs should remain in a liquid state during the extraction process. Hence, the operational temperature range at which DESs will remain in a liquid phase must be defined by determining their melting and freezing temperatures. Several experimental techniques have been used to test the freezing points in DESs, including the visual method, differential scanning calorimetry (DSC), and capillary melting point equipment [47, 107, 153]. However, up to this moment, the selection of liquid-state DESs is done by trial-and-error experiments toward a targeted application. To improve the screening process, thermodynamic models have been applied to obtain solid–liquid equilibrium (SLE) among HBAs and HBDs.

Silva et al. examined the Conductor-like Screening Model for Real Solvents (COSMO-RS) prediction of eutectic points and the SLE of sugar-based DESs with ChCl [154]. As influenced COSMO-RS parameters produced predictions that were in good agreement with actual data, the results showed that COSMO-RS could be a valuable tool in forecasting the phase state of DESs. Song and team [155] also presented a detailed assessment of the SLE predictions with the COSMO-RS model, for which the theoretically calculated results were compared with the experimentally established SLE data from eutectic systems. Their findings showed that though the COSMO-RS model provided reliable estimates of eutectic temperatures in many instances, its ability to predict the eutectic composition was relatively limited. However, COSMO-RS still shows promising prospects in assisting with selecting or designing DESs that will yield the optimal phase behavior in different applications.

6.4.2 *Approaches for Evaluating DESs for Separation Processes*

COSMO-RS is based on the quantum chemistry activity coefficient model applied as a tool for the identification of the thermodynamic equilibrium properties of pure substances and mixtures. Most of the studies used this liquid–liquid equilibrium (LLE), and other quick thermodynamic parameters can be computed using an activity coefficient model based on COSMO-RS, activity coefficients, solubility, and Henry's law constants [156, 157].

Since DESs involve more than one molecular species, the selection of the representation procedure of DES is crucial in the COSMO-RS framework to get reliable predictions. So far, three major methods have been followed in representing ILs: the ion pair approach, the electroneutral approach, and the metafile approach. The electroneutral method, in particular, has been widely adopted, wherein cations and anions are considered distinct molecules in an equimolar ratio for IL modeling [158]. This approach is also regarded as the most suitable for DESs in their liquid phase, with the cation, anion, and HBD being treated as separate species in their respective molar proportions [159].

Two concurrent studies described the effectiveness of glycerol-DESs as extraction media by utilizing the COSMO-RS-based model's predictive power [160, 161]. Their work exhibited a very good agreement between computed and experimental solute distribution coefficients and selectivity. Additionally, the model used DESs based on alkane diols to fractionate both the model and actual citrus essential oils. These, in turn, gave qualitative views on the COSMO-RS predictions for trends of distribution coefficients and selectivity of solutes, confirming the applicability for the estimation of the thermodynamic behavior of alkane diol-based DES systems in interaction with essential oils.

6.5 Mechanism of Separation Using DESs as Solvents

The effectiveness of DESs in extraction processes is notably affected by structural characteristics and their composition, which govern their interactions with target solutes [162]. It is essential to comprehend these relationships to elucidate the underlying separation mechanisms. In particular, DESs mainly interact with the conjugated π-electrons of aromatic rings through two important mechanisms when extracting aromatic compounds from alkanes: the CH–π and π–π interactions [163].

As can be observed, in the case of quaternary ammonium salts, the CH–π interactions are prevailing. For quaternary ammonium salts, longer alkyl chains decrease cation–anion interactions but increase CH–π interactions with aromatic molecules, which increases distribution coefficients and extraction efficiencies. Long-chain alkyl groups also increase van der Waals interactions with alkanes. Thus, selectivity decreases. In general, choline-based DESs have higher distribution coefficients compared to phosphonium DESs or long-chain quaternary ammonium but usually with lower selectivity.

Another critical factor influencing the properties of the HBD, which is important in extraction, is the HBA. Hou et al. [164] observed through Fourier-transform infrared spectroscopy (FT-IR) spectral analysis that in a tetrabutylphosphonium bromide (TBPB):levulinic acid-based DES in a molar ratio of 1:6, the C=O stretching vibration appeared at 1650–1740 cm^{-1} and significantly intensified upon extracting toluene into the DES. This suggests that the aromatic ring of toluene and levulinic acid has a strong hydrogen-bonding connection. In addition, the primary driving force behind DES-based extractions is the associative extraction of weakly polar solutes from an oil phase and the hydrogen bonding between organic salts and solutes. Additional elements that impact the effectiveness of DES-based separation methods include the chemical and functional group makeup of the solute, its molecular location, and the organic salt cation's composition and chain length.

Short-chain positive ions have strong cation–anion interactions, and hence, the solutes find it hard to disrupt the crystal lattice of the organic compounds to form DESs. On the contrary, cations with longer chains have weaker electrostatic interactions, and thus, the anion can combine with the HBD to form DESs more easily. Too long a chain in the cation may cause DESs to dissolve in the oil phase, leading to the creation of a homogenous and inseparable system. The polarity of the solute also influences DES interactions. Hydrophilic functional groups strengthen interactions with water to increase the compatibility of polar solutes with hydrophilic DESs. However, nonpolar solutes use van der Waals forces to interact with the hydrophobic DESs.

Verma et al. [165] used three hydrophobic DESs and molecular dynamics simulations to investigate the mechanism of alcohol extraction from water. The electrostatic contacts between DESs and alcohols were discovered to be stronger than the van der Waals interactions, according to their calculated interaction energies. The interaction energy of the HBA (menthol) with alcohol was higher than that of the HBD (lauric acid), except for interactions with water, which indicated that the HBA has a greater influence on DES hydrophobicity and alcohol extraction efficiency. Besides, the order of interaction energy ranking of DESs with alcohols was 1-butanol > 1-propanol > ethanol, which agreed well with the obtained trend in selectivity and distribution coefficient.

6.5.1 Application of DESs in Capillary Electrochromatography

A unique method for adjusting the selectivity of separation techniques has been introduced by the utilization of DESs in capillary electrochromatography (CEC) [166–168]. CEC blends the remarkable selectivity of high-performance liquid chromatography (HPLC) with the great separation effectiveness of capillary electrophoresis (CE) in a novel way [169, 170]. This method of in situ polymerization uses porogens, monomers, and radical initiators inside a glass capillary. According to multiple reports, the lengthy operating lifetime, quick mass transfer kinetics, and superior reproducibility of such a monolithic construction make it very appealing [171, 172].

Till now, organic solvents like toluene-isooctane-dimethyl sulfoxide and toluene-isooctane have been used as porogens for the preparation of monolithic columns [171]. However, the volatility, environmental problems, and toxicity associated with these solvents have driven research toward alternative porogenic systems [171, 173]. A common limitation of purely organic monolithic columns is their low surface area, attributed to inadequate pore formation and limited interaction sites. This negatively impacts separation efficiency, particularly for small molecules [171, 172]. For effective mass transfer, polymer monolithic columns need to have the highest column permeability to realize high flow rates and low pressure [174].

Furthermore, such complications have been partly overcome by newly developed hybrid monolithic columns, which have involved state-of-the-art nanomaterials, including hydroxyapatite, graphene oxide (GO), carbon nanotubes (CNTs), metal oxides, metal-organic frameworks (MOFs), and mesoporous molecular sieves. These materials favor surface area and allow a well-distributed porous network inside the polymer matrix [174]. With the increased interest in multifunctional polymer nanocomposites, studies have also been performed on the utilization of room-temperature ionic liquids (RTILs) as suspension media and porogens [172–175].

The combination of single-walled carbon nanotubes (SWCNTs) in a binary porogenic system consisting of ChCl:ethylene glycol and $[HMIM^+][BF_4^-]$ (1:2) was shown by Zhang and team [174]. The SWCNT-based hybrid monolithic columns demonstrated good thermal stability up to 270 °C, up to 251,000 plates per meter in CEC, and a much higher separation efficiency as compared to monoliths without SWCNTs.

Recently, Wang and team [169] described a novel approach where, for the first time, a DES acted as a functional monomer instead of a porogen. The formation of monoliths was made possible by a binary porogenic system that included isopropanol and PEG-400. At the same time, the DES, which was composed of ChCl and itaconic acid, was utilized in conjunction with the cross-linker of ethylene glycol dimethacrylate. This section briefly introduced the preparation methodology for the synthesis of a hybrid monolithic column [171, 172, 174, 175].

The monolithic column based on polymeric DESs demonstrated efficient separation of various analytes, including nucleosides, phenols, alkaloids, nucleotide bases, and toluidines [169]. Notably, the relative standard deviation (RSD) of migration time was 3.08% for columns prepared across different batches and 4.26% for columns within the same batch, indicating excellent column-to-column reproducibility and consistent performance.

Recent research has investigated the utilization of DESs as auxiliary additives to β-cyclodextrins (β-CD), which are commonly employed as chiral selectors in electrochromatography, thereby expanding the applicability of DESs in chiral separations [176]. Moufawad and colleagues [177] demonstrated that when DESs were utilized as cosolvents, inclusion complexes formed between analytes and β-CD. Mixtures of β-CD with four different ChCl-based DESs, such as ChCl:urea, ChCl:ethylene glycol, ChCl:propylene glycol, and ChCl:butylene glycol, were applied to pretreated fused-silica capillaries. DESs functioned as auxiliary additives to facilitate the enantiomeric separation of chiral pharmaceuticals, including racemic zopiclone (rac-Zop), racemic salbutamol (rac-Sal), and racemic amlodipine (rac-Aml). The data graph, depicted in Fig. 6.6, shows that the addition of ChCl-based DESs significantly improved the separation of all three racemic drugs. The highest resolution was obtained with 1.0% (v/v) of ChCl:urea-based DES, which also provided the longest retention time among the DESs tested [176]. Further testing of columns prepared with different concentrations of urea-based DESs (up to 2.0% v/v) demonstrated increased resolution and retention, with increasing DES concentration, though higher concentrations than 2.0% v/v were not evaluated owing to the potential for improved Joule heating [176].

The formation of monolithic columns with DESs to immobilize enzyme microreactors' utilization of a thiol-ene click reaction was also reported by Zhao et al. [173]. Using a binary combination of $[BMIM^+][BF_4^-]$ and ChCl:EG (1:3), trypsin was immobilized in a polymer matrix of poly(butyl methacrylate-*co*-α-methacrylic acid-*co*-EG dimethacrylate). The resultant immobilized enzyme reactor (IMER) showed excellent performance and digested bovine serum albumin (BSA) within less than 50 s, which is 864 times faster compared to the traditional in-solution digestion, which requires 12 h [173]. Further, fragmentation of the protein extract from rat liver, upon running a comparison against a protein database, led to the identification of 1034 unique protein groups.

6.5.2 Separation of Aromatics and Aliphatic Mixtures

Aromatic hydrocarbons are obtained mostly from fuel and coal processing and are generally combined with aliphatic hydrocarbons. The separation of aromatics from alkanes is an exciting task because their boiling points are very close, and some mixtures form azeotropes. Traditional methods of separation by liquid–liquid extraction are based on organic solvents like dimethyl sulfoxide, sulfolane, *N*-formylmorpholine, and *N*-methylpyrrolidone. These solvents exhibit certain disadvantages: They are usually toxic and inflammable and partly soluble in the raffinate phase, which reduces extraction efficiency. On the other hand, the

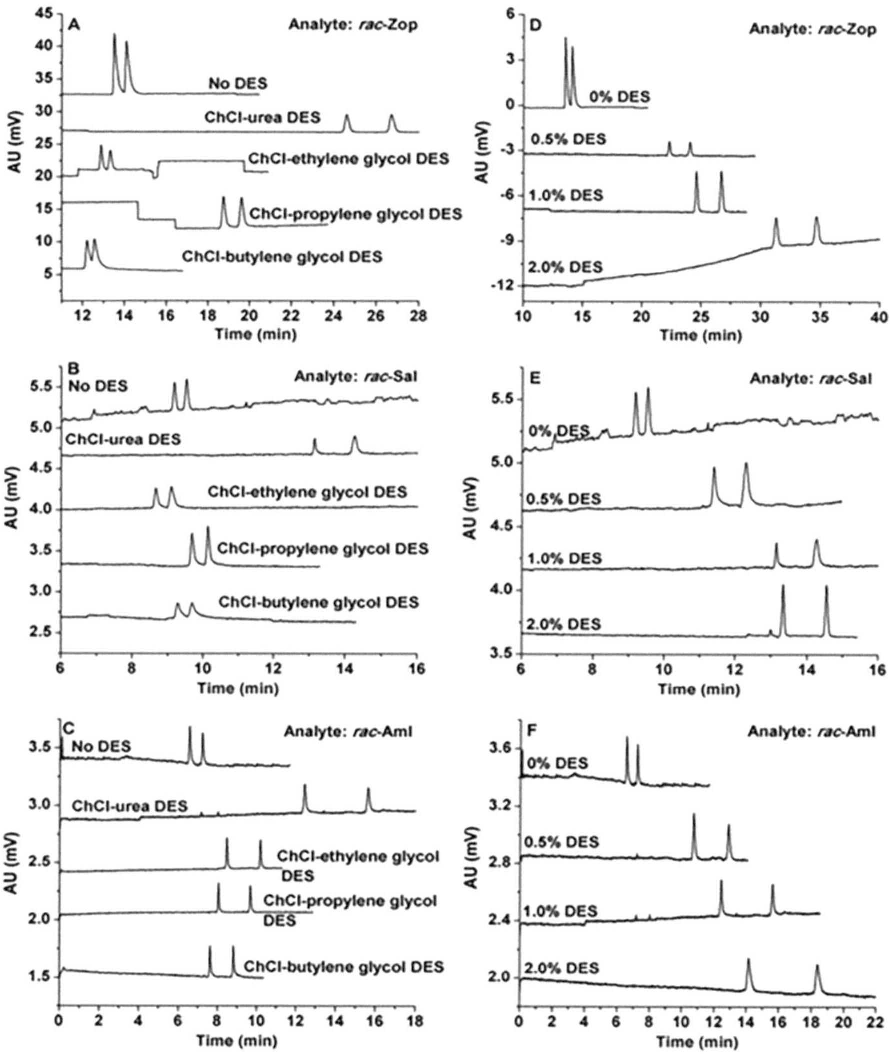

Fig. 6.6 Panels (**a–c**) display electropherograms illustrating the impact of different DESs used as additives, whereas panels (**d–f**) present electropherograms corresponding to varying concentrations of ChCl–urea DES as a chiral additive. The experiments were conducted using an uncoated fused-silica capillary (50.0/42.0 cm, 50 μm I.D.) at a controlled temperature of 20 °C. (**a**) Enantioseparation was performed in a 100-mM Tris-H_3PO_4 buffer (pH 2.5) supplemented with 15 mM β-CD and 1.0% v/v DES. Detection was carried out at 203 nm under an applied voltage of 15 kV. (**b**) A 30-mM borate-H_3PO_4 buffer (pH 4.0) containing 7 mM β-CD and 1.0% v/v DES was used, with UV detection at 254 nm and an applied voltage of 20 kV. (**c**) Enantioseparation was conducted in a 50-mM Tris-H_3PO_4 buffer (pH 2.5) with 10 mM β-CD and 1.0% v/v DES, using UV detection at 210 nm and an applied voltage of 25 kV. Panel (**d**) shows the effect of varying DES concentrations on the enantioseparation of rac-Zop, performed under the same conditions as in panel (**a**). (**e**) rac-Sal, with conditions identical to panel (**b**). (**f**) rac-Aml, following the experimental setup described in panel (**c**). (Reproduced with the license granted by **Wiley** from [176], Copyright **2019**)

development of alternative solvents has been increasingly investigated. Given their tunable properties, DESs have become a promising alternative to the separation of aromatics from aliphatic/aromatic mixtures.

Kareem and team [178] reported the first-time use of DESs in the extraction of aromatic hydrocarbons from aromatic/aliphatic mixtures. The study found that the DES consisting of TBPB and EG could separate benzene from hexane; therein, ethylene glycol acted as an HBD, while TBPB was used as the acceptor. However, high selectivity could not be achieved with high extraction efficiency. The same research group studied the LLE behavior of ternary systems containing heptane and toluene with TBPB-sulfolane and TBPB-EG DESs at 40, 50, and 60 °C in the follow-up study. These studies have further established the potential of these DESs as extraction media.

Kroon et al. [179] prepared tetrahexylammonium bromide (THAB) based DESs with either ethylene glycol or glycerol in a molar ratio of 1:2 and applied these to separate aromatic and aliphatic compounds. At 25.0 and 35.0 °C, they examined LLE data for hexane-benzene-DES systems and discovered that these DESs showed promise for the extraction of naphtha streams with low aromatic content. The same group developed DESs by combining polyols like EG and Gly as HBDs with quaternary ammonium salts as HBAs. The COSMO-RS model was utilized to correlate the LLE data in hexane-benzene-based DES systems at room temperature, which provided valuable information for the design of DES solvents.

Another research group recently performed work that demonstrated that the DES made from TBPB and levulinic acid effectively separated aromatic hydrocarbons from the aromatic or aliphatic mixture. The selectivity and extraction of toluene effectiveness were influenced by variables such as the molar ratio of levulinic acid to TBPB, the DES-to-toluene ratio, the mole fraction of toluene, and the extraction temperature [164]. The molar ratios of levulinic acid to TBPB (6:1) and DES to toluene (6.4:1) at room temperature are the ideal extraction conditions. DES regeneration is performed with distilled toluene at 100 °C under reduced pressure.

Currently, DESs are recognized as one of the most efficient extractants for the separation of aromatic compounds from aliphatic mixtures. In applications, the use of DESs demonstrates a greener extraction technique compared to the conventional processes by minimizing dependence on toxic organic solvents yet remaining highly efficient.

6.5.3 Separation of Alcohol and Water Mixtures

Certain alcohol–water mixtures form azeotropes, making their separation via conventional distillation challenging. To address this issue, DESs have been investigated as solvents to disrupt azeotropic behavior and enhance separation efficiency. Rodríguez et al. [180] demonstrated that the ethanol–water azeotrope could be effectively disrupted using DESs such as malonic acid (MA) with ChCl at a molar ratio of 1:1 and glycolic acid with ChCl at molar ratios of 1:1 and 3:1. Furthermore, the azeotropic composition was shifted toward the pure ethanol side when lactic acid and ChCl were combined in a molar ratio of 2:1, highlighting the tunability of DESs in azeotrope separation.

Two DESs, namely ChCl:urea (molar ratio of 1:2) and ChCl:triethylene glycol (molar ratio of 1:3), were synthesized by Gjineci et al. [181], and their effectiveness as entrainers for the separation of ethanol–water azeotropic mixtures was assessed. The azeotropic point was displaced in both conditions as a result of the addition of DESs, which raised the relative volatility of ethanol. The efficiency of DESs in aiding ethanol purification was demonstrated by the fact that, depending on the particular DES utilized, concentrations between 5.5% and 9% (w/w) were adequate to remove the azeotrope.

6.5.4 Separation of Mixtures of Alcohols and Esters

The tunable nature of DESs has also facilitated their application in the separation of alcohols from esters. Maugeri and team [182] demonstrated that DESs exhibit selective solubility for HBD-containing molecules, such as alcohols, while esters remain in a separate phase. A DES composed of Gly:ChCl (molar ratio of 2:1) was employed as an extractant to effectively separate alcohols (e.g., *n*-butanol and benzyl alcohol) from ester compounds (e.g., benzyl butyrate, benzyl acetate, and butyl acetate). The alcohols preferentially dissolved in the DES phase, whereas the esters remained undissolved, leading to efficient phase separation. This approach presents a feasible substitute to conventional chromatographic separation techniques, offering a bio-based and environmentally friendly method for alcohol–ester separation without the use of hazardous solvents.

6.5.5 Radioactive Nuclear Contaminant Separation Using DESs

The effective elimination and storage of radioactive nuclear pollutants are crucial for the advancement of nuclear power applications. A novel approach that utilizes DESs for the capture and storage of iodine (I_2), which serves as a model compound for radioactive nuclear pollutants, was introduced by Li [183]. Several DES formulations demonstrated superior I_2 removal efficiencies compared to past reported articles. One of them, the choline iodide (ChI)–methyl urea DES, exhibited the highest iodine uptake efficiency, achieving nearly 100% removal within 5 h. The exceptional productivity of DESs in iodine capture is primarily attributed to halogen-bonding interactions between the I_2 molecules and DES components. This research broadens the important applications of DESs in nuclear waste management and environmental remediation.

6.6 Conclusion and Future Prospects

DESs have gained attention as a promising alternative to conventional solvents in separation and extraction processes. Their unique properties—biodegradability, cost-effectiveness, recyclability, and environmental friendliness—position them as essential in advancing sustainable chemistry. Given their growing applications across diverse fields, such as pharmaceuticals, food processing, environmental remediation, and biofuel production, the potential for DESs in industrial-scale operations is substantial.

The future of DESs in extraction and separation processes is strongly tied to their scalability, environmental impact, and integration with emerging technologies. While laboratory-scale studies have demonstrated the efficiency of DESs, industrial-scale implementation remains a challenge due to concerns regarding solvent recovery, operational costs, and process stability. Addressing these issues will require the development of optimized large-scale extraction methods that ensure economic viability. Additionally, the environmental and health impacts of DESs, particularly those with synthetic components, must be thoroughly evaluated. The use of NADESs presents a greener alternative, but further research on their toxicity and biodegradability is essential to ensure safe and sustainable applications. One of the major benefits of DESs is their recyclability; however, efficient solvent regeneration methods must be developed to minimize waste and enhance long-term usability. The tunability of DESs enables selective extraction of target compounds, and continued exploration of novel formulations with tailored physicochemical properties will further improve extraction efficiency and specificity. Moreover, the integration of DESs with advanced technologies, such as MAE, UAE, and SFE, has already shown promising results, but further optimization and standardization are needed for widespread industrial adoption. Furthermore, with the increasing global emphasis on sustainability, DESs hold great potential for circular economy applications by facilitating the recovery of valuable bioactive materials from food waste, agro-industrial residues, and electronic waste. Expanding their role in waste valorization will not only support sustainable resource management but also drive innovation in green chemistry and industrial processing.

The utilization of DESs in extraction and separation processes represents a major shift toward greener and more sustainable chemical practices. Their unique ability to selectively dissolve and recover bioactive compounds, metals, and organic pollutants makes them indispensable in various industrial applications. However, for DESs to fully replace conventional solvents, challenges such as process scalability, solvent recovery, and environmental impact must be systematically addressed. As research continues to expand, DESs are expected to contribute significantly to the future of sustainable extraction and separation technologies, aligning with global efforts to reduce chemical waste, lower energy consumption, and promote environmentally friendly industrial processes. By integrating DESs with green chemistry principles and advanced extraction methodologies, these solvents have the potential to revolutionize multiple sectors, paving the way for a cleaner and more sustainable future.

Acknowledgments The authors would like to acknowledge The Maharaja Sayajirao University of Baroda, Indus University, and Parul University for their support of this work.

Conflict of Interest Statement The authors declare that they have no conflict of interest.

Declaration of Generative AI and AI-Assisted Technologies in the Writing Process While preparing this work, the authors utilized the ChatGPT tool to enhance the language and readability of several statements in the manuscript. After using this tool/service, the authors reviewed and edited the content as needed, and they take full responsibility for the publication content.

References

1. Belwal T, Ezzat SM, Rastrelli L et al (2018) A critical analysis of extraction techniques used for botanicals: trends, priorities, industrial uses and optimization strategies. TrAC Trends Anal Chem 100:82–102. https://doi.org/10.1016/j.trac.2017.12.018
2. Mbous YP, Hayyan M, Hayyan A et al (2017) Applications of deep eutectic solvents in biotechnology and bioengineering—promises and challenges. Biotechnol Adv 35:105–134. https://doi.org/10.1016/j.biotechadv.2016.11.006
3. Li G, Chen D (2024) Comparison of different extraction methods of active ingredients of Chinese medicine and natural products. J Sep Sci 47:e2300712. https://doi.org/10.1002/jssc.202300712
4. Bart H (2011) Extraction of natural products from plants—an introduction. In: Industrial scale natural products extraction. Wiley, New York, pp 1–25
5. Ventura SPM, Silva FAE, Quental MV et al (2017) Ionic-liquid-mediated extraction and separation processes for bioactive compounds: past, present, and future trends. Chem Rev 117:6984–7052. https://doi.org/10.1021/acs.chemrev.6b00550
6. Gil-Martín E, Forbes-Hernández T, Romero A et al (2022) Influence of the extraction method on the recovery of bioactive phenolic compounds from food industry by-products. Food Chem 378:131918. https://doi.org/10.1016/j.foodchem.2021.131918
7. Smith EL, Abbott AP, Ryder KS (2014) Deep eutectic solvents (DESs) and their applications. Chem Rev 114:11060–11082. https://doi.org/10.1021/cr300162p
8. Forsyth SA, Pringle JM, MacFarlane DR (2004) Ionic liquids—an overview. Aust J Chem 57:113. https://doi.org/10.1071/CH03231
9. MacFarlane DR, Seddon KR (2007) Ionic liquids—progress on the fundamental issues. Aust J Chem 60:3. https://doi.org/10.1071/CH06478
10. Krishnan A, Gopinath KP, Vo D-VN et al (2020) Ionic liquids, deep eutectic solvents and liquid polymers as green solvents in carbon capture technologies: a review. Environ Chem Lett 18:2031–2054. https://doi.org/10.1007/s10311-020-01057-y
11. Kareem MA, Mjalli FS, Hashim MA, AlNashef IM (2010) Phosphonium-based ionic liquids analogues and their physical properties. J Chem Eng Data 55:4632–4637. https://doi.org/10.1021/je100104v
12. Tang B, Zhang H, Row KH (2015) Application of deep eutectic solvents in the extraction and separation of target compounds from various samples. J Sep Sci 38:1053–1064. https://doi.org/10.1002/jssc.201401347
13. van Osch DJGP, Zubeir LF, van den Bruinhorst A et al (2015) Hydrophobic deep eutectic solvents as water-immiscible extractants. Green Chem 17:4518–4521. https://doi.org/10.1039/C5GC01451D
14. Abbott AP, Capper G, Davies DL et al (2002) Quaternary ammonium zinc- or tin-containing ionic liquids: water insensitive, recyclable catalysts for Diels–Alder reactions. Green Chem 4:24–26. https://doi.org/10.1039/b108431c

15. Dai Y, Rozema E, Verpoorte R, Choi YH (2016) Application of natural deep eutectic solvents to the extraction of anthocyanins from Catharanthus roseus with high extractability and stability replacing conventional organic solvents. J Chromatogr A 1434:50–56. https://doi.org/10.1016/j.chroma.2016.01.037
16. Choi YH, van Spronsen J, Dai Y et al (2011) Are natural deep eutectic solvents the missing link in understanding cellular metabolism and physiology? Plant Physiol 156:1701–1705. https://doi.org/10.1104/pp.111.178426
17. Dai Y, van Spronsen J, Witkamp G-J et al (2013) Natural deep eutectic solvents as new potential media for green technology. Anal Chim Acta 766:61–68. https://doi.org/10.1016/j.aca.2012.12.019
18. Mbous YP, Hayyan M, Wong WF et al (2017) Unraveling the cytotoxicity and metabolic pathways of binary natural deep eutectic solvent systems. Sci Rep 7:41257. https://doi.org/10.1038/srep41257
19. Mushtaq M, Butt FW, Akram S et al (2024) Deep eutectic liquids as tailorable extraction solvents: a review of opportunities and challenges. Crit Rev Anal Chem 54:1634–1660. https://doi.org/10.1080/10408347.2022.2125284
20. Fu M, Zhang H, Bai J et al (2025) Application of deep eutectic solvents with modern extraction techniques for the recovery of natural products: a review. ACS Food Sci Technol 5:444. https://doi.org/10.1021/acsfoodscitech.4c00973
21. Duan L, Dou L-L, Guo L et al (2016) Comprehensive evaluation of deep eutectic solvents in extraction of bioactive natural products. ACS Sustain Chem Eng 4:2405–2411. https://doi.org/10.1021/acssuschemeng.6b00091
22. Li X, Row KH (2016) Development of deep eutectic solvents applied in extraction and separation. J Sep Sci 39:3505–3520. https://doi.org/10.1002/jssc.201600633
23. Bhadange YA, Carpenter J, Saharan VK (2024) A comprehensive review on advanced extraction techniques for retrieving bioactive components from natural sources. ACS Omega 9:31274–31297. https://doi.org/10.1021/acsomega.4c02718
24. Mukhopadhyay S, Mukherjee S, Adnan NF et al (2016) Ammonium-based deep eutectic solvents as novel soil washing agent for lead removal. Chem Eng J 294:316–322. https://doi.org/10.1016/j.cej.2016.02.030
25. Rashid SN, Hizaddin HF, Hayyan A et al (2024) A kinetic study of ex-situ soil remediation by nickel extraction using natural deep eutectic solvent. Environ Technol 45:4820–4833. https://doi.org/10.1080/09593330.2023.2283093
26. Poole CF, Poole SK (2011) Ionic liquid stationary phases for gas chromatography. J Sep Sci 34:888–900. https://doi.org/10.1002/jssc.201000724
27. Tang B, Bi W, Row KH (2013) Using poly([1-vinyl-3-hexylimidazolium][bis(trifluoromethylsulfonyl)imide]) to adsorb bio-ethanol from a Chamaecyparis obtuse leaves fermentation broth. Bioresour Technol 137:25–32. https://doi.org/10.1016/j.biortech.2013.03.113
28. Abo-Hamad A, Hayyan M, AlSaadi MA, Hashim MA (2015) Potential applications of deep eutectic solvents in nanotechnology. Chem Eng J 273:551–567. https://doi.org/10.1016/j.cej.2015.03.091
29. Alqahtani AS (2024) Indisputable roles of different ionic liquids, deep eutectic solvents and nanomaterials in green chemistry for sustainable organic synthesis. J Mol Liq 399:124469. https://doi.org/10.1016/j.molliq.2024.124469
30. Gore S, Baskaran S, Koenig B (2011) Efficient synthesis of 3,4-dihydropyrimidin-2-ones in low melting tartaric acid–urea mixtures. Green Chem 13:1009. https://doi.org/10.1039/c1gc00009h
31. Chavda V, Patel B, Singh S et al (2023) A sustainable approach for the adsorption of methylene blue from an aqueous background: an adsorbent based on DES/CGS modified GO@ZrO_2. RSC Sustain 1:2038–2057. https://doi.org/10.1039/D3SU00236E
32. Shah PA, Chavda V, Hirpara D et al (2024) Advances in deep eutectic solvent-based synthesis of nanomaterials for environmental remediation. J Mol Liq 416:126482. https://doi.org/10.1016/J.MOLLIQ.2024.126482

33. Chavda V, Hirpara D, Rao V, Kumar S (2024) Deep eutectic solvent/surfactant-engineered GO@TiO_2 advanced material for sustained and ultrafast adsorption of hazardous dye from contaminated aqueous environment. ACS Appl Eng Mater 2:604–617. https://doi.org/10.1021/acsaenm.3c00696
34. Shah PA, Shrivastav PS, Sharma VS, Chavda V (2023) Uncovering the green frontier: harnessing deep eutectic solvents for sustainable bioanalysis. Bioanalysis 15:815–821. https://doi.org/10.4155/bio-2023-0088
35. Shah PA, Chavda V, Hirpara D et al (2023) Exploring the potential of deep eutectic solvents in pharmaceuticals: challenges and opportunities. J Mol Liq 390:123171. https://doi.org/10.1016/j.molliq.2023.123171
36. Chavda V, Hirpara D, Kumar S (2022) GO/ionic surfactant inspired photophysical modulation of rhodamine B in reline with or without additives. J Mol Liq 368:120614. https://doi.org/10.1016/j.molliq.2022.120614
37. Hirpara D, Patel B, Chavda V et al (2022) Micellization and clouding behaviour of an ionic surfactant in a deep eutectic solvent: a case of the reline-water mixture. J Mol Liq 364:119991. https://doi.org/10.1016/J.MOLLIQ.2022.119991
38. Hirpara D, Patel B, Chavda V, Kumar S (2022) Micellization of conventional and gemini surfactants in aquoline: a case of exclusively water based deep eutectic solvent. J Mol Liq 362:119672. https://doi.org/10.1016/J.MOLLIQ.2022.119672
39. Banjare RK, Banjare MK, Behera K et al (2021) Deep eutectic solvents as modulator on the micellization behaviour of cationic surfactants and potential application in human serum albumin aggregation. J Mol Liq 344:117864. https://doi.org/10.1016/J.MOLLIQ.2021.117864
40. Chavda V, Patel HD, Patel AK et al (2024) Deep eutectic solvent engineered GO/ZrO_2 nanofiller for mixed matrix membranes: an efficient hydrogen gas separation. J Appl Polym Sci 141:e55938. https://doi.org/10.1002/app.55938
41. Hirpara D, Chavda V, Hirapara N, Kumar S (2024) Inorganic salt-induced micellar morphologies in deep eutectic solvent: structure and curcumin solubilization. J Mol Liq 411:125761. https://doi.org/10.1016/j.molliq.2024.125761
42. Ma Z, Mi Y, Han X et al (2020) Transformation of ginsenoside via deep eutectic solvents based on choline chloride as an enzymatic reaction medium. Bioprocess Biosyst Eng 43:1195–1208. https://doi.org/10.1007/s00449-020-02314-8
43. Buarque FS, de Souza CEC, Ferreira RM et al (2024) Dissolution and enzymatic hydrolysis of sugarcane bagasse using ionic liquids and deep eutectic solvents. Process Biochem 147:257–267. https://doi.org/10.1016/j.procbio.2024.08.024
44. Zainal-Abidin MH, Hayyan M, Hayyan A, Jayakumar NS (2017) New horizons in the extraction of bioactive compounds using deep eutectic solvents: a review. Anal Chim Acta 979:1–23. https://doi.org/10.1016/j.aca.2017.05.012
45. Vieira V, Prieto MA, Barros L et al (2018) Enhanced extraction of phenolic compounds using choline chloride based deep eutectic solvents from Juglans regia L. Ind Crop Prod 115:261–271. https://doi.org/10.1016/J.INDCROP.2018.02.029
46. Dai Y, Witkamp GJ, Verpoorte R, Choi YH (2015) Tailoring properties of natural deep eutectic solvents with water to facilitate their applications. Food Chem 187:14–19. https://doi.org/10.1016/J.FOODCHEM.2015.03.123
47. Zhang Q, De Oliveira Vigier K, Royer S, Jérôme F (2012) Deep eutectic solvents: syntheses, properties and applications. Chem Soc Rev 41:7108–7146. https://doi.org/10.1039/c2cs35178a
48. Ijardar SP, Singh V, Gardas RL (2022) Revisiting the physicochemical properties and applications of deep eutectic solvents. Molecules 27:1368. https://doi.org/10.3390/molecules27041368
49. Hansen BB, Spittle S, Chen B et al (2021) Deep eutectic solvents: a review of fundamentals and applications. Chem Rev 121:1232–1285. https://doi.org/10.1021/acs.chemrev.0c00385
50. Mandal S, Lahiri S (2022) A review on extraction, preconcentration and speciation of metal ions by sustainable cloud point extraction. Microchem J 175:107150. https://doi.org/10.1016/j.microc.2021.107150

51. Pandey A, Rai R, Pal M, Pandey S (2014) How polar are choline chloride-based deep eutectic solvents? Phys Chem Chem Phys 16:1559–1568. https://doi.org/10.1039/c3cp53456a
52. Huang Y, Feng F, Jiang J et al (2017) Green and efficient extraction of rutin from tartary buckwheat hull by using natural deep eutectic solvents. Food Chem 221:1400–1405. https://doi.org/10.1016/J.FOODCHEM.2016.11.013
53. Wang M, Wang J, Zhang Y et al (2016) Fast environment-friendly ball mill-assisted deep eutectic solvent-based extraction of natural products. J Chromatogr A 1443:262–266. https://doi.org/10.1016/J.CHROMA.2016.03.061
54. Pandey A, Pandey S (2014) Solvatochromic probe behavior within choline chloride-based deep eutectic solvents: effect of temperature and water. J Phys Chem B 118:14652–14661. https://doi.org/10.1021/jp510420h
55. Makoś P, Słupek E, Gębicki J (2020) Hydrophobic deep eutectic solvents in microextraction techniques—a review. Microchem J 152:104384
56. Wang J, Song Z, Cheng H et al (2018) Computer-aided design of ionic liquids as absorbent for gas separation exemplified by CO_2 capture cases. ACS Sustain Chem Eng 6:12025–12035. https://doi.org/10.1021/acssuschemeng.8b02321
57. Qin H, Song Z, Zeng Q et al (2019) Bifunctional imidazole-PTSA deep eutectic solvent for synthesizing long-chain ester IBIBE in reactive extraction. AICHE J 65:675–683. https://doi.org/10.1002/aic.16472
58. Vargas-Serna CL, Ochoa-Martínez CI, Vélez-Pasos C (2022) Microwave-assisted extraction of phenolic compounds from pineapple peel using deep eutectic solvents. Horticulturae 8:791. https://doi.org/10.3390/horticulturae8090791
59. Ma W, Liu M, Zhang X et al (2023) An efficient and precipitant-free approach to selectively recover lithium cobalt oxide made for cathode materials using a microwave-assisted deep eutectic solvent. Energy Fuel 37:724–734. https://doi.org/10.1021/acs.energyfuels.2c03242
60. Duru KC, Slesarev GP, Aboushanab SA et al (2022) An eco-friendly approach to enhance the extraction and recovery efficiency of isoflavones from kudzu roots and soy molasses wastes using ultrasound-assisted extraction with natural deep eutectic solvents (NADES). Ind Crop Prod 182:114886. https://doi.org/10.1016/J.INDCROP.2022.114886
61. Zhang L, Wang M (2017) Optimization of deep eutectic solvent-based ultrasound-assisted extraction of polysaccharides from Dioscorea opposita Thunb. Int J Biol Macromol 95:675–681. https://doi.org/10.1016/J.IJBIOMAC.2016.11.096
62. Hayyan M, Mjalli FS, Hashim MA, AlNashef IM (2010) A novel technique for separating glycerine from palm oil-based biodiesel using ionic liquids. Fuel Process Technol 91:116–120. https://doi.org/10.1016/j.fuproc.2009.09.002
63. Vicente FA, Malpiedi LP, e Silva FA et al (2014) Design of novel aqueous micellar two-phase systems using ionic liquids as co-surfactants for the selective extraction of (bio)molecules. Sep Purif Technol 135:259–267. https://doi.org/10.1016/J.SEPPUR.2014.06.045
64. Domínguez-Rodríguez G, Amador-Luna VM, Castro-Puyana M et al (2025) Sustainable strategies to obtain bioactive compounds from citrus peels by supercritical fluid extraction, ultrasound-assisted extraction, and natural deep eutectic solvents. Food Res Int 202:115713. https://doi.org/10.1016/J.FOODRES.2025.115713
65. Smith KB, Bridson RH, Leeke GA (2014) Crystallisation control of paracetamol from ionic liquids. CrystEngComm 16:10797–10803. https://doi.org/10.1039/C4CE01796J
66. Cheng H, Cui Y, Ge Z et al (2021) Insight into the mechanism of tuned extractive desulfurization by aqueous tetrabutylphosphonium bromide. Sep Purif Technol 262:118342. https://doi.org/10.1016/J.SEPPUR.2021.118342
67. Rodríguez García SL, Raghavan V (2022) Green extraction techniques from fruit and vegetable waste to obtain bioactive compounds—a review. Crit Rev Food Sci Nutr 62:6446–6466. https://doi.org/10.1080/10408398.2021.1901651
68. Saini A, Panesar PS, Bera MB (2019) Valorization of fruits and vegetables waste through green extraction of bioactive compounds and their nanoemulsions-based delivery system. Bioresour Bioprocess 6:26. https://doi.org/10.1186/s40643-019-0261-9

69. Socas-Rodríguez B, Torres-Cornejo MV, Álvarez-Rivera G, Mendiola JA (2021) Deep eutectic solvents for the extraction of bioactive compounds from natural sources and agricultural by-products. Appl Sci 11:4897. https://doi.org/10.3390/app11114897
70. Benvenutti L, Zielinski AAF, Ferreira SRS (2019) Which is the best food emerging solvent: IL, DES or NADES? Trends Food Sci Technol 90:133–146. https://doi.org/10.1016/J.TIFS.2019.06.003
71. Samtiya M, Aluko RE, Dhewa T, Moreno-Rojas JM (2021) Potential health benefits of plant food-derived bioactive components: an overview. Foods 10:839. https://doi.org/10.3390/foods10040839
72. Saini A, Kumar A, Panesar PS, Thakur A (2022) Potential of deep eutectic solvents in the extraction of value-added compounds from agro-industrial by-products. Appl Food Res 2:100211. https://doi.org/10.1016/J.AFRES.2022.100211
73. Bonacci S, Di Gioia ML, Costanzo P et al (2020) Natural deep eutectic solvent as extraction media for the main phenolic compounds from olive oil processing wastes. Antioxidants 9:513. https://doi.org/10.3390/antiox9060513
74. Popovic BM, Micic N, Potkonjak A et al (2022) Novel extraction of polyphenols from sour cherry pomace using natural deep eutectic solvents—ultrafast microwave-assisted NADES preparation and extraction. Food Chem 366:130562. https://doi.org/10.1016/J.FOODCHEM.2021.130562
75. Vorobyova V, Skiba M, Miliar Y, Frolenkova S (2021) Enhanced phenolic compounds extraction from apricot pomace by natural deep eutectic solvent combined with ultrasonic-assisted extraction. J Chem Technol Metall 56:919–931
76. Krisanti EA, Saputra K, Arif MM, Mulia K (2019) Formulation and characterization of betaine-based deep eutectic solvent for extraction phenolic compound from spent coffee grounds. AIP Conf Proc 2175(1):020040
77. Vázquez-González M, Fernández-Prior Á, Bermúdez Oria A et al (2020) Utilization of strawberry and raspberry waste for the extraction of bioactive compounds by deep eutectic solvents. LWT 130:109645. https://doi.org/10.1016/J.LWT.2020.109645
78. Pal CBT, Jadeja GC (2020) Microwave-assisted extraction for recovery of polyphenolic antioxidants from ripe mango (*Mangifera indica* L.) peel using lactic acid/sodium acetate deep eutectic mixtures. Food Sci Technol Int 26:78–92. https://doi.org/10.1177/1082013219870010
79. Alibade A, Lakka A, Bozinou E et al (2021) Development of a green methodology for simultaneous extraction of polyphenols and pigments from red winemaking solid wastes (pomace) using a novel glycerol-sodium benzoate deep eutectic solvent and ultrasonication pretreatment. Environments 8:90. https://doi.org/10.3390/environments8090090
80. Hernández-Aguirre OA, Muro C, Hernández-Acosta E et al (2021) Extraction and stabilization of betalains from beetroot (Beta vulgaris) wastes using deep eutectic solvents. Molecules 26:6342. https://doi.org/10.3390/molecules26216342
81. Vasyliev G, Lyudmyla K, Hladun K et al (2022) Valorization of tomato pomace: extraction of value-added components by deep eutectic solvents and their application in the formulation of cosmetic emulsions. Biomass Convers Biorefinery 12:95–111. https://doi.org/10.1007/s13399-022-02337-z
82. Tzani A, Pitterou I, Divani F et al (2022) Green extraction of Greek Propolis using natural deep eutectic solvents (NADES) and incorporation of the NADES-extracts in cosmetic formulation. Sustain Chem 4:8–25. https://doi.org/10.3390/suschem4010002
83. Moore KE, Mangos DN, Slattery AD et al (2016) Wool deconstruction using a benign eutectic melt. RSC Adv 6:20095–20101. https://doi.org/10.1039/C5RA26516A
84. Fatima SR, Ahmed D (2025) RSM-based green extraction of valuable bioactives from chili peppers using choline chloride-urea deep eutectic solvent coupled with microwave heating. Biomass Convers Biorefinery 15:20165. https://doi.org/10.1007/s13399-025-06568-8
85. Mulia K, Fauzia F, Krisanti EA (2019) Polyalcohols as hydrogen-bonding donors in choline chloride-based deep eutectic solvents for extraction of Xanthones from the pericarp of Garcinia mangostana L. Molecules 24:636. https://doi.org/10.3390/molecules24030636

86. Popović BM, Uka D, Boublia A et al (2024) Solubility and extractability enhancement of the main food flavonoids by using choline chloride-based natural deep eutectic solvents. J Mol Liq 415:126333. https://doi.org/10.1016/J.MOLLIQ.2024.126333
87. Zainal-Abidin MH, Hayyan M, Ngoh GC et al (2019) Extraction of bioactive compounds. In: Deep eutectic solvents. Wiley, New York, pp 217–233
88. Meng Z, Zhao J, Duan H et al (2018) Green and efficient extraction of four bioactive flavonoids from Pollen Typhae by ultrasound-assisted deep eutectic solvents extraction. J Pharm Biomed Anal 161:246–253. https://doi.org/10.1016/J.JPBA.2018.08.048
89. Peng X, Duan MH, Yao XH et al (2016) Green extraction of five target phenolic acids from Lonicerae japonicae Flos with deep eutectic solvent. Sep Purif Technol 157:249–257. https://doi.org/10.1016/J.SEPPUR.2015.10.065
90. Yao XH, Zhang DY, Duan MH et al (2015) Preparation and determination of phenolic compounds from Pyrola incarnata Fisch. with a green polyols based-deep eutectic solvent. Sep Purif Technol 149:116–123. https://doi.org/10.1016/J.SEPPUR.2015.03.037
91. Cui Q, Peng X, Yao XH et al (2015) Deep eutectic solvent-based microwave-assisted extraction of genistin, genistein and apigenin from pigeon pea roots. Sep Purif Technol 150:63–72. https://doi.org/10.1016/J.SEPPUR.2015.06.026
92. Zhao B-Y, Xu P, Yang F-X et al (2015) Biocompatible deep eutectic solvents based on choline chloride: characterization and application to the extraction of rutin from *Sophora japonica*. ACS Sustain Chem Eng 3:2746–2755. https://doi.org/10.1021/acssuschemeng.5b00619
93. Rashid R, Mohd Wani S, Manzoor S et al (2023) Green extraction of bioactive compounds from apple pomace by ultrasound assisted natural deep eutectic solvent extraction: optimisation, comparison and bioactivity. Food Chem 398:133871. https://doi.org/10.1016/J.FOODCHEM.2022.133871
94. Xia B, Yan D, Bai Y et al (2015) Determination of phenolic acids in Prunella vulgaris L.: a safe and green extraction method using alcohol-based deep eutectic solvents. Anal Methods 7:9354–9364. https://doi.org/10.1039/C5AY02035B
95. Dheyab AS, Abu Bakar MF, AlOmar M et al (2021) Deep eutectic solvents (DESs) as green extraction media of beneficial bioactive phytochemicals. Separations 8:176. https://doi.org/10.3390/separations8100176
96. Sakti AS, Saputri FC, Mun'im A (2019) Optimization of choline chloride-glycerol based natural deep eutectic solvent for extraction bioactive substances from Cinnamomum burmannii barks and Caesalpinia sappan heartwoods. Heliyon 5:e02915. https://doi.org/10.1016/j.heliyon.2019.e02915
97. Macchioni V, Carbone K, Cataldo A et al (2021) Lactic acid-based deep natural eutectic solvents for the extraction of bioactive metabolites of Humulus lupulus L.: supramolecular organization, phytochemical profiling and biological activity. Sep Purif Technol 264:118039. https://doi.org/10.1016/J.SEPPUR.2020.118039
98. Fanali C, Gallo V, Della Posta S et al (2021) Choline chloride–lactic acid-based NADES as an extraction medium in a response surface methodology-optimized method for the extraction of phenolic compounds from hazelnut skin. Molecules 26:2652. https://doi.org/10.3390/molecules26092652
99. Ivanović M, Grujić D, Cerar J et al (2022) Extraction of bioactive metabolites from Achillea millefolium L. with choline chloride based natural deep eutectic solvents: a study of the antioxidant and antimicrobial activity. Antioxidants 11:724. https://doi.org/10.3390/antiox11040724
100. Bakirtzi C, Triantafyllidou K, Makris DP (2016) Novel lactic acid-based natural deep eutectic solvents: efficiency in the ultrasound-assisted extraction of antioxidant polyphenols from common native Greek medicinal plants. J Appl Res Med Aromat Plants 3:120–127. https://doi.org/10.1016/J.JARMAP.2016.03.003
101. García A, Rodríguez-Juan E, Rodríguez-Gutiérrez G et al (2016) Extraction of phenolic compounds from virgin olive oil by deep eutectic solvents (DESs). Food Chem 197:554–561. https://doi.org/10.1016/J.FOODCHEM.2015.10.131

102. Abdul Hadi N, Ng MH, Choo YM et al (2015) Performance of choline-based deep eutectic solvents in the extraction of tocols from crude palm oil. J Am Oil Chem Soc 92:1709–1716. https://doi.org/10.1007/s11746-015-2720-6
103. Wei Z, Qi X, Li T et al (2015) Application of natural deep eutectic solvents for extraction and determination of phenolics in Cajanus cajan leaves by ultra performance liquid chromatography. Sep Purif Technol 149:237–244. https://doi.org/10.1016/J.SEPPUR.2015.05.015
104. Dai Y, Witkamp G-J, Verpoorte R, Choi YH (2013) Natural deep eutectic solvents as a new extraction media for phenolic metabolites in Carthamus tinctorius L. Anal Chem 85:6272–6278. https://doi.org/10.1021/ac400432p
105. Nam MW, Zhao J, Lee MS et al (2015) Enhanced extraction of bioactive natural products using tailor-made deep eutectic solvents: application to flavonoid extraction from Flos sophorae. Green Chem 17:1718–1727. https://doi.org/10.1039/C4GC01556H
106. Tien NNT, Le NL, Khoi TT, Richel A (2022) Optimization of microwave-ultrasound-assisted extraction (MUAE) of pectin from dragon fruit peels using natural deep eutectic solvents (NADES). J Food Process Preserv 46. https://doi.org/10.1111/jfpp.16117
107. Liu Y, Friesen JB, McAlpine JB et al (2018) Natural deep eutectic solvents: properties, applications, and perspectives. J Nat Prod 81:679–690. https://doi.org/10.1021/acs.jnatprod.7b00945
108. Elgharbawy A, Hayyan A, Hayyan M et al (2019) Natural deep eutectic solvent-assisted pectin extraction from pomelo peel using sonoreactor: experimental optimization approach. Processes 7:416. https://doi.org/10.3390/pr7070416
109. Chen M, Lahaye M (2021) Natural deep eutectic solvents pretreatment as an aid for pectin extraction from apple pomace. Food Hydrocoll 115:106601. https://doi.org/10.1016/J.FOODHYD.2021.106601
110. Castro VIB, Craveiro R, Silva JM et al (2018) Natural deep eutectic systems as alternative nontoxic cryoprotective agents. Cryobiology 83:15–26. https://doi.org/10.1016/J.CRYOBIOL.2018.06.010
111. Zhao Y, Wang P, Zheng W et al (2019) Three-stage microwave extraction of cumin (Cuminum cyminum L.) seed essential oil with natural deep eutectic solvents. Ind Crops Prod 140:111660. https://doi.org/10.1016/J.INDCROP.2019.111660
112. Torres-Vega J, Gómez-Alonso S, Pérez-Navarro J, Pastene-Navarrete E (2020) Green extraction of alkaloids and polyphenols from Peumus boldus leaves with natural deep eutectic solvents and profiling by HPLC-PDA-IT-MS/MS and HPLC-QTOF-MS/MS. Plants 9:242. https://doi.org/10.3390/plants9020242
113. Pavlić B, Mrkonjić Ž, Teslić N et al (2022) Natural deep eutectic solvent (NADES) extraction improves polyphenol yield and antioxidant activity of wild thyme (Thymus serpyllum L.) extracts. Molecules 27:1508. https://doi.org/10.3390/molecules27051508
114. Hewage A, Olatunde OO, Nimalaratne C et al (2024) Improved protein extraction technology using deep eutectic solvent system for producing high purity fava bean protein isolates at mild conditions. Food Hydrocoll 147:109283. https://doi.org/10.1016/J.FOODHYD.2023.109283
115. van den Bruinhorst A, Raes S, Maesara SA et al (2019) Hydrophobic eutectic mixtures as volatile fatty acid extractants. Sep Purif Technol 216:147–157. https://doi.org/10.1016/J.SEPPUR.2018.12.087
116. Abbott AP, Capper G, Davies DL et al (2005) Selective extraction of metals from mixed oxide matrixes using choline-based ionic liquids. Inorg Chem 44:6497–6499. https://doi.org/10.1021/ic0505450
117. van Osch DJGP, Parmentier D, Dietz CHJT et al (2016) Removal of alkali and transition metal ions from water with hydrophobic deep eutectic solvents. Chem Commun 52:11987–11990. https://doi.org/10.1039/C6CC06105B
118. Shahbaz K, Mjalli FS, Hashim MA, AlNashef IM (2010) Using deep eutectic solvents for the removal of glycerol from palm oil-based biodiesel. J Appl Sci 10:3349–3354. https://doi.org/10.3923/jas.2010.3349.3354

119. Shahbaz K, Mjalli FS, Hashim MA, AlNashef IM (2011) Using deep eutectic solvents based on methyl triphenyl phosphunium bromide for the removal of glycerol from palm-oil-based biodiesel. Energy Fuel 25:2671–2678. https://doi.org/10.1021/ef2004943
120. Shahbaz K, Mjalli FS, Hashim MA, AlNashef IM (2011) Eutectic solvents for the removal of residual palm oil-based biodiesel catalyst. Sep Purif Technol 81:216–222. https://doi.org/10.1016/J.SEPPUR.2011.07.032
121. Shahbaz K, Baroutian S, Mjalli FS et al (2012) Prediction of glycerol removal from biodiesel using ammonium and phosphunium based deep eutectic solvents using artificial intelligence techniques. Chemom Intell Lab Syst 118:193–199. https://doi.org/10.1016/J.CHEMOLAB.2012.06.005
122. Sanad MMS, Toghan A (2024) Industrial perspectives for electrochemical extraction of metals from primary and secondary resources. Inorg Chem Commun 167:112710. https://doi.org/10.1016/J.INOCHE.2024.112710
123. Panda P, Mishra S (2023) Deep eutectic solvents: physico-chemical properties and their use for recovery of metal values from waste products. J Mol Liq 390:123070. https://doi.org/10.1016/J.MOLLIQ.2023.123070
124. Guo M, Deng R, Gao M et al (2024) Sustainable recovery of metals from e-waste using deep eutectic solvents: advances, challenges, and perspectives. Curr Opin Green Sustain Chem 47:100913. https://doi.org/10.1016/j.cogsc.2024.100913
125. Fan Y, Kong Y, Jiang P et al (2023) Development and challenges of deep eutectic solvents for cathode recycling of end-of-life lithium-ion batteries. Chem Eng J 463:142278. https://doi.org/10.1016/J.CEJ.2023.142278
126. Zhu A, Bian X, Han W et al (2023) The application of deep eutectic solvents in lithium-ion battery recycling: a comprehensive review. Resour Conserv Recycl 188:106690. https://doi.org/10.1016/J.RESCONREC.2022.106690
127. Peeters N, Janssens K, de Vos D et al (2022) Choline chloride–ethylene glycol based deep-eutectic solvents as lixiviants for cobalt recovery from lithium-ion battery cathode materials: are these solvents really green in high-temperature processes? Green Chem 24:6685–6695. https://doi.org/10.1039/D2GC02075K
128. Schiavi PG, Altimari P, Sturabotti E et al (2022) Decomposition of deep eutectic solvent aids metals extraction in lithium-ion batteries recycling. ChemSusChem 15:e202200966. https://doi.org/10.1002/cssc.202200966
129. Li T, Xiong Y, Yan X et al (2022) Closed-loop cobalt recycling from spent lithium-ion batteries based on a deep eutectic solvent (DES) with easy solvent recovery. J Energy Chem 72:532–538. https://doi.org/10.1016/J.JECHEM.2022.05.008
130. Mirshokraee SA, Muhyuddin M, Morina R et al (2023) Upcycling of waste lithium-cobalt-oxide from spent batteries into electrocatalysts for hydrogen evolution reaction and oxygen reduction reaction: a strategy to turn the trash into treasure. J Power Sources 557:232571. https://doi.org/10.1016/J.JPOWSOUR.2022.232571
131. Tang S, Yang Z, Zhang M, Guo M (2023) A simple green method for in-situ selective extraction of Li from spent LiFePO4 batteries by synergistic effect of deep-eutectic solvent and ozone. Environ Res 239:117393. https://doi.org/10.1016/J.ENVRES.2023.117393
132. Carreira ARF, Nogueira A, Crema APS et al (2023) Super concentrated HCl in a deep eutectic solvent as media for the integrated leaching and separation of metals from end-of-life lithium-ion batteries. Chem Eng J 475:146374. https://doi.org/10.1016/J.CEJ.2023.146374
133. Liao Y, Gong S, Wang G et al (2022) A novel ternary deep eutectic solvent for efficient recovery of critical metals from spent lithium-ion batteries under mild conditions. J Environ Chem Eng 10:108627. https://doi.org/10.1016/J.JECE.2022.108627
134. Huang F, Li T, Yan X et al (2022) Ternary deep eutectic solvent (DES) with a regulated rate-determining step for efficient recycling of lithium cobalt oxide. ACS Omega 7:11452–11459. https://doi.org/10.1021/acsomega.2c00742
135. Zhang Y, Wang F, Zhang W et al (2024) Recyclable deep eutectic solvents for recycling LiCoO2 from spent lithium-ion batteries with high selectivity. Sep Purif Technol 330:125498. https://doi.org/10.1016/J.SEPPUR.2023.125498

136. Cao S, Ma Y, Yang L et al (2023) Designing low-cost, green, and recyclable deep eutectic solvents for selective separation and recovery of valuable metals from spent Li-ion batteries. ACS Sustain Chem Eng 11:16984–16994. https://doi.org/10.1021/acssuschemeng.3c04802
137. Kozhevnikova AV, Zinov'eva IV, Zakhodyaeva YA et al (2022) Application of hydrophobic deep eutectic solvents in extraction of metals from real solutions obtained by leaching cathodes from end-of-life Li-ion batteries. Processes 10:2671. https://doi.org/10.3390/pr10122671
138. Łukomska A, Wiśniewska A, Dąbrowski Z et al (2022) Recovery of metals from electronic waste-printed circuit boards by ionic liquids, DESs and organophosphorous-based acid extraction. Molecules 27:4984. https://doi.org/10.3390/molecules27154984
139. Zhao Q, Ma S, Ho W et al (2023) Simple and environmentally friendly metal recovery from waste printed circuit boards by using deep eutectic solvents. J Clean Prod 421:138508. https://doi.org/10.1016/J.JCLEPRO.2023.138508
140. Jacobson B, Li S, Marin Rivera R et al (2023) A mechanistic study identifying improved technology critical metal delamination from printed circuit boards at lower power sonications in a deep eutectic solvent. Ultrason Sonochem 101:106701. https://doi.org/10.1016/J.ULTSONCH.2023.106701
141. Hartley JM, Scott S, Marin Rivera R et al (2023) Tailoring lixiviant properties to optimise selectivity in E-waste recycling. RSC Sustain 1:107–116. https://doi.org/10.1039/D2SU00038E
142. Marin Rivera R, Zante G, Hartley JM et al (2022) Catalytic dissolution of metals from printed circuit boards using a calcium chloride–based deep eutectic solvent. Green Chem 24:3023–3034. https://doi.org/10.1039/D1GC04694B
143. Marin Rivera R, Elgar CE, Jacobson B et al (2024) Ultra-fast extraction of metals from a printed circuit board using high power ultrasound in a calcium chloride-based deep eutectic solvent. RSC Sustain 2:403–415. https://doi.org/10.1039/D3SU00147D
144. Liu K, Wang M, Zhang Q et al (2024) Negative-carbon recycling of copper from waste as secondary resources using deep eutectic solvents. J Hazard Mater 465:133258. https://doi.org/10.1016/J.JHAZMAT.2023.133258
145. Pramanik S, Kumari A, Sinha MK et al (2023) Valorization of phosphor powder of waste fluorescent tubes with an emphasis on the recovery of terbium oxide (Tb4O7). Sep Purif Technol 322:124332. https://doi.org/10.1016/J.SEPPUR.2023.124332
146. Dhawan N, Tanvar H (2022) A critical review of end-of-life fluorescent lamps recycling for recovery of rare earth values. Sustain Mater Technol 32:e00401. https://doi.org/10.1016/J.SUSMAT.2022.E00401
147. Jiang T, Singh S, Dunn KA, Liang Y (2022) Optimizing leaching of rare earth elements from red mud and spent fluorescent lamp phosphors using Levulinic acid. Sustainability 14:9682. https://doi.org/10.3390/su14159682
148. Binnemans K, Jones PT (2023) Ionic liquids and deep-eutectic solvents in extractive metallurgy: mismatch between academic research and industrial applicability. J Sustain Metall 9:423–438. https://doi.org/10.1007/s40831-023-00681-6
149. Varghese AT, Malar CG, Seenuvasan M, Jayapradha V (2024) Neodymium recovery from permanent magnets: a breakthrough approach using deep eutectic solvents and mica. J Mol Liq 394:123690. https://doi.org/10.1016/J.MOLLIQ.2023.123690
150. Yang Q, Li Y, Li B et al (2024) Selective leaching and recovery of neodymium from NdFeB carbonyl residues. Sep Purif Technol 329:125137. https://doi.org/10.1016/J.SEPPUR.2023.125137
151. Prusty S, Pradhan S, Mishra S (2023) Extraction and separation studies of Nd/Fe and Sm/Co by deep eutectic solvent containing Aliquat 336 and glycerol. J Chem Technol Biotechnol 98:1631–1641. https://doi.org/10.1002/jctb.7381
152. Yu G, Ni S, Gao Y et al (2024) Recovery of rare earth metal oxides from NdFeB magnet leachate by hydrophobic deep eutectic solvent extraction, oxalate stripping and calcination. Hydrometallurgy 223:106209. https://doi.org/10.1016/J.HYDROMET.2023.106209

153. Kollau LJBM, Vis M, van den Bruinhorst A et al (2018) Quantification of the liquid window of deep eutectic solvents. Chem Commun 54:13351–13354. https://doi.org/10.1039/C8CC05815F
154. Silva LP, Martins MAR, Abranches DO et al (2021) Solid-liquid phase behavior of eutectic solvents containing sugar alcohols. J Mol Liq 337:116392. https://doi.org/10.1016/J.MOLLIQ.2021.116392
155. Song Z, Wang J, Sundmacher K (2021) Evaluation of COSMO-RS for solid–liquid equilibria prediction of binary eutectic solvent systems. Green Energy Environ 6:371–379. https://doi.org/10.1016/J.GEE.2020.11.020
156. Wojeicchowski JP, Ferreira AM, Abranches DO et al (2020) Using COSMO-RS in the design of deep eutectic solvents for the extraction of antioxidants from rosemary. ACS Sustain Chem Eng 8:12132–12141. https://doi.org/10.1021/acssuschemeng.0c03553
157. Hadj-Kali MK, Hizaddin HF, Wazeer I et al (2017) Liquid-liquid separation of azeotropic mixtures of ethanol/alkanes using deep eutectic solvents: COSMO-RS prediction and experimental validation. Fluid Phase Equilib 448:105–115. https://doi.org/10.1016/J.FLUID.2017.05.021
158. Wang B, Jin C, Shao S et al (2023) Electron-deficient Cu site catalyzed acetylene hydrochlorination. Green Energy Environ 8:1128–1140. https://doi.org/10.1016/J.GEE.2022.01.005
159. Hayyan A, Hadj-Kali MK, Salleh MZM et al (2020) Characterization of tetraethylene glycol-based deep eutectic solvents and their potential application for dissolving unsaturated fatty acids. J Mol Liq 312:113284. https://doi.org/10.1016/J.MOLLIQ.2020.113284
160. Ozturk B, Parkinson C, Gonzalez-Miquel M (2018) Extraction of polyphenolic antioxidants from orange peel waste using deep eutectic solvents. Sep Purif Technol 206:1–13. https://doi.org/10.1016/J.SEPPUR.2018.05.052
161. Ozturk B, Esteban J, Gonzalez-Miquel M (2018) Deterpenation of citrus essential oils using glycerol-based deep eutectic solvents. J Chem Eng Data 63:2384–2393. https://doi.org/10.1021/acs.jced.7b00944
162. Perkins SL, Painter P, Colina CM (2014) Experimental and computational studies of choline chloride-based deep eutectic solvents. J Chem Eng Data 59:3652–3662. https://doi.org/10.1021/je500520h
163. Cheng H, Qi Z (2021) Applications of deep eutectic solvents for hard-to-separate liquid systems. Sep Purif Technol 274:119027. https://doi.org/10.1016/J.SEPPUR.2021.119027
164. Hou Y, Li Z, Ren S, Wu W (2015) Separation of toluene from toluene/alkane mixtures with phosphonium salt based deep eutectic solvents. Fuel Process Technol 135:99–104. https://doi.org/10.1016/J.FUPROC.2014.11.001
165. Verma R, Mohan M, Goud VV, Banerjee T (2018) Operational strategies and comprehensive evaluation of menthol based deep eutectic solvent for the extraction of lower alcohols from aqueous media. ACS Sustain Chem Eng 6:16920–16932. https://doi.org/10.1021/acssuschemeng.8b04255
166. Yu L, Li Z, Huang W et al (2024) Recovery and post-treatment processes for ionic liquids and deep eutectic solvents. J Mol Liq 402:124767. https://doi.org/10.1016/j.molliq.2024.124767
167. Emara S, Kamal M, Sallam IE et al (2024) Application of molecular imprinting approach for alkaloids analysis in food and nutraceuticals: review and perspective. Phytochem Rev 23:459–483. https://doi.org/10.1007/s11101-023-09893-w
168. Li Y, Miao S, Tan J et al (2024) Capillary electrophoresis: a three-year literature review. Anal Chem 96:7799–7816. https://doi.org/10.1021/acs.analchem.4c00857
169. Wang R, Mao Z, Chen Z (2018) Monolithic column with polymeric deep eutectic solvent as stationary phase for capillary electrochromatography. J Chromatogr A 1577:66–71. https://doi.org/10.1016/J.CHROMA.2018.09.046
170. Qin S, Jin F, Gao L et al (2019) Determination of sulfamerazine in aquatic products by molecularly imprinted capillary electrochromatography. R Soc Open Sci 6:190119. https://doi.org/10.1098/rsos.190119
171. Li XX, Zhang LS, Wang C et al (2018) Green synthesis of monolithic column incorporated with graphene oxide using room temperature ionic liquid and eutectic solvents for capillary electrochromatography. Talanta 178:763–771. https://doi.org/10.1016/J.TALANTA.2017.10.014

172. Zhang LS, Zhao QL, Li XX et al (2016) Green synthesis of mesoporous molecular sieve incorporated monoliths using room temperature ionic liquid and deep eutectic solvents. Talanta 161:660–667. https://doi.org/10.1016/J.TALANTA.2016.09.030
173. Zhao X, Fan PR, Mo CE et al (2020) Green synthesis of monolithic enzyme microreactor based on thiol-ene click reaction for enzymatic hydrolysis of protein. J Chromatogr A 1611:460618. https://doi.org/10.1016/J.CHROMA.2019.460618
174. Zhang LS, Gao SP, Huang YP, Liu ZS (2016) Green synthesis of polymer monoliths incorporated with carbon nanotubes in room temperature ionic liquid and deep eutectic solvents. Talanta 154:335–340. https://doi.org/10.1016/J.TALANTA.2016.03.088
175. Zhou X-J, Zhang L-S, Song W-F et al (2018) A polymer monolith incorporating stellate mesoporous silica nanospheres for use in capillary electrochromatography and solid phase microextraction of polycyclic aromatic hydrocarbons and organic small molecules. Microchim Acta 185:444. https://doi.org/10.1007/s00604-018-2964-6
176. Mu Y, Wu X, Huang Y, Liu Z (2019) Investigation of deep eutectic solvents as additives to β-CD for enantiomeric separations of Zopiclone, Salbutamol, and Amlodipine by CE. Electrophoresis 40:1992–1995. https://doi.org/10.1002/elps.201900067
177. Moufawad T, Moura L, Ferreira M et al (2019) First evidence of cyclodextrin inclusion complexes in a deep eutectic solvent. ACS Sustain Chem Eng 7:6345–6351. https://doi.org/10.1021/acssuschemeng.9b00044
178. Kareem MA, Mjalli FS, Hashim MA, AlNashef IM (2012) Liquid–liquid equilibria for the ternary system (phosphonium based deep eutectic solvent–benzene–hexane) at different temperatures: a new solvent introduced. Fluid Phase Equilib 314:52–59. https://doi.org/10.1016/j.fluid.2011.10.024
179. Rodriguez NR, Requejo PF, Kroon MC (2015) Aliphatic–aromatic separation using deep eutectic solvents as extracting agents. Ind Eng Chem Res 54:11404–11412. https://doi.org/10.1021/acs.iecr.5b02611
180. Rodríguez NR, González ASB, Tijssen PMA, Kroon MC (2015) Low transition temperature mixtures (LTTMs) as novel entrainers in extractive distillation. Fluid Phase Equilib 385:72–78. https://doi.org/10.1016/j.fluid.2014.10.044
181. Gjineci N, Boli E, Tzani A et al (2016) Separation of the ethanol/water azeotropic mixture using ionic liquids and deep eutectic solvents. Fluid Phase Equilib 424:1–7. https://doi.org/10.1016/j.fluid.2015.07.048
182. Maugeri Z, Leitner W, Domínguez de María P (2012) Practical separation of alcohol–ester mixtures using deep-eutectic-solvents. Tetrahedron Lett 53:6968–6971. https://doi.org/10.1016/j.tetlet.2012.10.044
183. Li G, Yan C, Cao B et al (2016) Highly efficient I_2 capture by simple and low-cost deep eutectic solvents. Green Chem 18:2522–2527. https://doi.org/10.1039/C5GC02691A

Chapter 7
Deep Eutectic Solvents: A Green Alternative Catalysis Approach for Chemical Transformation

Vishwajit Chavda, Priyanka A. Shah, Darshna Hirpara, Gandhi Mansi, Sanjeev Kumar, and Pranav S. Shrivastav

7.1 Introduction

The challenges related to the environmental catastrophes stemming from climate change have led us to establish more sustainable and efficient synthetic techniques in various fields of chemical sciences. The research community identifies many sustainable alternatives to traditional solvents for minimizing the negative impacts on humans and the environment [1–5].

In synthesis and catalysis, several media types, such as supercritical fluids, ionic liquids (ILs), water, and fluorous solvents, hold a high position [6]. Even though the above-mentioned next-generation solvents follow the trail laid by principles of

V. Chavda (✉)
Department of Chemistry, Institute of Science, Humanities and Liberal Studies, Indus University, Shilaj, Ahmedabad, Gujarat, India
e-mail: vishwajitchavda1998@gmail.com

P. A. Shah
School of Forensic Science, National Forensic Sciences University, Gandhinagar, Gujarat, India

D. Hirpara · S. Kumar
Department of Applied Chemistry, Faculty of Technology and Engineering, The Maharaja Sayajirao University of Baroda, Vadodara, Gujarat, India
e-mail: drksanjeev@gmail.com

G. Mansi
Department of Chemistry, Gujarat Technological University, Chandkheda, Ahmedabad, Gujarat, India

P. S. Shrivastav
Department of Chemistry, School of Sciences, Gujarat University, Ahmedabad, Gujarat, India
e-mail: pranav_shrivastav@yahoo.com

K. Kuperkar et al. (eds.), *Deep Eutectic Solvents*, Environmental Footprints and Eco-design of Products and Processes,
https://doi.org/10.1007/978-3-032-03801-2_7

green chemistry [7], there are many limitations. The simplest media, water, often leads to issues like limited solubility and substrate insolubility. Special cutting-edge instruments are required for supercritical fluids. However, in the case of some ionic liquids, higher toxicity and costs are major concerns. Hence, researchers are still seeking an optimal reaction medium suitable for particular circumstances and practical applicability.

Different domains, such as chemical processes, development of advanced materials, catalysis (including biocatalysis), extraction, and separation methods, utilize the ILs over other traditional solvents [8]. However, the latest investigations on ILs based on phosphonium, pyridinium, and imidazolium have found them less environmentally friendly than previously. Their toxic nature, especially in living organisms, can be traced to their higher chemical and thermal stability, lower biodegradability, and superior water solubility [9]. Even the synthesis processes for some ILs are usually not environmentally generous and require significant amounts of solvents and salts to exchange anions completely. Such limitations, adding to already complicated preparation and purification steps, increase the costs of biocompatible ILs, making it difficult to scale up the process at the industrial level.

Overcoming these issues, deep eutectic solvents (DESs) have emerged for the past few decades as a novel class of solvents owing to their unique physicochemical features, allowing researchers to explore them by sustainable green chemistry principles [10–13]. Following the groundbreaking research by Abbott et al. [14], numerous studies have investigated DESs for the synthesis routes, various characteristics, and potential applications [15–22]. These DES-based solvents can easily be prepared from renewable and sustainable precursors, providing a strong case for their use in eco-friendly solutions [23–27].

In synthetic chemistry, DESs, popularly identified as a unique subclass of ILs, offer an environmentally suitable alternative solution to traditional organic solvents. DESs exhibit a notable improvement in the effectiveness of various organic chemical transformations. DESs are low-melting blends of widely available, recyclable, cheap, and biodegradable constituents [28–34]. These solvents are generally obtained through the combination of quaternary ammonium (hydrogen bond acceptor; HBA) or a metal salt having a simple hydrogen bond donor (HBD) such as amides, acids, alcohols, or amines, as seen in Fig. 7.1 [35]. Compared to the initial constituents, the melting point of the mixed product is significantly lower, usually well below 100 °C. The delocalization of the charge due to the formation of

Fig. 7.1 Diagrammatic illustration of the DES synthesis process

hydrogen bonds between the hydrogen donor moiety and the halide anion is responsible for lowering the melting point of the mixture. Similar to ILs, a prominently advantageous feature of the DESs is the lower vapor pressure, which results in lower volatility and makes them suitable for green catalytic technological applications [35–37].

Different novel approaches have been investigated in green chemistry, such as using ILs [38], polyethylene glycol (PEG) [39, 40], water, supercritical carbon dioxide [41], and solvent-omitted conditions [42], along with the alternative novel energy input techniques including mechanochemical ball-mixing [43, 44], ultrasound [45], and microwave irradiation [46], as well as related combinations [47–49]. DESs have especially been found to pay significant attention to widely differing chemical processes as an effective medium for synthesizing nanomaterials [13]. These studies highlight the multifunctionality of DESs serving as a solvent and a catalyst in various organic reactions [50, 51], assisting in synthesizing biologically active molecules. This versatile nature of DESs has made a huge impact, not limited to the advancement of further scientific explorations but also extending to the optimization of different chemical processes at the industrial level. The DESs are also highly efficient as solvents involving cross-coupling reactions [52–55]. Arguably, the hydrogen bond catalysis accelerated by DES constituents has made several nucleophilic additions and substitution reactions possible, along with the heterocyclic-dehydration processes, for proceeding under milder and more eco-friendly conditions [56].

Organic reactions covering a large spectrum have been identified to have benefited from the addition of DESs, proving their versatile and efficient nature [57–59]. Lipase-catalyzed reactions [60], biotransformations [61], organocatalysis and the multistep processes that combine the organocatalysts and enzymes [62–64], synthesis of polymers and other related materials [65–67], metal-catalyzed [68–70], and metal-mediated transformations [70–72] are some of the applications. This chapter focuses on the potential uses of DESs, particularly as an environmentally sustainable catalyst for different chemical transformations.

7.1.1 Properties of the DESs

The conventional ILs and DESs have widely diverse material properties since they consist of almost comparable traits, especially in their capacity to be used as adaptable chemical solutions to achieve a desired type of chemistry; other useful attributes are lower vapor pressure, a wider solution reach, and nonflammability [73–76]. Viscosity, extremity, density, melting temperature, surface strain, and conductivity are some of the properties of DES that can be tailored according to the requirement. The following sections discuss heat capacity, thermal behavior, primary physicochemical properties, preparation methods, toxicity, thermochromic conduct, and other exceptional properties of DESs.

7.1.1.1 Freezing Point

Synthesis of DESs involves making a mixture of two suitable solids, resulting in a liquid phase due to hydrogen bonds without additives; the resultant liquid phase has a lower melting point than the original solid components [77–79]. One DES is a urea and choline chloride (ChCl) mixture with a 1:2 molar ratio. The melting points of solid precursors urea and ChCl are 132 and 301 °C, respectively. However, the DES solution made from the two has a melting point of 12 °C [70]. The negative ions' interaction with the substrate HBD segment, in the present instance, carbamide, is responsible for the key freezing point. The freezing points of all DESs that have positive evaluations are below 149 °C. However, most of them exhibit freezing points much lower than that, usually below 45 °C, being a positive attribute in many applications. The number of different DESs exhibiting a liquid phase at room temperature is strictly limited despite the large number of possible DESs, made using an expensive scale of amides and ChCl, with a freezing point of less than 90 °C. Owing to their ability to frame hydrogen bond interactions with the ChCl, a mixture of ChCl and 2,2,2-trifluoroacetamide produces liquid DESs at room temperature. The selection of HBD species becomes a crucial point in the synthesis of DES due to the presence of ChCl, which can offer a lower freezing point. The anion from choline-based salts further impacts the solidification point of the DES. For instance, the solidification points for choline salt-based DESs increase $Cl < BF_4 < F < NO_3$ when urea is used; this hints toward a relationship as a function of hydrogen bond quality. A further vital effect is introduced on the solidification point due to the natural molar ratio of salt-to-HBD, such as a mixture of ChCl and urea with the molar proportion 1:2 and 1:1.

7.1.1.2 Density

The density of a solvent is one of its primary characteristics. ChCl-HBD, ChCl-glycerol, and ChCl: urea are some of the basic DESs having densities of ~1.20, 1.18, and 1.24 g/cm^3, respectively. This large difference observed in the densities is attributed to the alternating subatomic association or packing of the DES or to the existence of holes/voids. Taking the example of zinc chloride mixed with urea, a small increment in the density in the resulting DES, as compared to pure urea, is observed due to the decrement of the hole radius [80].

7.1.1.3 Viscosity

Viscosity is also one of the important physical attributes for applying any liquid. Each part of DESs consists of an extensive hydrogen bond network, which decreases the transportability of the unregulated species that exist within. Therefore, compared to other ionic fluids, DESs have more acute conductance and higher homogeneity. Larger particle size and extremely small void volume, along with the Van der

Waals connections, may also contribute to the higher consistency of DESs. Such an alluring property is highly desirable for DESs from the standpoint of applications as green media. Generally, the viscosity of eutectic mixtures depends on the multifaceted segments present in a particular DES. A direct relation between the electrical energy of ionic liquids and their fluidity has been generally observed [81], which has also been observed for many DESs. The electrical energy corresponding to various DESs depends on the temperature [82]. The binding of ChCl to urea, glycerol, 1,3-ethanediol, and other such hydrogen bond donors is expected, due to which these liquids are observed to be dipolar [83].

7.2 DESs in Catalytic Reactions

DESs have recently been taking center stage in the field of solvents for catalytic reactions. This success is due to excellent physical characteristics, including a wide viscosity range, easy recovery, higher solubility, and lower toxicity observed in most DES solvents. Modern catalytic activity exhibited by DESs allows them to be potential candidates for sustainable replacements for traditional solvents and catalysts in various chemical reactions [71, 84]. Such eutectic blends are effective in organic synthesis and catalytic chemical processes, including amidation, cross-coupling, oxidation, esterification, condensation, and cycloaddition reactions. Notably, the capacity to reuse the DESs in homogeneous and heterogeneous catalytic systems makes them highly desirable in more sustainable and practical ventures. The facile synthesis procedures, such as affordable and sustainable precursors to reactants, decreased toxic effects, and the disintegrating nature of DESs, particularly the natural deep eutectic solvents (NADESs), make products attractive to conventional solvents and catalysts. Recycling and the capacity to be reused and recycled throughout different chemical reaction cycles further reinforce their outstanding position as green and sustainable alternatives [52].

7.2.1 Heterocyclic Compound Synthesis in DESs

7.2.1.1 Synthesis of Pyrroles and Furans

The Paal-Knorr reaction is a widely employed method of producing furans and pyrroles, which are important heteroaromatic compounds [85]. Handy et al. [86] have reported a Paal-Knorr synthesis process using DES, formed using ChCl and either urea or glycerol, which shows the potential of being utilized as a catalyst and a solvent for the reaction. An equimolar solution of amine and dione (1 M) is stirred in a DES at 80 °C, resulting in pyrrole derivatives according to their method (Scheme 7.1). This reaction can also be adopted for less nucleophilic anilines by extending the reaction time while still producing quantitative yields of pyrrole derivatives. The

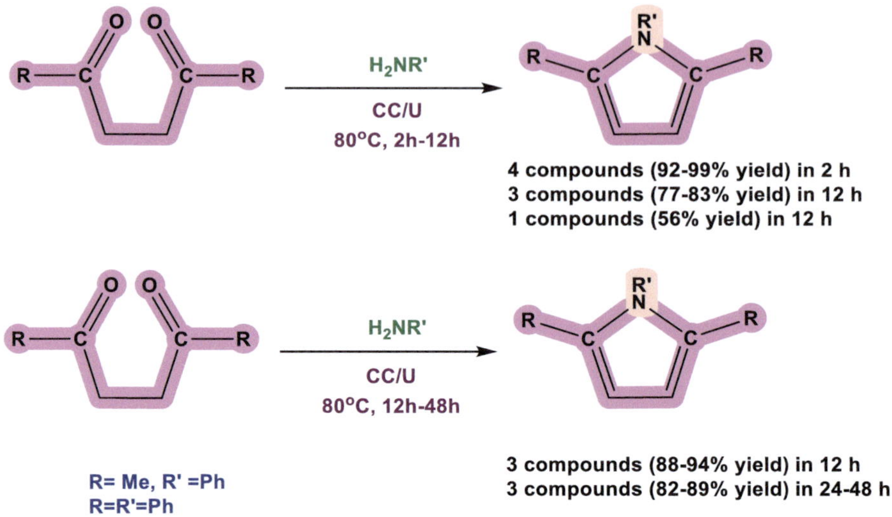

Scheme 7.1 Paal-Knorr synthesis of pyrrole

less reactive diones need reaction times of up to 24 h to enhance the product yield without losing the efficiency of the process. It is to be noted that additional Brønsted or Lewis acid catalysts are not required in this approach, which highlights the mild and environmentally friendly nature of the reaction conditions since DES is non-toxic and reusable. Moreover, furans have been obtained as side products while synthesizing pyrrole using nucleophilic amines. In the amine-free conditions, the furan yield was much higher comparatively (Scheme 7.2). Hence, due to this dual utility, the versatile nature of this green synthetic strategy can be confirmed.

In 2015, a one-pot, four-component synthesis of pyrroles with DES as a catalyst and nitromethane, 1,3-dicarbonyl, amines, and aldehyde compounds as reactants was developed by Hu et al. [87]. Scheme 7.3 represents the reaction schematically. DES utilized in this process consists of ChCl and malonic acid (MA). Among the different available systems, ChCl/MA is the most effective catalytic medium for the reactions involving aniline, 4-chlorobenzaldehyde, nitromethane, and acetylacetone in the presence of various hydrogen bond-donating species and ChCl. Only the aromatic amines with ortho-substituents show a lowering of reactivity among the wide range of amines, suggesting a relatively large steric hindrance effect. The ChCl/MA system is found to improve the electrophilic nature of acetylacetone through hydrogen bond interactions with the respective carbonyl group, accelerating the nucleophilic attack through aniline for the formation of the enaminone intermediate (A) as represented in Scheme 7.4.

The Knoevenagel condensation process has been shown to be facilitated by the hydrogen bond that exists between the carbonyl groups from 4-chlorobenzaldehyde and the ChCl/MA-based DES whenever nitromethane is used in the intermediates. A critical aspect is the presence of intermolecular hydrogen bonding to generate an additional attractive force among the molecules. This approach is further improved by the higher catalytic efficiency and the superior recyclability of the ChCl/MA

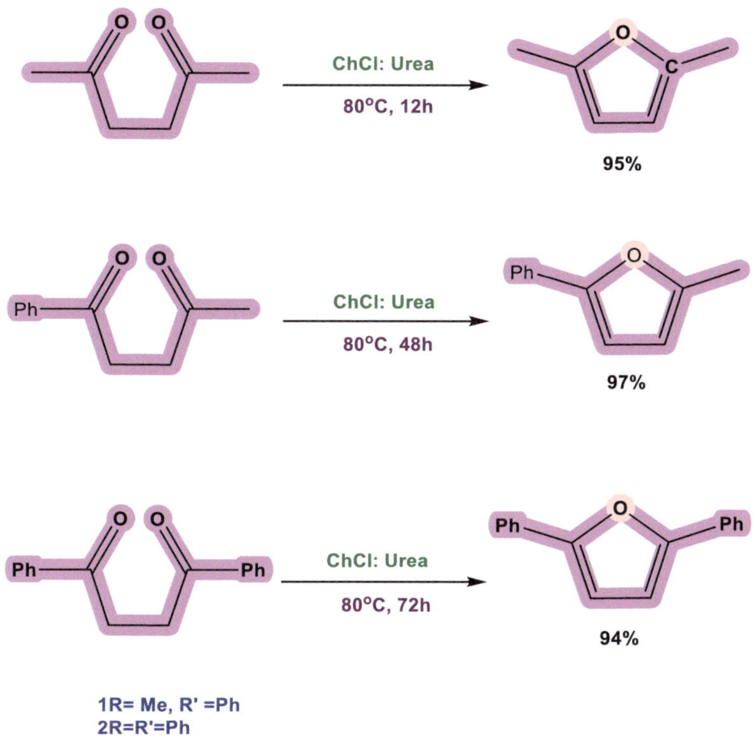

Scheme 7.2 Paal-Knorr synthesis of furan

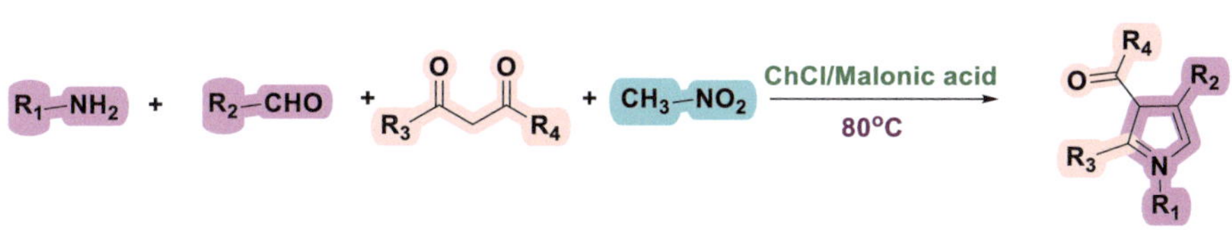

Scheme 7.3 Diagrammatic illustration of the synthesis of pyrrole using DES based on ChCl/MA

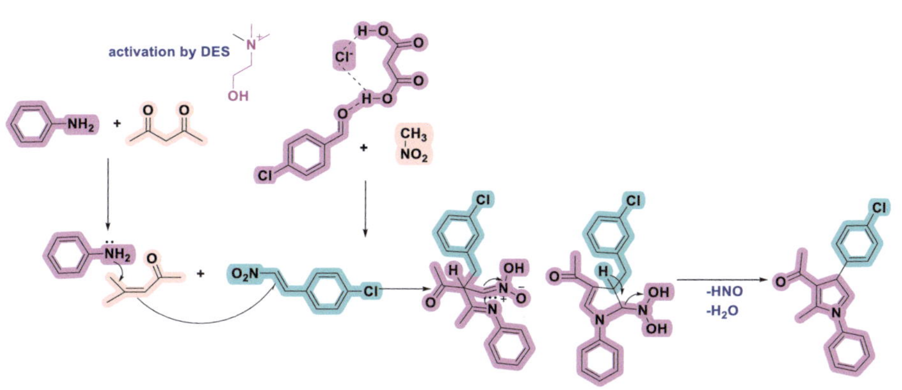

Scheme 7.4 A graphical illustration of a potential method for pyrrole synthesis

DES, making it a better approach than the conventionally used techniques for the large-scale production of multi-substituted pyrroles.

7.2.1.2 Synthesis of Thiophene Derivatives

Mancuso et al. [88] have investigated the synthesis process for thiophene derivatives using DES. S-heterocyclization and iodocyclization of the 1-mercapto-3-yn-2-ols in the presence of a ChCl/glycerol mixture (ChCl/Gly, 1:2 molar ratio) were carried out as an unconventional green solvent at room temperature, as depicted in Scheme 7.5. This process yields eight 3-iodothiophene derivatives with aryl or alkyl substituents, reaching 62–79% yields. The iodocyclization of methylthiophenylacetylenes into benzothiophene derivatives has also been successfully achieved via a DES made from ChCl and urea with a molar ratio of 1:2 (Scheme 7.5) [89]. The optimal efficiency of the reaction was achieved with I_2/KI at 60 °C, whereas the yield started decreasing with further increments in the temperature of the reaction. A wide range of substituted derivatives involving the iodobenzothiazole framework have been successfully synthesized, producing several different compounds with an efficiency ranging from 62% to 89%.

7.2.1.3 Synthesis of 1,4-Dihydropyridines

Famous for their therapeutic potential, 1,4-Dihydropyridines (1,4-DHPs) are highly regarded for their bioactivity. Such molecules are generally known for their capacity to modulate calcium channels and also find applications in hypertension treatment

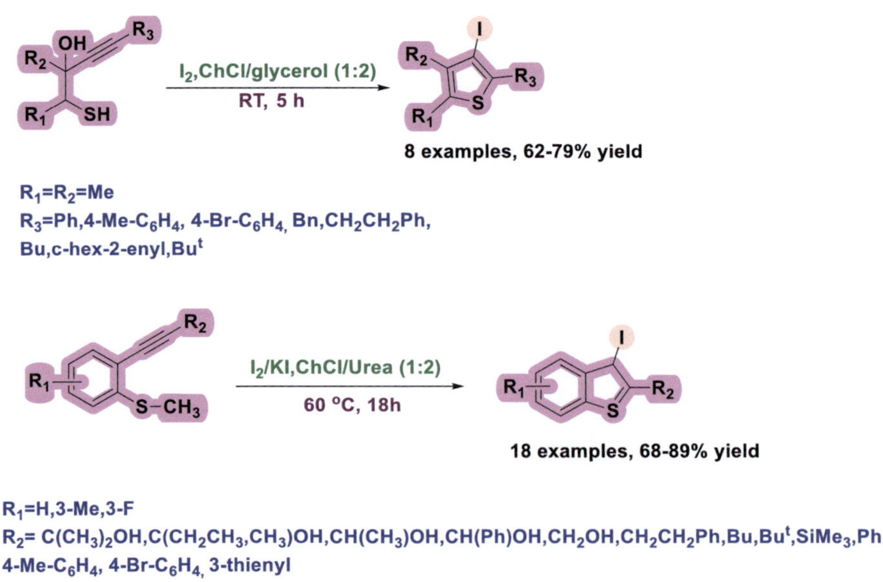

Scheme 7.5 Thiophene derivatives are synthesized using DESs as solvents

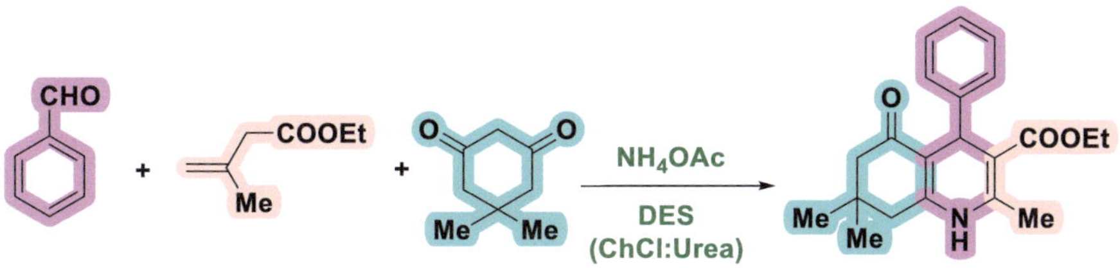

Scheme 7.6 Synthesis of 1,4-dihydropyridines derivatives

[90]. As shown in Scheme 7.6, Pednekar et al. [91] reported on inventing a one-pot, multicomponent approach for making 1,4-dihydropyridine derivatives via combining dimedone, ammonium acetate, aldehydes, and ethyl acetate with biocompatible DESs composed of ChCl and urea. This study systematically investigated water, different polar/nonpolar traditional organic solvents, and DESs to optimize reaction conditions. In aiding the synthesis of 1,4-DHPs, the results show that DESs perform better than traditional organic solvents. This green synthetic method has shown higher product yields and more efficient recyclability, resulting in insignificant losses in catalytic efficiency up to five cycles.

7.2.1.4 Synthesis of Dihydropyrimidinones

The synthesis process of 3,4-dihydropyrimidin-2(1*H*)-one derivative through the Biginelli reaction, which involves a chemical reaction among aromatic and aliphatic aldehydes, urea, and 1,3-dicarbonyl compounds, has been documented by Azizi et al. [92] as depicted in Scheme 7.7. The reaction uses a simple tin (II) chloride-based DES that acts as a catalyst and the necessary medium. Five different ChCl-based DESs were initially checked under various conditions to optimize the reaction conditions. Larger yields of the products that are required have been obtained from the ChCl:$SnCl_2$ DES, which proved to be the most effective of all. This green synthesis approach's specific conditions permitted the preparation of a wide range of molecules, which comprise urea, 1,3-dicarbonyl compounds, and aliphatic and aromatic aldehydes, yielding dihydropyrimidinones in high amounts with an increased rate of reactions and quality.

7.2.1.5 Trisubstituted and Tetrasubstituted Imidazoles Synthesis

Azizi et al. [93] have also demonstrated an efficient but straightforward one-pot, four-component reaction for preparing tetrasubstituted imidazoles. The reaction uses aldehydes, ammonium acetate, 1,2-diphenylethane-1,2-dione, and amines with DES, which was stabilized with nanoparticles of iron oxide at 60 °C. The reaction yielded different imidazole derivatives in moderate to good amounts (Scheme 7.8). Other competing methods involve solvent-free conditions and solvents and fail to produce sufficient outcomes. Interestingly, in the absence of

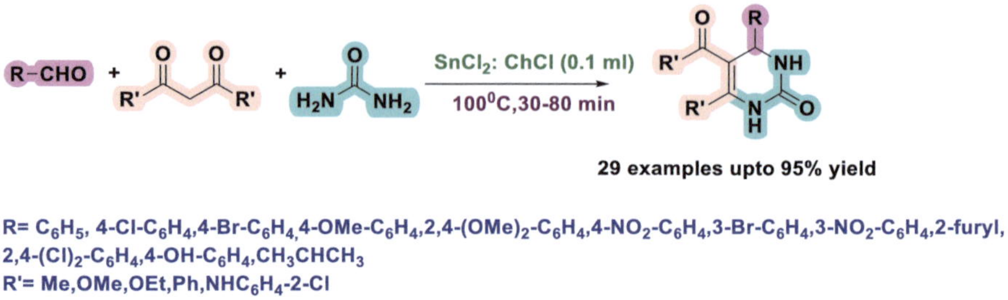

Scheme 7.7 Synthesis of dihydropyrimidinones

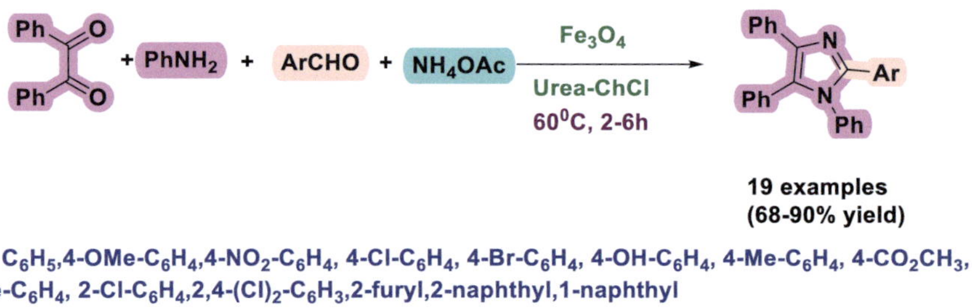

Scheme 7.8 1,2,4,5-Tetrasubstituted imidazoles synthesis

ferrofluids, the same DESs produced only 25% of the yield compared to when ferrofluid was added. The multifunctional role of DES was highlighted in this reaction, viz., of the reaction medium, of a hydrogen bond catalyst, and of stabilizing Fe_3O_4 nanoparticles.

Using a Brønsted acidic DES consisting of ChCl and *p*-toluenesulfonic acid, the researchers Wang and colleagues [94] explored an approach to make 1,2,4,5-tetrasubstituted and 2,4,5-trisubstituted imidazoles in a single pot with higher yields. The identical compounds were established by Amiri et al. via a condensation reaction involving benzil, ammonium acetate/aniline, and aldehydes, with DES functioning as a catalyst at elevated temperatures of 100 °C without the use of solvents. The ChCl·2ZnCl$_2$ DES was discovered to be the most effective catalyst for the preparation of 2,4,5-triphenyl imidazole, following a series of initial investigations that were carried out to optimize the crucial conditions for the reaction [95]. Later, another one-pot four-component condensation reaction involving benzil (1 mmol), ammonium acetate (1.5 mmol), aniline (1 mmol), and benzaldehyde (1 mmol) was investigated for synthesizing tetrasubstituted imidazole. This investigation also indicated that ChCl·2ZnCl$_2$ ionic liquid performs significantly better than other catalysts for this specific reaction. Under the optimized conditions, 1,2,4,5-tetrasubstituted and 2,4,5-trisubstituted imidazoles have been reported to be synthesized via three-component and four-component reactions (Scheme 7.9).

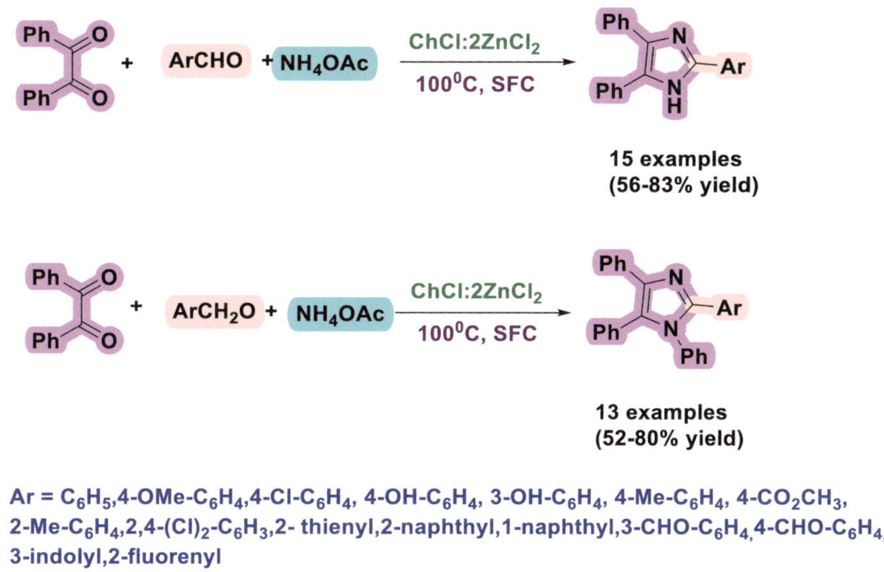

Scheme 7.9 2,4,5-Trisubstituted and 1,2,4,5-tetrasubstituted imidazoles synthesis

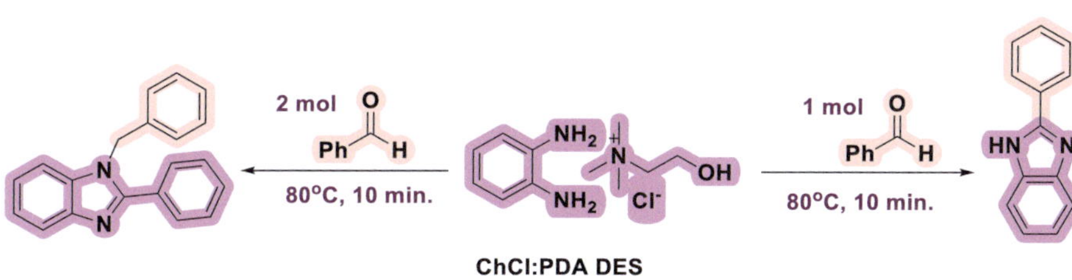

Scheme 7.10 Synthesis of benzimidazoles in DES

7.2.1.6 Synthesis of Benzimidazoles and Imidazole-Fused Heterocycles

Di Gioia and team reported an effective and environmentally suitable method for preparing 1,2-disubstituted or 2-substituted benzimidazoles in 2019 [96]. Using *o*-phenylenediamine and different aldehydes as initial materials, this method employs DES as a medium and a reagent (Scheme 7.10). Such an approach offers improved yields and streamlined work-up procedures, which are significantly more advantageous than traditional methods.

The synthesis process for structurally diverse small molecules such as chromenopyrroloimidazoles, imidazopyrrolopyrimidines, imidazoindoles, and indenopyrroloimidazoles utilizing guanidinium chloride and urea-composed DES was reported by Rushell et al. [97], showing the dual-purpose nature of DESs. The first time this DES was utilized as a reusable reaction medium and a catalyst for synthesizing nitrogen heterocyclic compounds via a one-pot, three-component reaction

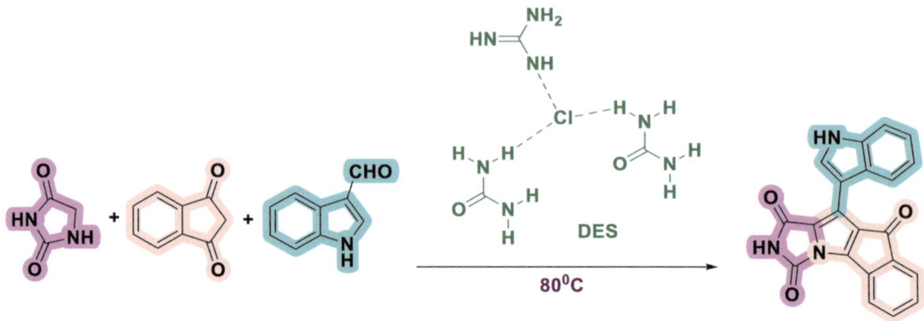

Scheme 7.11 Synthesis of imidazole-fused heterocycles in DES

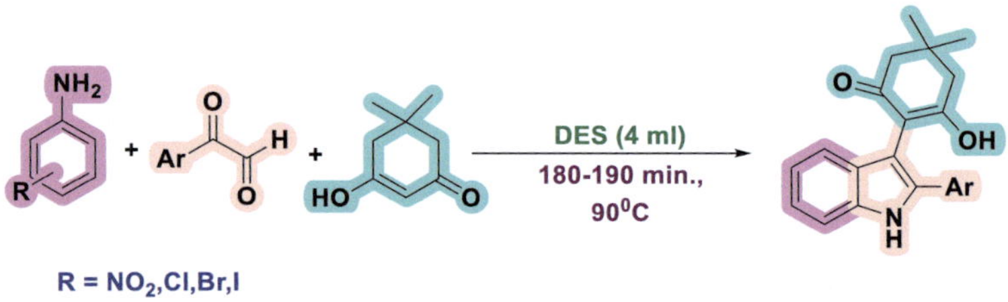

Scheme 7.12 Synthesis of indole derivatives

involving 1,3-diketones, aldehydes, and hydantoin. The significant advantages are mild reaction conditions, higher selectivity, excellent yields, higher atom economy, and shorter reaction times, making it suitable for more similar systems. The schematic of the reaction is presented in Scheme 7.11.

7.2.1.7 Synthesis of Indole Derivatives

An extremely efficient technique to make modified derivatives of indole was recently presented in a 2022 study. This involves interactions including anilines, arylglyoxals, and 1,3-dicarbonyl groups in the form of DES (ChCl and oxalic acid), as shown in acids (Scheme 7.12). This approach utilizes biodegradable components and demonstrates quicker reaction durations with higher yields of the products, making it an eco-friendly and more sustainable approach. The DES catalyst also exhibited superior recyclability, with activity lasting up to at least four reaction cycles [98].

7.2.1.8 Naphthyridines Synthesis

Functionally modified naphthyridines and their derivative compounds are usually found in marine sources. They are widely known for their excellent biological processes, including anti-HCV and antiproliferative effects [99, 100]. Hooshmand

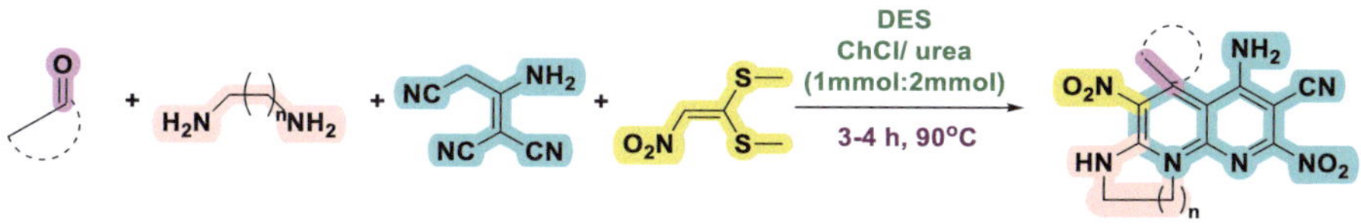

Scheme 7.13 Naphthyridine scaffold synthesis in urea and ChCl

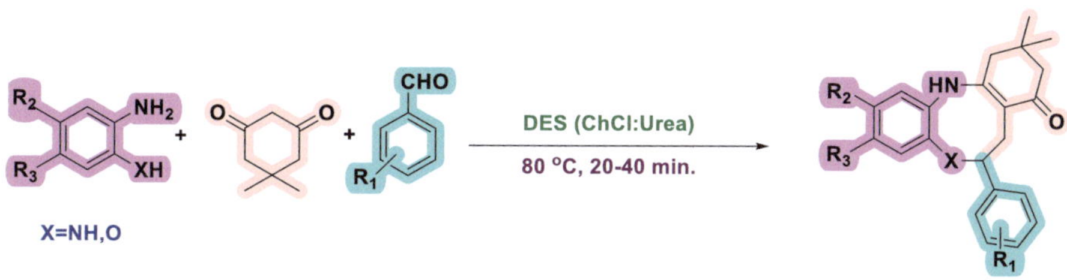

Scheme 7.14 DES synthesizes 1,4-benzodiazepines and 1,4-benzoxazepine derivatives

et al. claimed that they had been able to make naphthyridine derivatives immediately by applying a new domino four-component procedure. The strategy involves blending a 2-aminoprop-1-ene-1,1,3-tricarbonitrile, diamine, 1,1-bis(methylthio)-2-nitroethylene, and other carbonyl compounds with a ChCl-derived DES. A new category of polyfunctionalized fused naphthyridines was produced, which were characterized by their simultaneous formation of six π-bonds and three new rings. The requirement for the additional base catalyst was also eliminated by this approach when conducted under mild conditions (Scheme 7.13) [101].

7.2.1.9 Synthesis of Benzodiazepines and Benzoxazepines

With two heteroatoms at positions 1 and 4, seven-membered heterocycles like 1,4-benzoxazepines and 1,4-benzodiazepines have attracted a lot of attention recently due to their significant biological impacts. DES composed of ChCl:urea has been utilized for synthesizing seven-membered heterocycles bonded with benzo, such as 1,4-benzoxazepines and 1,4-benzodiazepines, through a one-pot, three-component process [102]. Such a green synthetic process involving reactions among phenylenediamine or 2-aminophenol, dimedone, and various benzaldehydes under mild reaction conditions effectively eliminates the requirement for metal catalysts, organic solvents, and acidic reagents (Scheme 7.14).

7.2.1.10 Oxindole and Oxazolidinones Synthesis

Kureshy and coworkers have investigated the synthesis process and enhancement of a novel class of spirocyclic carbonates [103]. The team has developed a very effective approach using cycloaddition of spiro epoxy oxindoles in the presence of CO_2

Scheme 7.15 Spirocyclic carbonate synthesis in DES

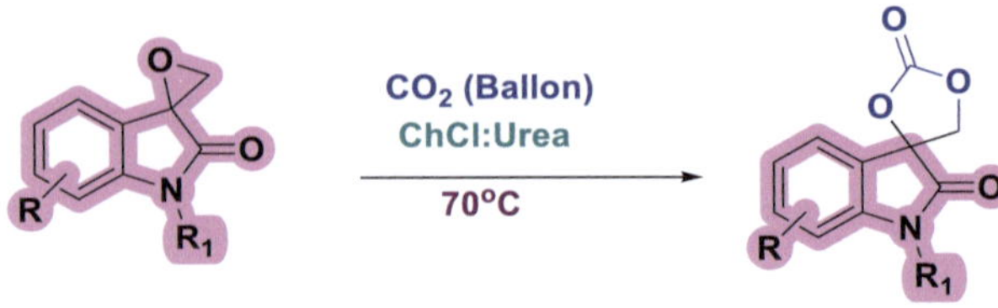

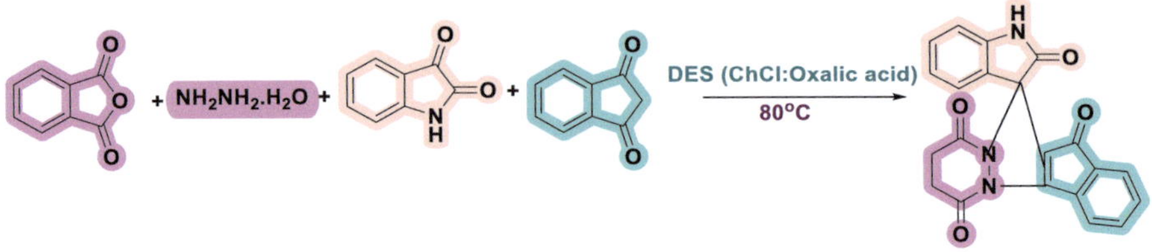

Scheme 7.16 Synthesis of spirooxindoles

via DES. This reaction occurred at mild conditions, yielding spirocyclic carbonates in yields that are fair to outstanding (~98%). The reaction products could be easily separated, with good recyclability of the DES sustaining up to four reaction cycles while maintaining the necessary activity level (Scheme 7.15).

Using a ternary, eco-friendly DES with glycerol, aluminum chloride hexahydrate, and ChCl (Glycerol: $AlCl_3 \cdot 6H_2O$: ChCl) as a catalyst and reaction media, an environmentally friendly strategy was proposed in 2022. The same DES has been applied in a three-component process involving epoxides, dimethyl carbonate, and amines for synthesizing different 3,5-disubstituted oxazolidinones. This approach proves the functional and efficient nature of the reaction with straightforward conditions and biodegradable reagents [104].

Kumar and team [105] studied an effective, eco-friendly, one-pot, four-component domino reaction to synthesize different spirooxindoles. The reaction reported in the study involves a combination of isatins, phthalic anhydride, cyclic diketones or diamides, and anhydride in a DES constituted of ChCl and urea (molar ratio of 1:2). Several solvents were checked initially, leading to the best outcomes in the case of DES, which exhibits a great potential for producing desired products with high yields and shorter reaction durations. Moreover, the DES catalyst offered higher reusability by maintaining its catalytic efficiency after four reaction cycles without losing the yield (Scheme 7.16).

A more efficient and environmentally friendly approach for synthesizing different kinds of structurally varying spiroxindoles, spiroannulated with indenopyridopyrimidines, chromenopyridopyrimidines, and pyranopyridopyrimidines, has been explored by Kumar and coworkers [106]. This approach involved a three-component reaction of isatins, aminouracils, and cyclic carbonyl compounds in the presence of DES (ChCl and oxalic acid with a 1:1 molar ratio). The catalytic protocol extending to 6-aminouracil with various isatins and carbonyl compounds produced spirooxindoles with excellent yields (Scheme 7.17).

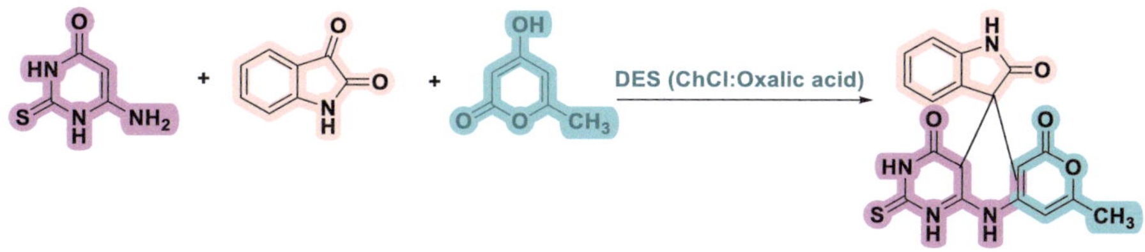

Scheme 7.17 Synthesis of spirooxindoles

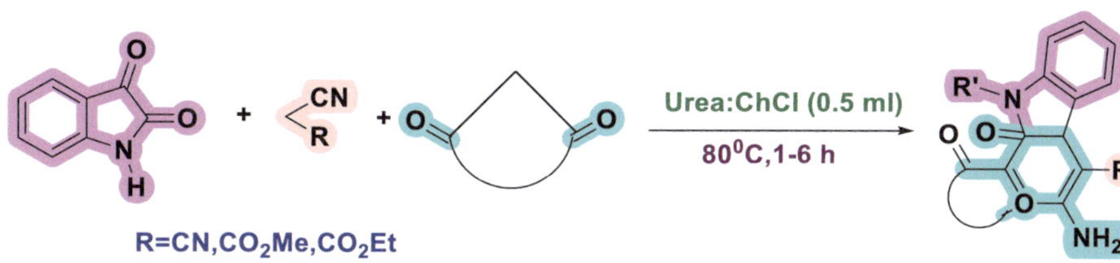

Scheme 7.18 Synthesis of spirooxindole derivatives

In a separate article by Azizi and a researcher [107], a multicomponent synthesis process for spirooxindole derivatives was reported involving isatin or acenaphthoquinone, cyanoacetic ester or malononitrile, 1,3-dicarbonyl compounds, 4-hydroxycoumarin, and naphthol, utilizing a biodegradable ChCl-based DES. At the beginning of the investigation, several different ChCl-based DESs were tested at optimized temperature conditions to evaluate the efficiency of the catalytic reaction. From all the solvents tested in the investigation, the urea-ChCl blend was the most suitable, providing a large yield of 95%. In contrast, the remaining DESs showed poor catalytic response. These optimized protocols were later applied to several other 1,3-dicarbonyl derivatives, cyanoacetic ester or malononitrile, and isatin, which further improved the effectiveness of this reaction medium (Scheme 7.18).

7.2.1.11 Quinoline and Quinazoline Compound Synthesis

The ring quinoline framework is a key structural element observed in numerous natural alkaloids, therapeutic agents, and synthetic analogs, which exhibit notable biological activity [108, 109]. Zhang and team [110] have reported a synthesis process for biologically important quinoline derivatives via the Friedländer heteroannulation reaction, which involves α-methylene ketones and 2-aminoaryl ketones reacting with each other (Scheme 7.19). Low-melting mixtures of L-(+)-tartaric acid and urea derivatives were used in this process as an affordable, biodegradable, and nontoxic reaction medium. By using the melted substance as both a catalyst and reaction media, high yields of quinolines are obtained.

Quinazoline is an important nitrogen-containing heterocycle in many natural chemicals, pharmaceutical chemicals, and various functional materials [111–113].

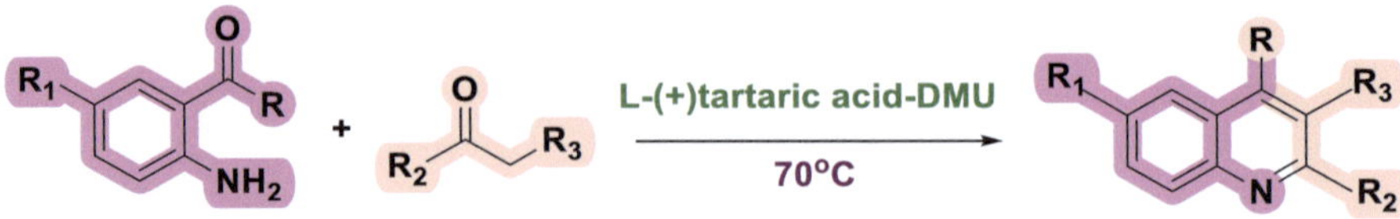

Scheme 7.19 Synthesis of quinoline derivatives

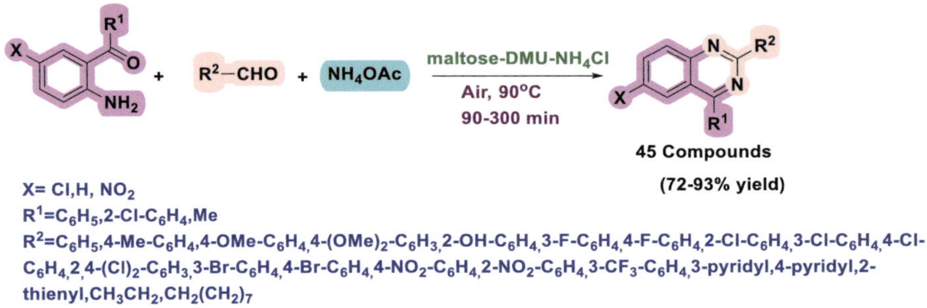

Scheme 7.20 Quinazolines synthesis

Zhang et al. [114] have used a DES made up of maltose, ammonium chloride (NH_4Cl), and dimethylurea (DMU) as a medium for carrying out a catalyst-free reaction for synthesizing quinazoline derivatives. The reaction among 4-nitrobenzaldehyde, ammonium acetate, and 2-aminobenzophenone is carried out in the above-mentioned DES, as depicted in Scheme 7.20. The same approach was successfully applied to different types of 2-amino aryl ketones and aldehydes under aerobic oxidation conditions, producing quinazoline derivatives in high-quality yields. The simple process, milder reaction conditions, and good yields make this approach a good candidate for synthesizing a broad range of biologically important quinazolines.

7.2.1.12 Synthesis of Thiazole and Thiazolone Derivatives

An interesting application of DESs is the S-cyclization of α-bromoacetylferrocene in the presence of different thioamides using a choline chloride/glycine blend with a 1:2 molar ratio at 85 °C (Scheme 7.21). This reaction is an environmentally friendly version of the Hantzsch synthesis. The mild reaction conditions, impact on sustainability, recyclability of solvent, simple experimental setup, and easy isolation process characterize this approach. Under optimized conditions, this approach efficiently produces substituted thiazoles with large yields (74–91%) [115].

DESs also demonstrate significant catalytic potential. A direct ammonium-based DES, composed of ChCl and urea, has been shown to assist in the preparation of 2-methyl- and 2-aminothiazole compounds via the chemical reaction involving substituted phenacyl bromides with thioacetamide or thiourea derivatives (Scheme 7.22). This process of catalysis, exploiting the properties of DES, enables quicker

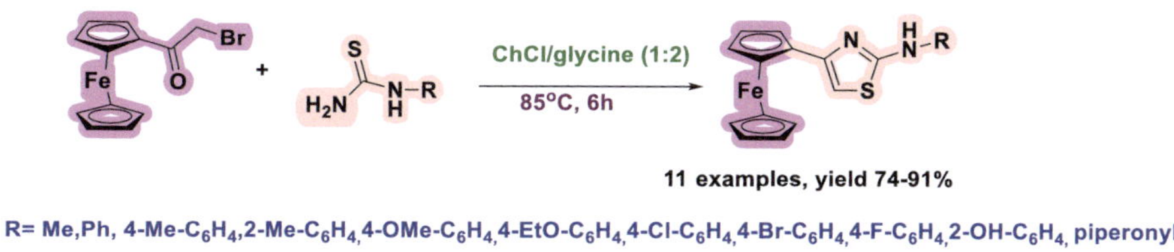

Scheme 7.21 Synthesis of thiazole-derivatives of ferrocene

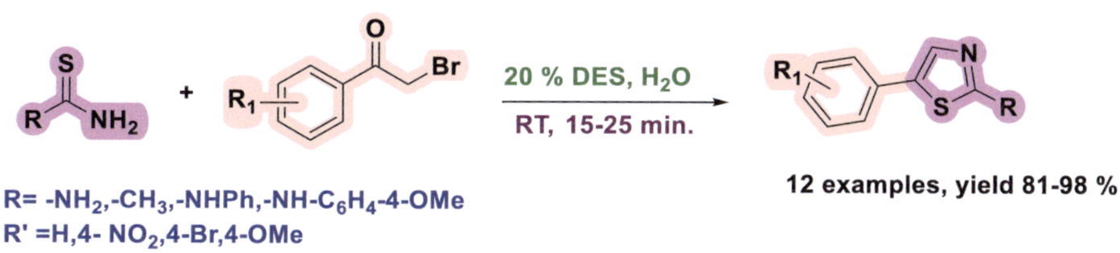

Scheme 7.22 Synthesis of substituted thiazole

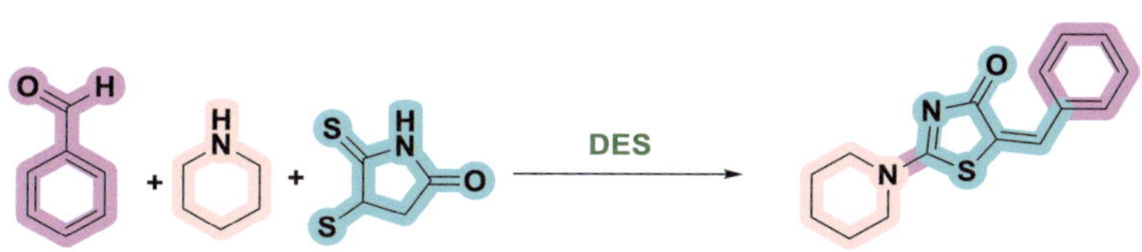

Scheme 7.23 Preparation of 2-amino-5-alkylidenethiazol-4-one compounds

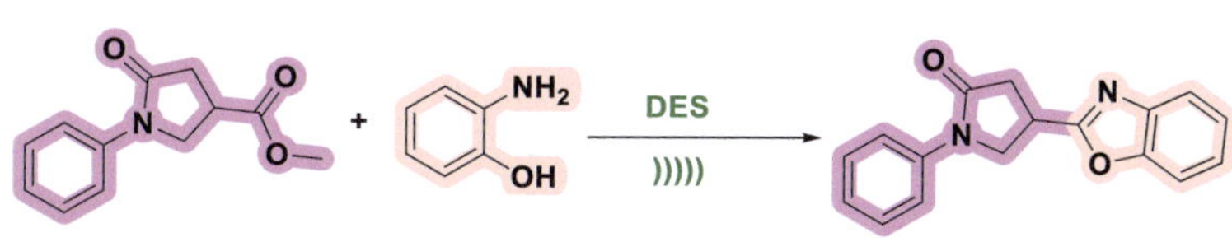

Scheme 7.24 Synthesis of new benzoxazole-linked pyrrolidinone compounds

production of the desired products within 15–25 min with an excellent yield ratio between 81% and 98% [116].

A recently published report presents a one-pot synthesis of 2-amino-5-alkylidenethiazol-4-one from nitrogen, rhodamine, and carbonyl compounds to be applied with ChCl-based DES as both a catalyst and dual solvent, obtaining superior results promptly (Scheme 7.23) [117].

7.2.1.13 Benzoxazole-Linked Pyrrolidin-2-One Heterocycle Synthesis

As shown in Scheme 7.24, a novel metal-free DES has been utilized in creating a new benzoxazole-linked pyrrolidinone heterocyclic series using 2-aminophenol and substituted benzylamines [118]. This DES was obtained via the combination of glycine and a urea-derived synthetic IL. This approach offers multiple advantages, such as improved productivity, quicker reactions, and direct processing. The DES also proved highly recyclable, with the potential for easy isolation for subsequent reactions.

7.2.2 Other Chemical Transformations in DESs

7.2.2.1 Reaction of Cycloloaddition with CO_2

Using the potential advantages of greater CO_2 solubility in DESs and the hydrogen-bonded characteristics of DES components, Lu and coworkers designed an improved catalyst material in 2017 that allows the cycloaddition of epoxides with CO_2. When $ZnBr_2$ is used as a cocatalyst, different DESs based on dicarboxylic acids and amino acids (AAs) have been used as a reaction medium for the catalytic cycloaddition reaction of CO_2 and propylene oxide (PO). The suitable reactivities and selection for propylene carbonate (PC) have been investigated in every aspect of studies [119]. The highest selectivity and reactivity were observed for the DES formed by combining L-proline and propanedioic acid (L-pro/PA). The ideal reaction parameters are 150 °C temperature and 1.2 MPa pressure. While aromatic styrene oxide and aliphatic acyclic epoxides yielded significant yields of cyclic carbonate via cycloaddition of CO_2, cyclohexene oxide exhibited small yields despite steric interference. But while we wait for an improved method of making cyclic carbonate from CO_2 and PO via a combination of halide anions, AAs, and dicarboxylic acid, the DESs developed using that presented positives within their capacity for recycling with decreasing functionality.

Subsequently, the same research team established procedures for the CO_2 cycloaddition process with PO (as shown in Scheme 7.25), whereby DES (polyethylene glycol/choline chloride) served as a catalyst and no further cocatalyst compounds would be required because of the solvent- and metal-free reaction conditions. Table 7.1 summarizes the numerous reaction conditions for the explored cycloaddition of CO_2 with PO. When PEG200 and ChCl were used alone (Table 7.1: entries 1 and 2), the yields were lower; however, with the DES formed using the combination of the two (entries 3–6 in Table 7.1), the results improved significantly. The reduction in the yields of PC was observed with the increasing PEG chain length due to the limitations in mass transportation. The optimum

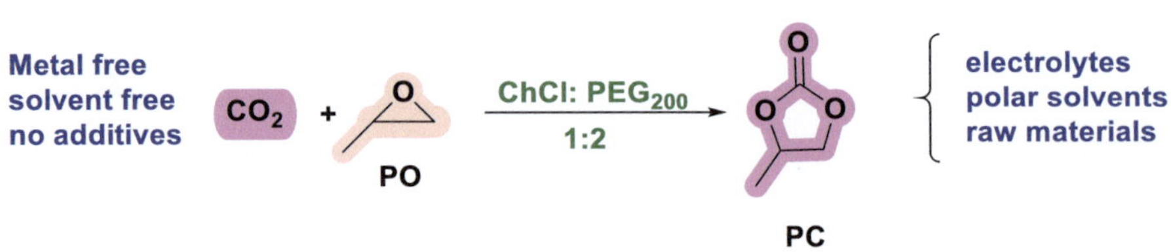

Scheme 7.25 CO_2 cycloaddition reaction in DES

Table 7.1 Effectiveness of several catalysts in CO_2 cycloaddition

Run	HBA	HBD	PC selectivity	PO conversion	PC yield
1[a]	ChCl	–	99.6%	36.8%	36.6%
2	–	PEG_{200}	98.1%	13.9%	13.6%
3	ChCl	PEG_{200}	99.6%	94.9%	94.6%
4	ChCl	PEG_{400}	99.4%	89.9%	89.4%
5	ChCl	PEG_{600}	99.6%	85.4%	85.1%
6	ChCl	PEG_{1000}	99.4%	67.5%	67.1%
7[b]	ChOH	PEG_{200}	99.8%	33.0%	32.9%
8[c]	ChOH	PEG_{200}	99.7%	87.8%	87.6%

Conditions involve the following: (a) PO (71.5 mmol), catalyst (1.5 mmol), a 1:2 HBA:HBD ratio, 130 °C, 1.2 MPa CO_2 pressure, with a 5-h duration. (b) Temperature set at 150 °C, CO_2 pressure maintained at 0.8 MPa. (c) Similar to (b) with an additional 1.5 mmol of KCl

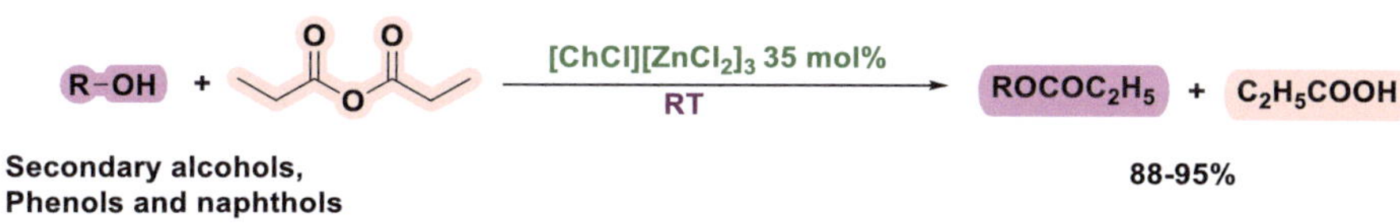

Scheme 7.26 Acylation reaction in DES based on $[ChCl_2]$ $[ZnCl_2]_3$

temperature for the best PC yield was 150 °C. Further increments in temperature result in decreased PO conversion since the solubility of CO_2 falls as the temperature rises. The optimum pressure of 0.8 MPa was observed for the maximum PC yield. However, increasing the CO_2 pressure further decreased the PC yield drastically due to the dilution of PO concentration at excessive carbon dioxide pressure. As presented in entries 7 and 8, significant yields were observed with the addition of KCL, with chloride ions from ChCl:PEG-based DES playing a major part in the cycloaddition reaction [120]. This approach is highly stable, sustainable in the long term, and biocompatible.

7.2.2.2 Acylation Reaction

A key step in the synthesis of physiologically and medicinally active substances involves the acylation of naphthols, phenols, and alcohols, among several additional acylation procedures. For this procedure, Nguyen and coworkers [121] suggested applying $[ChCl_2][ZnCl_2]_3$-based DES as a powerful catalyst. The main shortcomings of these supplies are their respective positions, the massive amounts of disposal, the recrystallization from the harmful solvent, and their costly demand for donor solvents. Additionally, they have longer reaction times, and neither are they reusable or recoverable. Up to four uses of the $ChCl_2$- and $(ZnCl_2)_3$-based DES are feasible without an apparent reduction in enzymatic activity (Scheme 7.26).

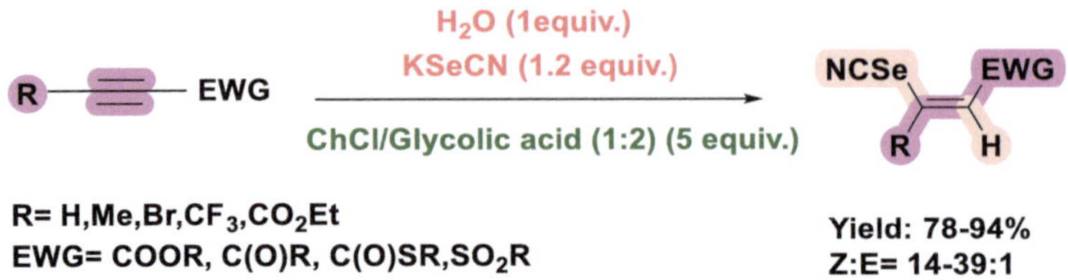

Scheme 7.27 Selenocyanation reactions in DES based on ChCl/glycolic acid

7.2.2.3 Selenocyanation Reactions

Organo-selenocyanate compounds and other selenocyanation products have shown a superior capacity for synthesizing biologically active chemicals and a strong prevalence in the pharmaceutical industry. However, these are versatile preparatory precursors used in the manufacturing of compounds with additional value [122]. Organic synthesis, advanced functional materials, and medicinal chemistry all benefit from the Se-containing chemical, having limited protocols for their synthesis [123–126]. The reaction between ethyl propiolate and KSeCN in a varied medium was shown by Wu and colleagues [127] (as shown in Scheme 7.27). In this study, the most efficient reaction environment has been determined to be ChCl: glycolic acid-based DES. In contrast, the components of the eutectic blend, ChCl and glycolic acid, resulted in poor yields. Using other DES blends offered lower yields, whereas weaker hydrogen bonding interactions provided a slight trace of yields and poor conversion. It is to be emphasized that the ultrasonic (US) irradiation treatment produced the largest yields due to the increased size and number of the functioning cavitation bubbles, allowing quicker reaction partners' micromixing and subsequently resulting in facilitating the reaction with different substrates. This DES showed reusability up to five reaction cycles without losing the effectiveness of catalysis.

7.2.2.4 Reactions of Bromination

In the opinion of Phadtare and Shankarling, DES acts as an affordable, safe catalyst to make high-yield brominated products via substituted 1-aminoanthra-9,10-quinone bromination (as shown in Scheme 7.28) [128]. For producing the DES used in the report, 1 mol of ChCl was mixed with 2 mol of urea, resulting in no unwanted by-products. A beneficial brominated product was successfully manufactured by brominating a range of 1-aminoanthra-9,10-quinone derivatives employing the ChCl:urea-based DES. The comparison of these results with those of other organic solvents, including methanol and chloroform, can be seen in Table 7.2. It was discovered that DES outperforms the solvents described above with respect to purity, yields, and time to response. The authors reported steady DES catalytic activity when recycled up to five times. This approach improves the directness, reaction

Scheme 7.28 Aminoanthraquinone's catalyzed bromination in DES based on ChCl:urea reaction

Table 7.2 Using 2.5 equivalents of bromine, the 1-aminoanthra-9,10-quinone molecule was brominated in an organic solvent between 20 and 30 °C

Input	Medium	Response time (h)	Percentage (%)	Purity (%)
1	Methanol	10–12	75	95
2	Chloroform	10–12	70	92
3	DES	2–3	95	97

time, applicability, yield amount, complexity of the reaction conditions, and eco-friendly nature of solvents over the previous approaches. Harmful organic solvents and catalysts are also avoided by following this approach; the isolation process is easy and complete with high purity, making this DES an ideal candidate for an industrial synthesis solvent.

7.2.2.5 Chlorination Reactions

Since chlorinated organic compounds are capable of moderate chemical transformations that produce different functional atoms, these constitute a key category of intermediates [129]. The β-ketoesters and α-chlorinated ketones are among the most versatile intermediates used in medicine and agriculture. Their increased reactivity enables them to react with a wide variety of nucleophiles to produce a variety of beneficial compounds [130]. A focus relevant nowadays is the novel, more selective halogenation techniques that meet the demand for sustainable green chemistry methods [131]. Despite these advances, developing selective mono- or dichlorination reactions is still challenging [132]. There are, however, some methods for overcoming some aspects of this challenge [133]. DCDMH was observed to be the most effective reagent among those reported, suitable for selective chlorination, with the disadvantage of producing acid sewage [134]. To decrease the acid sewage without losing the selectivity of the material, Chen Zou et al. [135] tested several catalysts. They observed that the silica gel added to a mixture of acetophenone, DCDMH, and methanol, and the α-chlorination performed better under reflux. The relation between the amount of silica and the produced yield was observed to be that a large amount of silica gel does not necessarily result in positive outcomes. An identical approach was also employed to prepare dechlorinated acetophenones. However, an

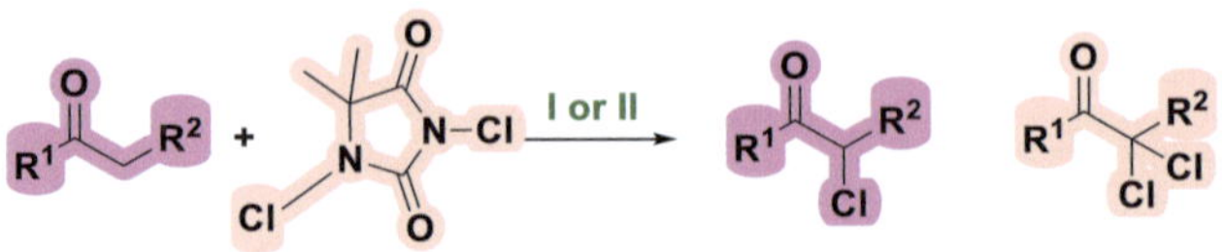

I = SiO_2,MeOH,reflux,1h,yield: monochlorinated products (86-98%)
II= DES, rt,45 min,yield: dichlorinated products (86-95%)

R_1= C_6H_5,4-Cl-C_6H_4,4-Br-C_6H_4, 4-MeC_6H_4, 4-$NO_2C_6H_4$,3-$NO_2C_6H_4$, 4-Br-biphenyl,4-OMe-C_6H_4, Me,benzoyl,4-NO_2-benzoyl, t-butyl
R_2= H,COOEt

Scheme 7.29 Synthesis of α-mono or α,α-dichloro ketones

amalgamation of α-monochloroacetophenone and α,α-dichloroacetophenone was collected. Hence, an effective approach for synthesizing α,α-dichlorinated ketones must be explored. Within this context, ChCl/p-TsOH-derived DES produced α,α-dichlorinated products with large yields ranging between 86% and 95% by simply whisking for 45 min at room temperature (Scheme 7.29).

7.2.2.6 Epoxide Ring-Opening

Several functionally modified ring-opening chemicals find different applications, such as biologically active compounds and therapeutic agents, which are produced due to the ring-opening of epoxide [136]. The ChCl:urea-based DES has been used to brominate a variety of 1-aminoanthra-9,10-quinone derivatives, producing a good, brominated product. As shown in Table 7.2, these results were compared with those of other organic solvents, such as methanol and chloroform. Compared to the solvents previously mentioned, DES has been demonstrated to possess higher purity, higher yields, and faster reaction times [137] (Scheme 7.30). A good yield of 40 mol% of pristine $SnCl_2$ is obtained from the epoxide via a ring-opening reaction. In contrast, it produces decent yields with trimethylsilyl azide ($TMSN_3$), trimethylsilyl cyanide (TMSCN), and methanol with significant stereoselectivity and regioselectivity. From the products produced from the ring-opening reactions of epoxides catalyzed with DES, highly respectable chemoselectivities, stereoselectivities, and regioselectivities are anticipated.

7.2.2.7 Amides Synthesis

Amide synthesis is an organic transformation because it serves as a primary building element for peptides, several other human-engineered compounds, and some natural compounds [138]. Nagarkar and team [139] have reported the development of a suitable DES composed of ChCl and $ZnCl_2$ at a 2:1 molar ratio, displaying effective catalytic efficacy for the synthesis of the amides (Scheme 7.31). Different

Scheme 7.30 Epoxide ring-opening reactions in ChCl:$SnCl_2$

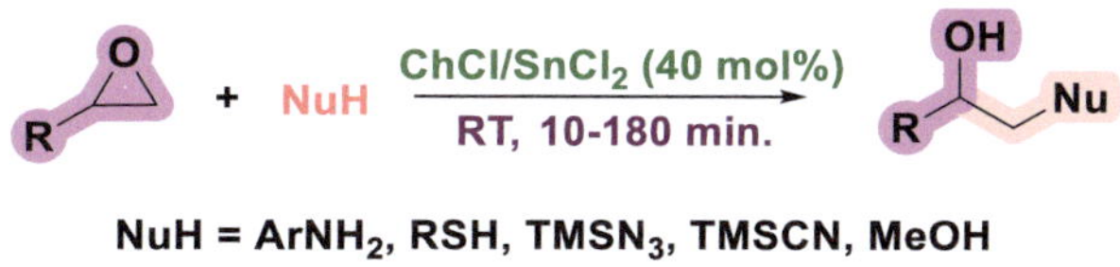

Scheme 7.31 Primary amides synthesis in ChCl:$ZnCl_2$-based DES

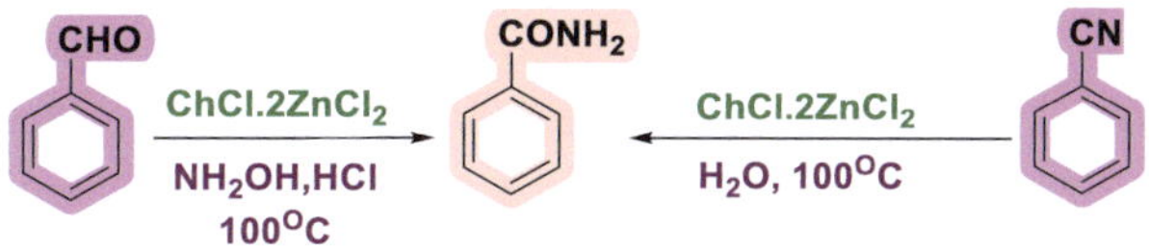

Scheme 7.32 Carbamates synthesis in [ChCl][$ZnCl_2$]$_2$-based DES

substrates (nitriles and aldehydes), such as aliphatic, conjugated, aromatic, and aromatic substrates having electron-attracting and electron-supplying groups, showed great yields as compared to the earlier literature. Once again, the strong aspects of DES-based catalytic systems, such as safety, recyclability, simplicity, eco-friendliness, and low cost, are cemented.

7.2.2.8 Carbamates Synthesis

Conventionally, the carbonylation with carbonating agents like carbon monoxide, carbon dioxide, and other carbonates is catalyzed with metal complexes, which require a higher molar ratio of carbonylating agents for longer reaction intervals and at higher temperatures when there are, when usually harsher safety conditions are needed for mixing of CO and O_2. As a result, carbonates can be prepared only on a smaller scale [140, 141]. In recent years, urea and its derivatives have become the center of attraction when serving as a carbonylating agent due to their nontoxic nature and lower prices [142]. A greener and more suitable catalyst for synthesizing carbamates from the reaction between urea and alcohol was reported in 2018 by Dindarloo Inaloo and Majnooni [143]. The catalysis was in the form of [ChCl] [$ZnCl_2$]$_2$ DES, which yielded the most significant amounts with the smallest side product formation compared to other DES catalysts. Notably, this was also a more reliable and recoverable catalyst exhibiting great potential for synthesizing biologically active chemicals (Scheme 7.32).

A three-component coupling technique known as [ChCl][$ZnCl_2$]$_2$-based DES has been shown to have been invented in 2019 by the same group. It employs amines, alkyl halides, and CO_2 in the same chemical reaction system and media for generating carbamates. Due to their high yields, the fundamental aliphatic amines and anilines have been shown to be superior nucleophiles and substrates to secondary and aromatic amines; in particular, benzyl halides and alkyl halides produced noticeably higher yields, while allyl halides underperformed [144].

7.2.2.9 Dithiocarbamates Derivatives Synthesis

Many important biological compounds constitute dithiocarbamates, which find utilization as vulcanization accelerators in rubber factories, biocides in some household products, and animal repellents [145, 146]. Dithiocarbamates have been utilized as effective ligands in nanomaterial and surface chemistry. Additionally, many biologically active heterocyclic chemicals are synthesized using dithiocarbamates [147–149]. An eco-friendly, reliable, mild, scalable, catalyst-free four-component preparation approach synthesizing pharmaceutically viable and chemically important dithiocarbamate compounds using a one-pot reaction was initially investigated by Azizi et al. [150]. This approach involves a reaction of different aliphatic amines, aromatic aldehydes, carbon disulfide, and ketones in the presence of NaOH, working as an additive in ChCl/urea DES or ethanol. Different solvents and additives have been investigated in a model reaction to optimize the reaction conditions involving acetophenone, pyrrolidine, benzaldehyde, and carbon disulfide. DES and ethanol work well as solvents, whereas NaOH works well as an additive, providing a high yield to synthesize dithiocarbamate (Scheme 7.33).

7.2.2.10 Cinnamic Acid and Its Derivatives Synthesis

Perfumery, cosmetics, agriculture, medicines, and polymers are some fields where cinnamic acid and related compounds are widely utilized [151, 152]. Traditionally, cinnamic acid and its derivatives were synthesized via a base-catalyzed Perkin

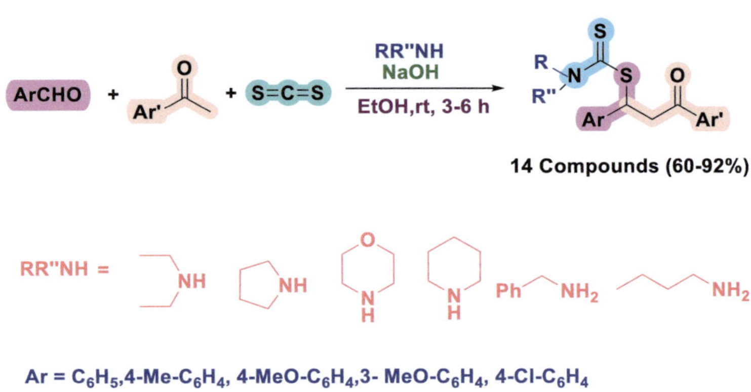

Scheme 7.33 Synthesis of dithiocarbamates

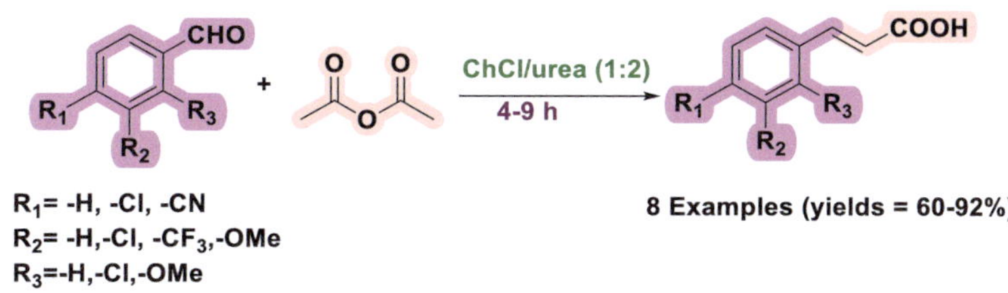

Scheme 7.34 Synthesis of cinnamic acid derivatives

reaction utilizing acetic anhydride and benzaldehyde in the presence of a base [153]. Cinnamic acid and its derivatives have also been synthesized via the Knoevenagel condensation method utilizing pyridine as a catalyst. Despite a large number of methods being available for synthesizing cinnamic acid and its derivatives, it is clear that organic solvents and bases that are generally utilized in these processes are harmful and hazardous due to their higher volatility, toxicity, and the amount used in these processes. Weng et al. [153] have proposed using ammonium ionic liquids to synthesize cinnamic acid to overcome these drawbacks. This eco-friendly approach requires a high temperature (120 °C) for a relatively longer duration (8 h) for this reaction to be carried out, making it highly energy-draining. Shankarling et al. [154] have recently used a biodegradable DES (ChCl/urea) as a reaction medium and a catalyst to synthesize cinnamic acid and its derivatives through the Perkin reaction. Under the mild reaction conditions, the reaction provided better yields without requiring any additional catalyst. This approach is proven to be environmentally benign and, overall, highly efficient because of the solvents' ease of recovery and reusability, which maintain their activity after several reaction cycles. This approach and a simple operational mechanism are more energy efficient (Scheme 7.34).

7.2.2.11 Phenol Synthesis

Phenol compounds can be discovered in many natural materials and act as important synthetic component parts for complex structures [155]. A sustainable gram-scale preparation approach for phenols using aryl/heteroaryl boronic acids and their compounds in ChCl/urea-based DES in water mixed with 30% aqueous H_2O_2 as an oxidant (for the model reaction, 5 min and 92% yield) has been reported by He et al. [156]. Tert-butyl hydroperoxide (TBHP) and ozone were also investigated for their suitability as oxidants. However, these oxidants resulted in an inferior yield for the products. The optimization of the reaction parameters revealed the optimal reaction temperature to be room temperature, with good product selectivity in a few minutes (5–15 min), exhibiting yields ranging from 81% to 96% and recyclability up to five times without losing any activity (Scheme 7.35). The activation of H_2O_2 through DES exhibiting polarization in the peroxide bonds via hydrogen bonding was suggested as the mechanism behind the reaction. In the next step, the activated peroxide quickly reacts with the boronic acid, followed by the phenyl migration, and ultimately, hydrolysis causes the creation of the end material (Scheme 7.36).

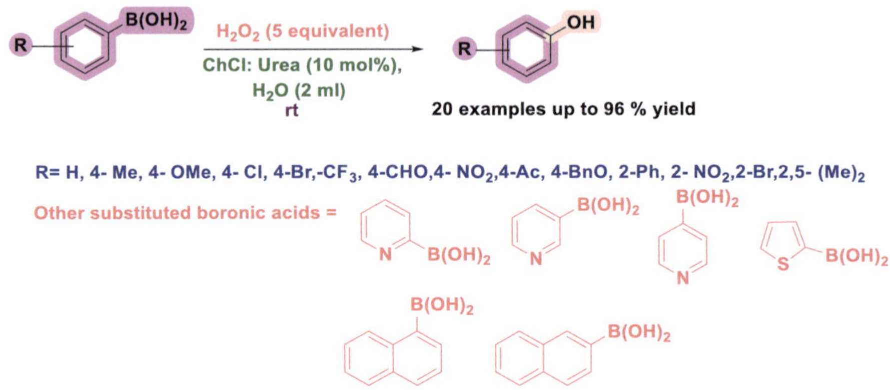

Scheme 7.35 Phenol synthesis

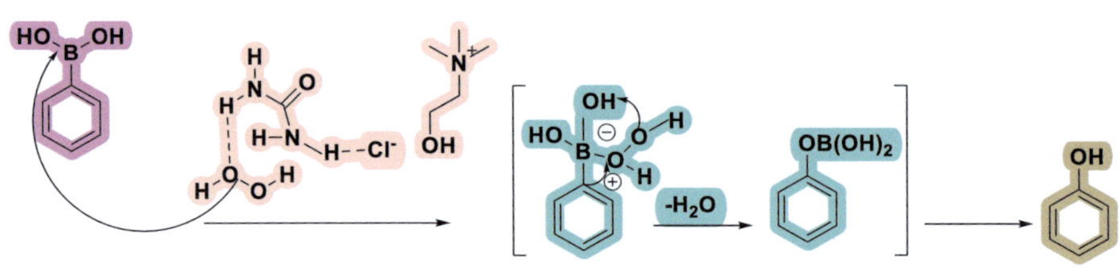

Scheme 7.36 Reaction mechanism of phenols

7.2.3 *Multicomponent Reactions Based on Isocyanide*

In designing different routes for the sustainable preparation of heterocyclic chemicals, drugs, biologically active molecules, and peptides, multicomponent reactions (MCRs) have attracted significant interest from the research community [157, 158]. MCRs are one-pot transformations achieved via three or more starting reagents leading to the formation of complex compounds as depicted in Fig. 7.2. Among the widely popular MCRs, isocyanide-based ones, such as Passerini, Groebke-Blackburn-Bienayme, and Ugi reactions, have taken up significant attention [159]. The main characteristics of MCRs are negative activation volumes and entropy decrement executed by the conversion of many components into a product. Notably, pressure assists and facilitates the reactions possessing the entropy-decreasing transformations [160, 161]. One of the key characteristics of DESs that assist in the formation of high-pressure cavities is their increased viscosity; reactions linked to entropy decrement and increased thermal stability make DESs an ideal reaction medium for organic chemical transformation at higher pressures and temperatures. Subsequently, the selection of DESs is also favorable for expedient and quick access to MCRs.

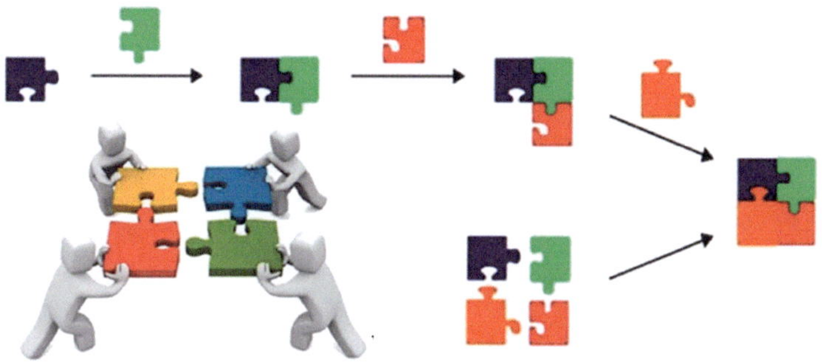

Fig. 7.2 Diagrammatic comparison of the MCR approach and traditional synthetic pathways. (Reproduced with the license granted by the Royal Society of Chemistry from [159], Copyright 2014)

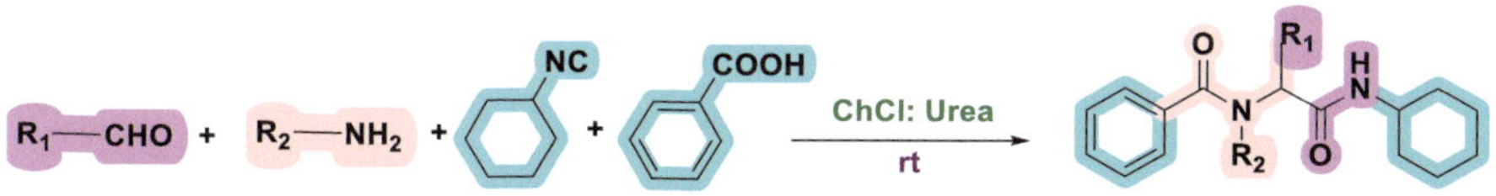

Scheme 7.37 In DES, Ugi four-component reactions

7.2.3.1 Ugi Reactions

Initially reported in 1959, MCRs have widely used the Ugi four-component reaction in producing peptide-like products using common reagents like carboxylic acids, isocyanides, carbonyl chemicals, and amines [162]. This Ugi-4CR reaction was explored by Azizi and team [163] in different reaction media, and it was observed that the ChCl/urea-based DESs offered relatively superior results over the conventional organic media like dichloromethane and acetonitrile. Reactions can be conducted at room temperature in between 2 and 5 h with these simple experimental setups. By using water to remove the DES, the reaction's by-products may be separated (Scheme 7.37).

7.2.3.2 Multicomponent Synthesis Via Groebke Blackburn-Bienayme (GBB) Reactions

Another popular reaction for synthesizing imidazo[1,2-a] pyridines is the Groebke Blackburn-Bienayme (GBB) reaction, which has been exploited as a synthetic pathway to prepare biologically active compounds, illustrated by the anxiolytic drugs Zolpidem and Alpidem [164]. Shaabani et al. claimed that 3-aminoimidazo-fused heterocycles are formed by a one-pot cascade reaction of 3-formyl chromone derivatives, 2-aminoheterocycles, benzaldehydes, and isocyanides with DES [62]. The process was aided by the strong catalytic activity of DES, which could be recycled in four subsequent reactions without losing its effectiveness (Scheme 7.38).

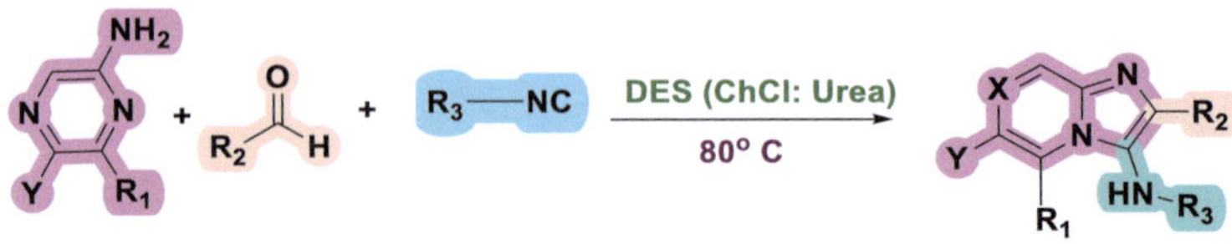

Scheme 7.38 Representation of Groebke–Blackburn–Bienayme reactions utilizing DES as the reaction medium

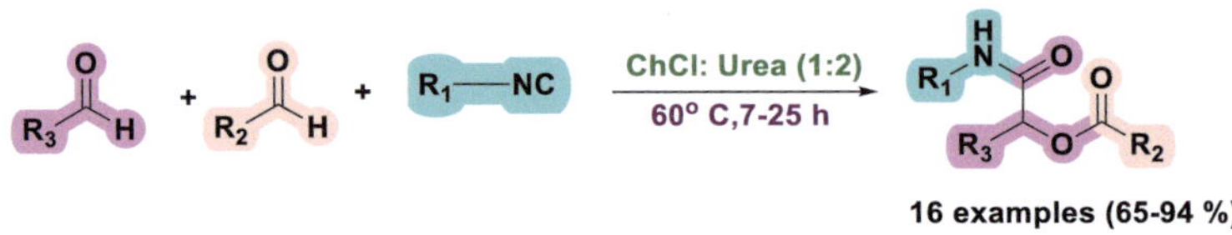

Scheme 7.39 Passerini three-component reactions using DES

7.2.3.3 Passerini Reactions

Following Mumm's rearrangement, isocyanides, carboxylic acids, and aldehydes condensed to produce acyloxy amide compounds in the first-ever multicomponent isocyanide-based reaction, as described by Mario Passerini. A synthetic method for synthesizing a-acyloxy amides is also explained by Shaabani and the group employing single-pot three-component cascade Passerini reactions within DES [165]. This reaction produced the related product with a yield percentage ranging from 65% to 94%. It involved groups of aromatic and aliphatic aldehydes, isatin, alloxsan, aliphatic isocyanides, and carboxylic acids in DES at 60 °C (Scheme 7.39).

7.2.3.4 MCR Based on Meldrum's Acid and Isocyanide

The reaction involving Meldrum's acid and isocyanide, in addition to various kinds of carbonyl compounds and nucleophiles, formed the conceptual framework for this novel MCR. In general, dichloromethane has been employed as a solvent to synthesize a wide range of physiologically active molecules, including benzodiazepines, iminofuropyranones, amidodiesters, succinimides, benzooxazepines, coumarins, and malonamides. In 2019, Balalaie and team investigated an environmentally sustainable protocol to directly synthesize amidoesters in a DES solvent composed of ChCl/urea in 1:2 proportions via a pseudo five-component reaction of Meldrum's acid, primary alcohols, isocyanides, and 3-formyl chromone [166]. Proceeding with this reaction with trifluoroethanol resulted in succinimide derivatives, as DES was found to be reusable even after four reaction cycles due to its improved solubility in the solvent (Scheme 7.40).

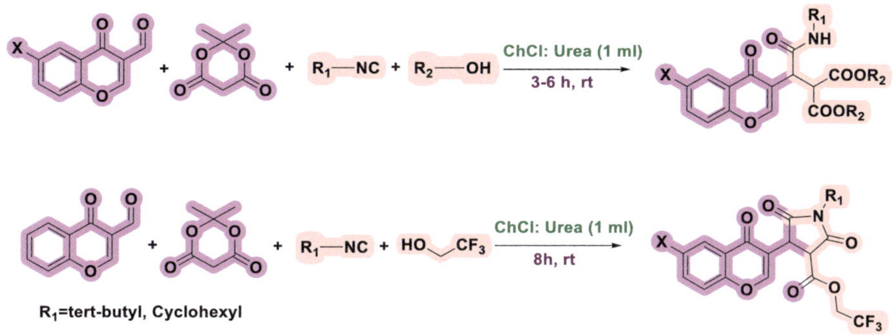

Scheme 7.40 Amidoesters and succinimides synthesis in DES

7.3 Future Perspective

Despite the catalytic effects of DES being well-known from antiquity, the fundamental understanding of the phenomenon was not clear till the late eighteenth or early nineteenth century. During these times, modifications in the previously known reactions were observed to be made possible by the addition of different chemical substances in small amounts—later termed "catalysts." Catalytic reactions significantly contributed to the economic side of the chemical industries, including refining crude oil and synthesizing basic chemical compounds. They play a vital role in the pharmaceutical industry by generating the molecular frameworks that comprise various chemicals, which are employed in pharmaceuticals. Since environmental problems have gained widespread attention, replacing hazardous materials with less harmful ones requires an enormous amount of work. Finding new catalytic reactions requires new technological and intellectual innovations, which is where this issue resides. The application of novel research methods that can take advantage of cutting-edge technologies is required to explore cheaper and energy-efficient catalysts and chemical reactions, which exhibit the potential to shorten the reaction duration and improve the recovery of the catalysts. DES has garnered the interest of research groups from different fields in opening up desired prospects in green chemistry due to the significant advantages over traditional IL. The DESs offer more straightforward synthesis procedures, cheaper raw chemicals, and better biodegradability than traditional solvents and catalysts. Discoveries in catalytic organic processes have been systematically summarized in this chapter, including the introduction of DESs as reaction media, catalysts, and advanced electrolytes, while emphasizing the fundamental understanding of the structural and compositional properties of DESs and their impact on the yield, highlighting their effectiveness as catalysts. Generally, the major drawbacks of DESs in catalysis are due to their higher viscosity, which alters other physicochemical characteristics, including transport characteristics like hygroscopic nature and conductivity. The lack of investigations focusing on the storage time of the eutectic blends in optimal conditions and the primary understanding of the intrinsic reactions among various components of the eutectic blends need to be tackled further. To overcome this problem in scientific articles, computer technology can be used relevantly, such as in the prediction of the molecular behavior of novel solvents and their constituent components. Before DESs become widely used as a better green alternative, it is equally, if not more, essential to carry out the necessary techno-economic analysis and solve the economic aspects of using them. This is because the primary purpose of using novel green solvents like DESs in organic chemical catalysis is to eliminate the traditional solvents and suggest new sustainable chemical reactions. To focus on this concern, upcoming efforts should not be limited to the scientific aspects of DES-accelerated catalysis. Additionally, they should use sustainable and economic analytical techniques to improve various catalytic processes in mild response environments while lowering the power costs that stimulate their use in industry.

Minimal attention in the field of techno-economics focusing on DESs and their industrial impact has been given so far, the majority of which focuses on the area of biorefineries [167, 168]; some of the reviews discussing the same topic can be a starting point for gathering deeper insights on the topic [169]. To completely comprehend the manner in which DESs affect catalytic reactions, however, additional investigation by academic institutions is required. Recent methods for generating organometallic structures from eutectic combinations, insights into the way DES affects enantioselectivity of reactions, the design of more efficient catalysts in a specific response media formed using eutectic mixtures, etc., could all be found to widen the scope of beneficial chemical transformations that are approved using the DESs.

7.4 Conclusion

This chapter validates the application of DESs as excellent catalysts and their primary function as solvents. Due to their remarkable physicochemical properties, the DES-based catalysts proved more efficient and exhibited reusability for numerous synthesis processes. Additionally, by optimizing the reaction conditions, the catalytic efficiency of DESs could be tuned, providing high yields in return. As illustrated in this work, DESs can be more advantageous in large-scale industrial synthesis processes since they are inexpensive and easily synthesized. The additional attractive features of DESs are their nontoxic nature and the readily available precursors for forming any DES. Hydrogen bonding among the components of DES reduces their reactivity in the end DES, despite the individual components being highly reactive. DES, which encompasses carboxylic acid, can potentially engage in acid-catalyzed preparation. DES, which contains metal chlorides, can act as a Lewis acid. Besides these, DESs can also be used as Brønsted acids for certain synthesis processes. DES, constituting a hydrogen bond donor such as urea, plays an important role in hydrogen bonding for the particular substituent of the reactants; this increases the electrophilicity and nucleophilicity of the reactant for product creation. The DESs have shown their potential to be an excellent medium for CO_2, which is important for cycloaddition reactions. In such reactions, DES functions as a dual-purpose chemical—first as a reaction medium and second as a catalyst. DESs can also be utilized as an alternative green solvent for synthesizing many different chemical compounds due to their good catalytic efficiency compared to other solvents reported in the scientific literature. In summary, DES displays remarkable recoverability and reusability while maintaining catalytic efficiency. They can be separated easily from the end chemical products, resulting in higher yields. Reactions catalyzed through DESs often occur at shorter durations than traditional catalysts or organic solvents. DESs require additional reagents in some synthesis processes to enhance the catalytic performance. As shown in this chapter, many relevant organic transformations clearly display the importance and requirement of DESs along with their advantages. Even after these known advancements made possible through using DESs, only a handful of them have, thus far, been investigated. The analysis presented in this work asks for more attention to developing novel DESs made from sustainable routes for specific tasks, further improving the organic reactions. Theoretical simulations can enhance our foundation of different reaction mechanisms and advance the field of novel DES catalysts. Moreover, it is especially important to establish inexpensive DESs as reaction media and reusable catalysts, improving the synthesis of commercially viable chemicals and drugs at the industrial scale for replacing the conventional media and catalysts. As the field of DES offers novel possibilities for the advancement of green organic synthesis, particularly applications of DES as an intermediate and catalyst, we believe that this chapter can function as an overview for sparking the interests of scholars who are willing to explore this topic. Observing the rapid development of green chemistry in recent times, DESs play an important role in the upcoming innovations in organic synthesis.

Acknowledgments The authors would like to acknowledge Indus University, National Forensic Sciences University, The Maharaja Sayajirao University of Baroda, Gujarat Technological University, and Gujarat University for their support of this work.

Conflict of Interest Statement The authors declare that they have no conflict of interest.

Declaration of Generative AI and AI-Assisted Technologies in the Writing Process While preparing this work, the authors utilized the ChatGPT tool to enhance the language and readability of several statements in the manuscript. After using this tool/service, the authors reviewed and edited the content as needed and take(s) full responsibility for the publication content.

References

1. Patel B, Singh S, Parikh K et al (2022) Composition triggered aggregation/solubilization behaviour of mixed counter charged gemini surfactants: a multi-technique investigations. J Mol Liq 359:119242. https://doi.org/10.1016/j.molliq.2022.119242
2. Patel B, Singh S, Parikh K et al (2022) Micro-environment mapping of mole fraction inspired contrasting charged aqueous gemini micelles: a drug solubilization/release study. J Mol Liq 363:119885. https://doi.org/10.1016/j.molliq.2022.119885

3. Mishra SR, Gadore V, Roy S et al (2024) Advancements in Ni-MOFs: unravelling the photocatalytic tapestry for organic contaminant degradation. Int J Environ Anal Chem:1–19. https://doi.org/10.1080/03067319.2024.2343106
4. Mishra SR, Chavda V, Roy S et al (2025) Covalent organic framework nanocomposites for superior lithium-ion storage: bridging molecular design and battery technology. J Energy Storage 108:115026. https://doi.org/10.1016/J.EST.2024.115026
5. Mishra SR, Gadore V, Chavda V et al (2025) Black phosphorene: a versatile allotrope revolutionizing environmental, energy, and biomedical applications. Coord Chem Rev 524:216345. https://doi.org/10.1016/J.CCR.2024.216345
6. Clarke CJ, Tu W-C, Levers O et al (2018) Green and sustainable solvents in chemical processes. Chem Rev 118:747–800. https://doi.org/10.1021/acs.chemrev.7b00571
7. Anastas P, Eghbali N (2010) Green chemistry: principles and practice. Chem Soc Rev 39:301–312. https://doi.org/10.1039/b918763b
8. Kaur G, Kumar H, Singla M (2022) Diverse applications of ionic liquids: a comprehensive review. J Mol Liq 351:118556. https://doi.org/10.1016/J.MOLLIQ.2022.118556
9. Gonçalves ARP, Paredes X, Cristino AF et al (2021) Ionic liquids—a review of their toxicity to living organisms. Int J Mol Sci 22:5612. https://doi.org/10.3390/ijms22115612
10. Hansen BB, Spittle S, Chen B et al (2021) Deep eutectic solvents: a review of fundamentals and applications. Chem Rev 121:1232–1285. https://doi.org/10.1021/acs.chemrev.0c00385
11. Shah PA, Chavda V, Hirpara D et al (2023) Exploring the potential of deep eutectic solvents in pharmaceuticals: challenges and opportunities. J Mol Liq 390:123171. https://doi.org/10.1016/j.molliq.2023.123171
12. Shah PA, Shrivastav PS, Sharma VS, Chavda V (2023) Uncovering the green frontier: harnessing deep eutectic solvents for sustainable bioanalysis. Bioanalysis 15:815–821. https://doi.org/10.4155/bio-2023-0088
13. Shah PA, Chavda V, Hirpara D et al (2024) Advances in deep eutectic solvent-based synthesis of nanomaterials for environmental remediation. J Mol Liq 416:126482. https://doi.org/10.1016/J.MOLLIQ.2024.126482
14. Abbott AP, Capper G, Davies DL et al (2003) Novel solvent properties of choline chloride/urea mixtures. Chem Commun:70–71. https://doi.org/10.1039/b210714g. Electronic supplementary information (ESI) available: spectroscopic data. See http://www.rsc.org/suppdata/cc/b2/b210714g/
15. Xu GC, Ding JC, Han RZ et al (2016) Enhancing cellulose accessibility of corn stover by deep eutectic solvent pretreatment for butanol fermentation. Bioresour Technol 203:364–369. https://doi.org/10.1016/J.BIORTECH.2015.11.002
16. Jangir AK, Dani U, Chaudhary VK (2025) Exploring the influence of composition on physicochemical and biological properties of choline chloride-based deep eutectic solvents: an experimental and theoretical endorsement. J Mol Liq 427:127419. https://doi.org/10.1016/j.molliq.2025.127419
17. Jangir AK, Bhawna, Verma G et al (2022) Design and thermophysical characterization of betaine hydrochloride-based deep eutectic solvents as a new platform for CO_2 capture. New J Chem 46:5332–5345. https://doi.org/10.1039/D1NJ05373F
18. Jangir AK, Nain AK, Kuperkar K (2021) Insight into structural properties and molecular interactions of maline (choline chloride + malonic acid) and 1,4-butanediol based pseudo-binary mixture: a thermophysical, spectral, and simulation portrayal. J Mol Liq 334:116050. https://doi.org/10.1016/j.molliq.2021.116050
19. Jangir AK, Sethy P, Verma G et al (2021) An inclusive thermophysical and rheology portrayal of deep eutectic solvents (DES) for metal oxides dissolution enhancement. J Mol Liq 332:115909. https://doi.org/10.1016/j.molliq.2021.115909
20. Jangir AK, Mandviwala H, Patel P et al (2020) Acumen into the effect of alcohols on choline chloride: L-lactic acid-based natural deep eutectic solvent (NADES): a spectral investigation unified with theoretical and thermophysical characterization. J Mol Liq 317:113923. https://doi.org/10.1016/j.molliq.2020.113923

21. Jangir AK, Lad B, Dani U et al (2020) *In vitro* toxicity assessment and enhanced drug solubility profile of green deep eutectic solvent derivatives (DESDs) combined with theoretical validation. RSC Adv 10:24063–24072. https://doi.org/10.1039/C9RA10320A
22. Jangir AK, Patel D, More R et al (2019) New insight into experimental and computational studies of choline chloride-based 'green' ternary deep eutectic solvent (TDES). J Mol Struct 1181:295–299. https://doi.org/10.1016/j.molstruc.2018.12.106
23. Nie Y, Zhou Y, Zhang Y et al (2025) Sustainable synthesis of functional materials assisted by deep eutectic solvents for biomedical, environmental, and energy applications. Adv Funct Mater 35:2418957. https://doi.org/10.1002/adfm.202418957
24. Alnajjar A, Onaizi SA (2025) CO2 capture and conversion using deep eutectic solvents: recent progress and outlooks. J Mol Liq 421:126832. https://doi.org/10.1016/J.MOLLIQ.2024.126832
25. Wang Z, Zhao X, Chen Y et al (2025) A review of designable deep eutectic solvents for green fabrication of advanced functional materials. RSC Sustain 3:738–756. https://doi.org/10.1039/D4SU00560K
26. Zdanowicz M, Paszkiewicz S, El Fray M (2025) Polyesters and deep eutectic solvents: from synthesis through modification to depolymerization. Prog Polym Sci 161:101930. https://doi.org/10.1016/J.PROGPOLYMSCI.2025.101930
27. Pasricha S, Gahlot P, Rangarajan TM et al (2025) Deep eutectic solvents (DESs): emerging viable solvent systems for transition-metal-catalyzed cross-coupling reactions. J Mol Liq 426:127287. https://doi.org/10.1016/J.MOLLIQ.2025.127287
28. Hirpara D, Chavda V, Hirapara N, Kumar S (2024) Inorganic salt-induced micellar morphologies in deep eutectic solvent: structure and curcumin solubilization. J Mol Liq 411:125761. https://doi.org/10.1016/j.molliq.2024.125761
29. Chavda V, Patel HD, Patel AK et al (2024) Deep eutectic solvent engineered GO/ZrO_2 nanofiller for mixed matrix membranes: an efficient hydrogen gas separation. J Appl Polym Sci 141:e55938. https://doi.org/10.1002/app.55938
30. Chavda V, Patel B, Singh S et al (2023) A sustainable approach for the adsorption of methylene blue from an aqueous background: an adsorbent based on DES/CGS modified GO@ZrO_2. RSC Sustain 1:2038–2057. https://doi.org/10.1039/D3SU00236E
31. Chavda V, Hirpara D, Kumar S (2022) GO/ionic surfactant inspired photophysical modulation of rhodamine B in reline with or without additives. J Mol Liq 368:120614. https://doi.org/10.1016/j.molliq.2022.120614
32. Chavda V, Hirpara D, Rao V, Kumar S (2024) Deep eutectic solvent/surfactant-engineered GO@TiO_2 advanced material for sustained and ultrafast adsorption of hazardous dye from contaminated aqueous environment. ACS Appl Eng Mater 2:604–617. https://doi.org/10.1021/acsaenm.3c00696
33. Hirpara D, Patel B, Chavda V, Kumar S (2022) Micellization of conventional and gemini surfactants in aquoline: a case of exclusively water based deep eutectic solvent. J Mol Liq 362:119672. https://doi.org/10.1016/J.MOLLIQ.2022.119672
34. Hirpara D, Patel B, Chavda V et al (2022) Micellization and clouding behaviour of an ionic surfactant in a deep eutectic solvent: a case of the reline-water mixture. J Mol Liq 364:119991. https://doi.org/10.1016/J.MOLLIQ.2022.119991
35. Smith EL, Abbott AP, Ryder KS (2014) Deep eutectic solvents (DESs) and their applications. Chem Rev 114:11060–11082. https://doi.org/10.1021/cr300162p
36. Zhang Q, De Oliveira VK, Royer S, Jérôme F (2012) Deep eutectic solvents: syntheses, properties and applications. Chem Soc Rev 41:7108–7146. https://doi.org/10.1039/c2cs35178a
37. Perna FM, Vitale P, Capriati V (2020) Deep eutectic solvents and their applications as green solvents. Curr Opin Green Sustain Chem 21:27–33. https://doi.org/10.1016/j.cogsc.2019.09.004
38. Kim YJ, Varma RS (2005) Tetrahaloindate(III)-based ionic liquids in the coupling reaction of carbon dioxide and epoxides to generate cyclic carbonates: H-bonding and mechanistic studies. J Org Chem 70:7882–7891. https://doi.org/10.1021/jo050699x

39. Halimehjani AZ, Dadras A, Ramezani M et al (2015) Synthesis of dithiocarbamates by Markovnikov addition reaction in PEG and their application in amidoalkylation of naphthols and indoles. J Braz Chem Soc 26(7). https://doi.org/10.5935/0103-5053.20150119
40. Namboodiri VV, Varma RS (2001) Microwave-accelerated Suzuki cross-coupling reaction in polyethylene glycol (PEG). Green Chem 3:146–148. https://doi.org/10.1039/b102337n
41. Han X, Poliakoff M (2012) Continuous reactions in supercritical carbon dioxide: problems, solutions and possible ways forward. Chem Soc Rev 41:1428. https://doi.org/10.1039/c2cs15314a
42. Gawande MB, Bonifácio VDB, Luque R et al (2014) Solvent-free and catalysts-free chemistry: a benign pathway to sustainability. ChemSusChem 7:24–44. https://doi.org/10.1002/cssc.201300485
43. Cagnetta G, Huang J, Yu G (2018) A mini-review on mechanochemical treatment of contaminated soil: from laboratory to large-scale. Crit Rev Environ Sci Technol 48:723–771. https://doi.org/10.1080/10643389.2018.1493336
44. Wang G-W (2013) Mechanochemical organic synthesis. Chem Soc Rev 42:7668. https://doi.org/10.1039/c3cs35526h
45. Baig RBN, Varma RS (2012) Alternative energy input: mechanochemical, microwave and ultrasound-assisted organic synthesis. Chem Soc Rev 41:1559–1584. https://doi.org/10.1039/C1CS15204A
46. Varma RS (2008) Chemical activation by mechanochemical mixing, microwave and ultrasonic irradiation. Green Chem 10:1129. https://doi.org/10.1039/b817559b
47. Vaddula BR, Varma RS, Leazer J (2013) Mixing with microwaves: solvent-free and catalyst-free synthesis of pyrazoles and diazepines. Tetrahedron Lett 54:1538–1541. https://doi.org/10.1016/J.TETLET.2013.01.029
48. Hooshmand SE, Ghadari R, Mohammadian R et al (2019) Rhodanine-furan Bis-heterocyclic frameworks synthesis via green one-pot sequential six-component reactions: a synthetic and computational study. ChemistrySelect 4:11893–11898. https://doi.org/10.1002/slct.201903361
49. Varma RS, Namboodiri VV (2001) An expeditious solvent-free route to ionic liquids using microwaves. Chem Commun 7(7):643–644. https://doi.org/10.1039/b101375k
50. Cicco L, Dilauro G, Perna FM et al (2021) Advances in deep eutectic solvents and water: applications in metal- and biocatalyzed processes, in the synthesis of APIs, and other biologically active compounds. Org Biomol Chem 19:2558–2577. https://doi.org/10.1039/D0OB02491K
51. Longo L Jr, Craveiro M (2018) Deep eutectic solvents as unconventional media for multicomponent reactions. J Braz Chem Soc 29(10). https://doi.org/10.21577/0103-5053.20180147
52. Hooshmand SE, Afshari R, Ramón DJ, Varma RS (2020) Deep eutectic solvents: cutting-edge applications in cross-coupling reactions. Green Chem 22:3668–3692. https://doi.org/10.1039/D0GC01494J
53. Saavedra B, González-Gallardo N, Meli A, Ramón DJ (2019) A bipyridine-palladium derivative as general pre-catalyst for cross-coupling reactions in deep eutectic solvents. Adv Synth Catal 361:3868–3879. https://doi.org/10.1002/ADSC.201900472
54. Heidari B, Heravi MM, Nabid MR et al (2019) Novel palladium nanoparticles supported on β-cyclodextrin@graphene oxide as magnetically recyclable catalyst for Suzuki–Miyaura cross-coupling reaction with two different approaches in bio-based solvents. Appl Organomet Chem 33:e4632. https://doi.org/10.1002/aoc.4632
55. Marset X, De Gea S, Guillena G, Ramón DJ (2018) NCN–pincer–Pd complex as catalyst for the Hiyama reaction in biomass-derived solvents. ACS Sustain Chem Eng 6:5743–5748. https://doi.org/10.1021/acssuschemeng.8b00598
56. García-Álvarez J (2015) Deep eutectic mixtures: promising sustainable solvents for metal-catalysed and metal-mediated organic reactions. Eur J Inorg Chem 2015:5147–5157. https://doi.org/10.1002/ejic.201500892

57. Hooshmand SE, Kumar S, Bahadur I et al (2023) Deep eutectic solvents as reusable catalysts and promoter for the greener syntheses of small molecules: recent advances. J Mol Liq 371:121013. https://doi.org/10.1016/J.MOLLIQ.2022.121013
58. Najaf-Abadi MK, Ghobadian B, Dehghani-Soufi M (2024) A review on application of deep eutectic solvents as green catalysts and co-solvents in biodiesel production and purification processes. Biomass Convers Biorefinery 14:3117–3134. https://doi.org/10.1007/s13399-022-02644-5
59. Ruan J, Chen L, Qi Z (2023) Deep eutectic solvents as a versatile platform toward CO_2 capture and utilization. Green Chem 25:8328–8348. https://doi.org/10.1039/D3GC02468G
60. Durand E, Lecomte J, Baréa B et al (2013) Evaluation of deep eutectic solvent–water binary mixtures for lipase-catalyzed lipophilization of phenolic acids. Green Chem 15:2275. https://doi.org/10.1039/c3gc40899j
61. Gorke JT, Srienc F, Kazlauskas RJ (2008) Hydrolase-catalyzed biotransformations in deep eutectic solvents. Chem Commun (10):1235. https://doi.org/10.1039/b716317g
62. Shaabani A, Hooshmand SE (2016) Choline chloride/urea as a deep eutectic solvent/organocatalyst promoted three-component synthesis of 3-aminoimidazo-fused heterocycles via Groebke–Blackburn–Bienayme process. Tetrahedron Lett 57:310–313. https://doi.org/10.1016/J.TETLET.2015.12.014
63. Guajardo N, Domínguez de María P, Ahumada K et al (2017) Water as cosolvent: nonviscous deep eutectic solvents for efficient lipase-catalyzed esterifications. ChemCatChem 9:1393–1396. https://doi.org/10.1002/cctc.201601575
64. Yadav N, Bhakuni K, Bisht M et al (2020) Expanding the potential role of deep eutectic solvents toward facilitating the structural and thermal stability of α-chymotrypsin. ACS Sustain Chem Eng 8:10151–10160. https://doi.org/10.1021/acssuschemeng.0c02213
65. García-Argüelles S, Serrano MC, Gutiérrez MC et al (2013) Deep eutectic solvent-assisted synthesis of biodegradable polyesters with antibacterial properties. Langmuir 29:9525–9534. https://doi.org/10.1021/la401353r
66. Iqbal J, Shah NS, Sayed M et al (2020) Deep eutectic solvent-mediated synthesis of ceria nanoparticles with the enhanced yield for photocatalytic degradation of flumequine under UV-C. J Water Process Eng 33:101012. https://doi.org/10.1016/j.jwpe.2019.101012
67. Iqbal J, Shah NS, Sayed M et al (2020) Synthesis of nitrogen-doped ceria nanoparticles in deep eutectic solvent for the degradation of sulfamethaxazole under solar irradiation and additional antibacterial activities. Chem Eng J 394:124869. https://doi.org/10.1016/j.cej.2020.124869
68. Marset X, Khoshnood A, Sotorríos L et al (2017) Deep eutectic solvent compatible metallic catalysts: cationic pyridiniophosphine ligands in palladium catalyzed cross-coupling reactions. ChemCatChem 9:1269–1275. https://doi.org/10.1002/cctc.201601544
69. Park T-J, Lee SH (2017) Deep eutectic solvent systems for $FeCl_3$-catalyzed oxidative polymerization of 3-octylthiophene. Green Chem 19:910–913. https://doi.org/10.1039/C6GC02789J
70. Azizi N, Rahimi Z, Alipour M (2015) A magnetic nanoparticle catalyzed eco-friendly synthesis of cyanohydrins in a deep eutectic solvent. RSC Adv 5:61191–61198. https://doi.org/10.1039/C5RA06176H
71. Khandelwal S, Tailor YK, Kumar M (2016) Deep eutectic solvents (DESs) as eco-friendly and sustainable solvent/catalyst systems in organic transformations. J Mol Liq 215:345–386. https://doi.org/10.1016/j.molliq.2015.12.015
72. Cicco L, Ríos-Lombardía N, Rodríguez-Álvarez MJ et al (2018) Programming cascade reactions interfacing biocatalysis with transition-metal catalysis in *deep eutectic solvents* as biorenewable reaction media. Green Chem 20:3468–3475. https://doi.org/10.1039/C8GC00861B
73. Shafique S, Belousov AS, Rashid R et al (2025) Deep eutectic solvents (DES): structure, properties, and cutting-edge applications in green catalysis. J Mol Liq 419:126769. https://doi.org/10.1016/J.MOLLIQ.2024.126769
74. Falcini C, de Gonzalo G (2024) Deep eutectic solvents as catalysts in the synthesis of active pharmaceutical ingredients and precursors. Catalysts 14:120. https://doi.org/10.3390/catal14020120

75. Sheikh Asadi AM, Cichocki Ł, Atamaleki A et al (2024) Catalysts for advanced oxidation processes: deep eutectic solvents-assisted synthesis—a review. Water Resour Ind 31:100251. https://doi.org/10.1016/J.WRI.2024.100251
76. Álvarez MS, Longo MA, Rodríguez A, Deive FJ (2024) The role of deep eutectic solvents in catalysis. A vision on their contribution to homogeneous, heterogeneous and electrocatalytic processes. J Ind Eng Chem 132:36–49. https://doi.org/10.1016/J.JIEC.2023.11.030
77. Paparella AN, Perrone S, Salomone A et al (2023) Use of deep eutectic solvents in plastic depolymerization. Catalysts 13:1035. https://doi.org/10.3390/catal13071035
78. Joarder S, Bansal D, Meena H et al (2023) Bioinspired green deep eutectic solvents: preparation, catalytic activity, and biocompatibility. J Mol Liq 376:121355. https://doi.org/10.1016/J.MOLLIQ.2023.121355
79. Rodríguez-Álvarez MJ, García-Garrido SE, Perrone S et al (2023) Deep eutectic solvents and heterogeneous catalysis with metallic nanoparticles: a powerful partnership in sustainable synthesis. Curr Opin Green Sustain Chem 39:100723. https://doi.org/10.1016/J.COGSC.2022.100723
80. Abbott AP, Barron JC, Ryder KS, Wilson D (2007) Eutectic-based ionic liquids with metal-containing anions and cations. Chem Eur J 13:6495–6501. https://doi.org/10.1002/chem.200601738
81. Singh SK, Savoy AW (2020) Ionic liquids synthesis and applications: an overview. J Mol Liq 297:112038. https://doi.org/10.1016/j.molliq.2019.112038
82. Bagh FSG, Shahbaz K, Mjalli FS et al (2013) Electrical conductivity of ammonium and phosphonium based deep eutectic solvents: measurements and artificial intelligence-based prediction. Fluid Phase Equilib 356:30–37. https://doi.org/10.1016/J.FLUID.2013.07.012
83. Pandey A, Rai R, Pal M, Pandey S (2014) How polar are choline chloride-based deep eutectic solvents? Phys Chem Chem Phys 16:1559–1568. https://doi.org/10.1039/c3cp53456a
84. Shaibuna M, Theresa LV, Sreekumar K (2022) Neoteric deep eutectic solvents: history, recent developments, and catalytic applications. Soft Matter 18:2695–2721. https://doi.org/10.1039/D1SM01797G
85. Khaghaninejad S, Heravi MM (2014) Paal–Knorr reaction in the synthesis of heterocyclic compounds. Adv Heterocycl Chem 111:95–146. https://doi.org/10.1016/B978-0-12-420160-6.00003-3
86. Handy S, Lavender K (2013) Organic synthesis in deep eutectic solvents: Paal–Knorr reactions. Tetrahedron Lett 54:4377–4379. https://doi.org/10.1016/J.TETLET.2013.05.122
87. Hu H-C, Liu Y-H, Li B-L et al (2015) Deep eutectic solvent based on choline chloride and malonic acid as an efficient and reusable catalytic system for one-pot synthesis of functionalized pyrroles. RSC Adv 5:7720–7728. https://doi.org/10.1039/C4RA13577F
88. Mancuso R, Maner A, Cicco L et al (2016) Synthesis of thiophenes in a deep eutectic solvent: heterocyclodehydration and iodocyclization of 1-mercapto-3-yn-2-ols in a choline chloride/glycerol medium. Tetrahedron 72:4239–4244. https://doi.org/10.1016/J.TET.2016.05.062
89. Mancuso R, Lettieri M, Strangis R et al (2022) Iodocyclization of 2-methylthiophenylacetylenes to 3-iodobenzothiophenes and their coupling reactions under more sustainable conditions. Asian J Org Chem 11:e202200353. https://doi.org/10.1002/ajoc.202200353
90. Safak C, Simsek R (2006) Fused 1,4-dihydropyridines as potential calcium modulatory compounds. Mini-Rev Med Chem 6:747–755. https://doi.org/10.2174/138955706777698606
91. Pednekar S, Bhalerao R, Ghadge N (2013) One-pot multi-component synthesis of 1,4-dihydropyridine derivatives in biocompatible deep eutectic solvents. J Chem Sci 125:615–621
92. Azizi N, Dezfuli S, Hahsemi MM (2012) Eutectic salt catalyzed environmentally benign and highly efficient Biginelli reaction. Sci World J 2012:1–6. https://doi.org/10.1100/2012/908702
93. Aziizi N, Manochehri Z, Nahayi A, Torkashvand S (2014) A facile one-pot synthesis of tetrasubstituted imidazoles catalyzed by eutectic mixture stabilized ferrofluid. J Mol Liq 196:153–158. https://doi.org/10.1016/J.MOLLIQ.2014.03.013

94. Wang L, Zhong X, Zhou M et al (2013) One-pot synthesis of polysubstituted imidazoles in a Brønsted acidic deep eutectic solvent. J Chem Res 37:236–238. https://doi.org/10.3184/174751913X13636339694414
95. Mobinikhaledi A, Amiri AK (2015) One-pot synthesis of tri- and tetrasubstituted imidazoles using eutectic salts as ionic liquid catalyst. Res Chem Intermed 41:2063–2070. https://doi.org/10.1007/s11164-013-1331-9
96. Di Gioia ML, Cassano R, Costanzo P et al (2019) Green synthesis of privileged benzimidazole scaffolds using active deep eutectic solvent. Molecules 24:2885. https://doi.org/10.3390/molecules24162885
97. Rushell E, Tailor YK, Khandewal S et al (2019) Deep eutectic solvent promoted synthesis of structurally diverse hybrid molecules with privileged heterocyclic substructures. New J Chem 43:12462–12467. https://doi.org/10.1039/C9NJ02694K
98. Berjis A, Mirza B, Anaraki-Ardakani H (2023) One-pot synthesis of highly functionalized indole in choline chloride/oxalic acid as a deep eutectic solvent. Polycycl Aromat Compd 43:4305–4312. https://doi.org/10.1080/10406638.2022.2089172
99. Huang S, Qing J, Wang S et al (2014) Design and synthesis of imidazo[1,2-α][1,8]naphthyridine derivatives as anti-HCV agents via direct C–H arylation. Org Biomol Chem 12:2344–2348. https://doi.org/10.1039/C3OB42525H
100. Rudys S, Ríos-Luci C, Pérez-Roth E et al (2010) Antiproliferative activity of novel benzo[b][1,6]naphthyridines in human solid tumor cell lines. Bioorg Med Chem Lett 20:1504–1506. https://doi.org/10.1016/J.BMCL.2010.01.112
101. Shaabani A, Hooshmand SE, Tabatabaei AT (2016) Synthesis of fully substituted naphthyridines: a novel domino four-component reaction in a deep eutectic solvent system based on choline chloride/urea. Tetrahedron Lett 57:351–353. https://doi.org/10.1016/J.TETLET.2015.12.017
102. Shaabani A, Hooshmand SE, Nazeri MT et al (2016) Deep eutectic solvent as a highly efficient reaction media for the one-pot synthesis of benzo-fused seven-membered heterocycles. Tetrahedron Lett 57:3727–3730. https://doi.org/10.1016/J.TETLET.2016.07.005
103. Tak RK, Patel P, Subramanian S et al (2018) Cycloaddition reaction of spiro-epoxy oxindole with CO_2 at atmospheric pressure using deep eutectic solvent. ACS Sustain Chem Eng 6:11200–11205. https://doi.org/10.1021/acssuschemeng.8b02566
104. Ahmadi F, Nejad MJ, Shariatipour M, Heydari A (2022) A greener approach to oxazolidinone synthesis using a ternary deep eutectic solvent. ChemistrySelect 7:e202200068. https://doi.org/10.1002/slct.202200068
105. Rajawat A, Khandelwal S, Kumar M (2014) Deep eutectic solvent promoted efficient and environmentally benign four-component domino protocol for synthesis of spirooxindoles. RSC Adv 4:5105. https://doi.org/10.1039/c3ra44600j
106. Khandelwal S, Rajawat A, Tailor Y, Kumar M (2014) A simple, efficient and environmentally benign synthetic protocol for the synthesis of spirooxindoles using choline chloride-oxalic acid eutectic mixture as catalyst/solvent system. Comb Chem High Throughput Screen 17:763–769. https://doi.org/10.2174/1386207317666141016125218
107. Azizi N, Dezfooli S, Mahmoudi Hashemi M (2014) Greener synthesis of spirooxindole in deep eutectic solvent. J Mol Liq 194:62–67. https://doi.org/10.1016/J.MOLLIQ.2014.01.009
108. Ferretti MD, Neto AT, Morel AF et al (2014) Synthesis of symmetrically substituted 3,3-dibenzyl-4-hydroxy-3,4-dihydro-1H-quinolin-2-ones, as novel quinoline derivatives with antibacterial activity. Eur J Med Chem 81:253–266. https://doi.org/10.1016/J.EJMECH.2014.05.024
109. Michael JP (2008) Quinoline, quinazoline and acridonealkaloids. Nat Prod Rep 25:166–187. https://doi.org/10.1039/B612168N
110. Ma F-P, Cheng G-T, He Z-G, Zhang Z-H (2012) A new and efficient procedure for Friedländer synthesis of quinolines in low melting tartaric acid-urea mixtures. Aust J Chem 65:409. https://doi.org/10.1071/CH12025
111. Alanazi AM, Abdel-Aziz AAM, Al-Suwaidan IA et al (2014) Design, synthesis and biological evaluation of some novel substituted quinazolines as antitumor agents. Eur J Med Chem 79:446–454. https://doi.org/10.1016/J.EJMECH.2014.04.029

112. Ugale VG, Bari SB (2014) Quinazolines: new horizons in anticonvulsant therapy. Eur J Med Chem 80:447–501. https://doi.org/10.1016/J.EJMECH.2014.04.072
113. Ghoneim MM, Abdelgawad MA, Elkanzi NAA et al (2024) A literature review on pharmacological aspects, docking studies, and synthetic approaches of quinazoline and quinazolinone derivatives. Arch Pharm (Weinheim) 357:e2400057. https://doi.org/10.1002/ardp.202400057
114. Zhang Z-H, Zhang X-N, Mo L-P et al (2012) Catalyst-free synthesis of quinazoline derivatives using low melting sugar–urea–salt mixture as a solvent. Green Chem 14:1502. https://doi.org/10.1039/c2gc35258c
115. Zhao D, Liu Y, Li Y, Chen Y (2022) A green synthesis and antibacterial activity of ferrocene-based thiazole derivatives in choline chloride/glycerol eutectic solvent. RSC Adv 12:22054–22059. https://doi.org/10.1039/D2RA04587G
116. Lobo HR, Singh BS, Shankarling GS (2012) Lipase and deep eutectic mixture catalyzed efficient synthesis of thiazoles in water at room temperature. Catal Lett 142:1369–1375. https://doi.org/10.1007/s10562-012-0902-5
117. Azizi N, Qomi M, Asghari M, Farhadi E (2021) Deep eutectic solvent mediated rapid and selective one-pot synthesis of 5-alkylidene-thiazolones. Sustain Chem Pharm 22:100457. https://doi.org/10.1016/J.SCP.2021.100457
118. Riadi Y, Ouerghi O, Geesi MH et al (2021) Efficient novel eutectic-mixture-mediated synthesis of benzoxazole-linked pyrrolidin-2-one heterocycles. J Mol Liq 323:115011. https://doi.org/10.1016/J.MOLLIQ.2020.115011
119. Lü H, Wu K, Zhao Y et al (2017) Synthesis of cyclic carbonates from CO2 and propylene oxide (PO) with deep eutectic solvents (DESs) based on amino acids (AAs) and dicarboxylic acids. J CO2 Util 22:400–406. https://doi.org/10.1016/J.JCOU.2017.10.024
120. Wu K, Su T, Hao D et al (2018) Choline chloride-based deep eutectic solvents for efficient cycloaddition of CO_2 with propylene oxide. Chem Commun 54:9579–9582. https://doi.org/10.1039/C8CC04412K
121. Nguyen HT, Tran PH (2016) An extremely efficient and green method for the acylation of secondary alcohols, phenols and naphthols with a deep eutectic solvent as the catalyst. RSC Adv 6:98365–98368. https://doi.org/10.1039/C6RA22757K
122. Nikolaienko P, Rueping M (2016) Trifluoromethylselenolation of aryldiazonium salts: a mild and convenient copper-catalyzed procedure for the introduction of the $SeCF_3$ group. Chemistry 22:2620–2623. https://doi.org/10.1002/chem.201504601
123. Shao L, Li Y, Lu J, Jiang X (2019) Recent progress in selenium-catalyzed organic reactions. Org Chem Front 6:2999–3041. https://doi.org/10.1039/C9QO00620F
124. Hoover GC, Seferos DS (2019) Photoactivity and optical applications of organic materials containing selenium and tellurium. Chem Sci 10:9182–9188. https://doi.org/10.1039/C9SC04279B
125. Muniraj N, Dhineshkumar J, Prabhu KR (2016) *N*-iodosuccinimide catalyzed oxidative selenocyanation and thiocyanation of electron rich arenes. ChemistrySelect 1:1033–1038. https://doi.org/10.1002/slct.201600292
126. Sun K, Chen XL, Li SJ et al (2018) Copper-catalyzed radical cascade cyclization to access 3-sulfonated indenones with the AIE phenomenon. J Org Chem 83:14419–14430. https://doi.org/10.1021/ACS.JOC.8B02175
127. Wu C, Xiao H-J, Wang S-W et al (2019) Natural deep eutectic solvent-catalyzed selenocyanation of activated alkynes via an intermolecular H-bonding activation process. ACS Sustain Chem Eng 7:2169–2175. https://doi.org/10.1021/acssuschemeng.8b04877
128. Phadtare SB, Shankarling GS (2010) Halogenation reactions in biodegradable solvent: efficient bromination of substituted 1-aminoanthra-9,10-quinone in deep eutectic solvent (choline chloride:urea). Green Chem 12:458. https://doi.org/10.1039/b923589b
129. Chiappe C, Leandri E, Tebano M (2006) [Hmim][NO_3]—an efficient solvent and promoter in the oxidative aromatic chlorination. Green Chem 8:742–745. https://doi.org/10.1039/B606258J

130. De Kimpe N, Verh R (1988) α-Haloketones, α-haloaldehydes and α-haloimines. John Wiley & Sons, Inc., Chichester
131. Pravst I, Zupan M, Stavber S (2006) Solvent-free bromination of 1,3-diketones and β-keto esters with NBS. Green Chem 8:1001–1005. https://doi.org/10.1039/B608446J
132. Hoffman RV, Weiner WS, Maslouh N (2001) Highly stereoselective synthesis of *anti*-N-protected-α-amino epoxides. J Org Chem 66:5790–5795. https://doi.org/10.1021/jo010319h
133. Meshram HM, Reddy PN, Sadashiv K, Yadav JS (2005) Amberlyst-15®-promoted efficient 2-halogenation of 1,3-keto-esters and cyclic ketones using N-halosuccinimides. Tetrahedron Lett 46:623–626. https://doi.org/10.1016/J.TETLET.2004.11.140
134. Xu Z, Zhang D, Zou X (2006) α-Chlorination of acetophenones using 1,3-dichloro-5,5-dimethylhydantoin. Synth Commun 36:255–258. https://doi.org/10.1080/00397910500334637
135. Chen Z, Zhou B, Cai H et al (2009) Simple and efficient methods for selective preparation of α-mono or α,α-dichloro ketones and β-ketoesters by using DCDMH. Green Chem 11:275–278. https://doi.org/10.1039/B815169E
136. Halimehjani AZ, Hooshmand SE, Shamiri EV (2014) Synthesis of α-phthalimido-α′-dithiocarbamato propan-2-ols via a one-pot, three-component epoxide ring-opening in water. Tetrahedron Lett 55:5454–5457. https://doi.org/10.1016/J.TETLET.2014.08.017
137. Azizi N, Batebi E (2012) Highly efficient deep eutectic solvent catalyzed ring opening of epoxides. Cat Sci Technol 2:2445. https://doi.org/10.1039/c2cy20456h
138. Pattabiraman VR, Bode JW (2011) Rethinking amide bond synthesis. Nature 480:471–479. https://doi.org/10.1038/nature10702
139. Patil UB, Singh AS, Nagarkar JM (2014) Choline chloride based eutectic solvent: an efficient and reusable solvent system for the synthesis of primary amides from aldehydes and from nitriles. RSC Adv 4:1102–1106. https://doi.org/10.1039/C3RA43552K
140. Guo W, Laserna V, Rintjema J, Kleij AW (2016) Catalytic one-pot oxetane to carbamate conversions: formal synthesis of drug relevant molecules. Adv Synth Catal 358:1602–1607. https://doi.org/10.1002/ADSC.201500895
141. Ren L, Jiao N (2014) PdCl2 catalyzed efficient assembly of organic azides, CO, and alcohols under mild conditions: a direct approach to synthesize carbamates. Chem Commun 50:3706. https://doi.org/10.1039/c4cc00538d
142. Manjunathan P, Ravishankar R, Shanbhag GV (2016) Novel bifunctional Zn–Sn composite oxide catalyst for the selective synthesis of glycerol carbonate by carbonylation of glycerol with urea. ChemCatChem 8:631–639. https://doi.org/10.1002/cctc.201501088
143. Dindarloo Inaloo I, Majnooni S (2018) Ureas as safe carbonyl sources for the synthesis of carbamates with deep eutectic solvents (DESs) as efficient and recyclable solvent/catalyst systems. New J Chem 42:13249–13255. https://doi.org/10.1039/C8NJ02624F
144. Dindarloo Inaloo I, Majnooni S (2019) Carbon dioxide utilization in the efficient synthesis of carbamates by deep eutectic solvents (DES) as green and attractive solvent/catalyst systems. New J Chem 43:11275–11281. https://doi.org/10.1039/C9NJ02810B
145. Hogarth G (2005) Transition metal dithiocarbamates: 1978–2003. In: Progress in inorganic chemistry. Wiley, New York, pp 71–561
146. Ronconi L, Marzano C, Zanello P et al (2006) Gold(III) dithiocarbamate derivatives for the treatment of cancer: solution chemistry, DNA binding, and hemolytic properties. J Med Chem 49:1648–1657. https://doi.org/10.1021/jm0509288
147. Qian Y, Ma GY, Yang Y et al (2010) Synthesis, molecular modeling and biological evaluation of dithiocarbamates as novel antitubulin agents. Bioorg Med Chem 18:4310–4316. https://doi.org/10.1016/J.BMC.2010.04.091
148. Alizadeh A, Rostamnia S, Zohreh N, Hosseinpour R (2009) A simple and effective approach to the synthesis of rhodanine derivatives via three-component reactions in water. Tetrahedron Lett 50:1533–1535. https://doi.org/10.1016/J.TETLET.2008.12.107
149. Ranu BC, Saha A, Banerjee S (2008) Catalysis by ionic liquids: significant rate acceleration with the use of [pmIm]Br in the three-component synthesis of dithiocarbamates. Eur J Org Chem 2008:519–523. https://doi.org/10.1002/ejoc.200700842

150. Azizi N, Khajeh M, Hasani M, Dezfooli S (2013) An efficient four-component synthesis of dithiocarbamate derivatives. Tetrahedron Lett 54:5407–5410. https://doi.org/10.1016/J.TETLET.2013.07.124
151. Shi Y, Chen QX, Wang Q et al (2005) Inhibitory effects of cinnamic acid and its derivatives on the diphenolase activity of mushroom (Agaricus bisporus) tyrosinase. Food Chem 92:707–712. https://doi.org/10.1016/J.FOODCHEM.2004.08.031
152. Aoyagi Y, Masuko N, Ohkubo S et al (2005) A novel cinnamic acid derivative that inhibits Cdc25 dual-specificity phosphatase activity. Cancer Sci 96:614–619. https://doi.org/10.1111/j.1349-7006.2005.00086.x
153. Weng J, Wang C, Li H, Wang Y (2006) Novel quaternary ammonium ionic liquids and their use as dual solvent-catalysts in the hydrolytic reaction. Green Chem 8:96–99. https://doi.org/10.1039/B508325G
154. Pawar PM, Jarag KJ, Shankarling GS (2011) Environmentally benign and energy efficient methodology for condensation: an interesting facet to the classical Perkin reaction. Green Chem 13:2130. https://doi.org/10.1039/c0gc00712a
155. Lattanzio V (2013) Phenolic compounds: introduction. In: Natural products. Springer, Berlin, pp 1543–1580
156. Wang L, Dai D, Chen Q, He M (2013) Rapid, sustainable, and gram-scale synthesis of phenols catalyzed by a biodegradable deep eutectic mixture in water. Asian J Org Chem 2:1040–1043. https://doi.org/10.1002/ajoc.201300192
157. Graebin CS, Ribeiro FV, Rogério KR, Kümmerle AE (2019) Multicomponent reactions for the synthesis of bioactive compounds: a review. Curr Org Synth 16:855–899. https://doi.org/10.2174/1570179416666190718153703
158. Shaabani A, Hooshmand SE (2018) Diversity-oriented catalyst-free synthesis of pseudo-peptides containing rhodanine scaffolds via a one-pot sequential isocyanide-based six-component reactions in water using ultrasound irradiation. Ultrason Sonochem 40:84–90. https://doi.org/10.1016/J.ULTSONCH.2017.06.030
159. Alvim HGO, da Silva Júnior EN, Neto BAD (2014) What do we know about multicomponent reactions? Mechanisms and trends for the Biginelli, Hantzsch, Mannich, Passerini and Ugi MCRs. RSC Adv 4:54282–54299. https://doi.org/10.1039/C4RA10651B
160. Morrison HG, Sun CC, Neervannan S (2009) Characterization of thermal behavior of deep eutectic solvents and their potential as drug solubilization vehicles. Int J Pharm 378:136–139. https://doi.org/10.1016/j.ijpharm.2009.05.039
161. Shaabani A, Hooshmand SE, Afshari R et al (2018) Direct construction of diverse metallophthalocyanines by manifold substrates in a deep eutectic solvent. J Solid State Chem 258:536–542. https://doi.org/10.1016/J.JSSC.2017.11.022
162. Lei J, Meng J-P, Tang D-Y et al (2018) Recent advances in the development of polycyclic skeletons via Ugi reaction cascades. Mol Divers 22:503–516. https://doi.org/10.1007/s11030-017-9811-2
163. Azizi N, Dezfooli S, Hashemi MM (2013) A sustainable approach to the Ugi reaction in deep eutectic solvent. C R Chim 16:1098–1102. https://doi.org/10.1016/J.CRCI.2013.05.013
164. Bagdi AK, Santra S, Monir K, Hajra A (2015) Synthesis of imidazo[1,2-a]pyridines: a decade update. Chem Commun 51:1555–1575. https://doi.org/10.1039/C4CC08495K
165. Shaabani A, Afshari R, Hooshmand SE (2016) Passerini three-component cascade reactions in deep eutectic solvent: an environmentally benign and rapid system for the synthesis of α-acyloxyamides. Res Chem Intermed 42:5607–5616. https://doi.org/10.1007/s11164-015-2390-x
166. Jamaati H, Balalaie S, Miraki MK et al (2019) Choline chloride/urea as mild media for the synthesis of the chromonyl amidodiester fragments and succinimide derivatives. ChemistrySelect 4:9074–9078. https://doi.org/10.1002/slct.201901893
167. Kumar AK, Sharma S, Dixit G et al (2020) Techno-economic evaluation of a natural deep eutectic solvent-based biorefinery: exploring different design scenarios. Biofuels Bioprod Biorefin 14:746–763. https://doi.org/10.1002/bbb.2110

168. Peng J, Xu H, Wang W et al (2021) Techno-economic analysis of bioethanol preparation process via deep eutectic solvent pretreatment. Ind Crop Prod 172:114036. https://doi.org/10.1016/J.INDCROP.2021.114036
169. Arnodo D, Maffeis E, Marra F et al (2023) Combination of enzymes and deep eutectic solvents as powerful toolbox for organic synthesis. Molecules 28:516. https://doi.org/10.3390/molecules28020516

Chapter 8
Deep Eutectic Solvents for Inorganic Synthesis Applications

Piyush Gupta, Vaibhav Sharma, and Pallavi Jain

8.1 Introduction

Deep eutectic solvents (DES) are a category of solvents that have arisen as viable and economical substitutes for conventional ionic liquids (ILs) in numerous chemical applications, such as inorganic synthesis. DES are usually synthesized by mixing a quaternary ammonium or phosphonium salt with urea or glycerol. When combined, these components form a mixture that melts at a much lower temperature than either of the individual substances on their own. These mixes generate liquids under mild conditions without requiring filtration, rendering them appealing from both economic and environmental perspectives. Similar to ILs, DES demonstrate low vapor pressure, elevated polarity, and adjustable chemical reactivity, and they are progressively used in the preparation of inorganic compounds as a result of their capacity to dissolve a diverse array of precursors and affect nucleation and particle growth. Ionothermal procedures utilize ILs or DES as both solvents and active agents in the reaction. Their distinctive microheterogeneous structure, particularly in imidazolium-based ILs with extended alkyl chains, facilitates the regulation of material shape, dimensions, and growth rate. Moreover, ionic liquids have remarkable thermal and chemical stability. The multifunctional characteristics of ILs and DES as solvents, templates, and reactants provide innovative, sustainable approaches for the design and synthesis of novel inorganic

P. Gupta (✉) · P. Jain
Department of Chemistry, Faculty of Engineering and Technology, SRM Institute of Science and Technology, NCR Campus, Delhi-NCR Campus, Modinagar, Ghaziabad, Uttar Pradesh, India

V. Sharma
Department of Chemistry, SRM Institute of Science and Technology, NCR Campus, Delhi-NCR Campus, Ghaziabad, Uttar Pradesh, India

K. Kuperkar et al. (eds.), *Deep Eutectic Solvents*, Environmental Footprints and Eco-design of Products and Processes,
https://doi.org/10.1007/978-3-032-03801-2_8

materials under softer and more controlled circumstances. DES have remained relevant and advanced since their introduction in 2003. They underwent significant advancements in 2008 as DES were first incorporated into the preparation of metal nanoparticles (NPs) for use in nanotechnology and catalysis. The concept advanced further with Natural Deep Eutectic Solvents (NDES) in 2011, where NDES are made of natural, biodegradable materials appealing to biological and pharmaceutical use. In 2015, Therapeutic Deep Eutectic Solvents (THEDES) emerged as incorporating active ingredients of the drugs for enhanced solubility, stability, and delivery of the drug. Recently, in 2020, Supramolecular Deep Eutectic Solvents (SUPRADES) were introduced, which are utilized as advanced materials in materials science and drug formulation [1–3]. The advancement in the DES over time is shown in Fig. 8.1.

This chapter explores how DES can be customized, allowing their physical and chemical properties to achieve properties such as low volatility, biodegradability, and tunable polarity. Special emphasis is placed on the application of DES in inorganic synthesis, where they serve as green solvents for nanoparticle and nanocrystal synthesis, demonstrating their versatility and growing importance in sustainable chemistry.

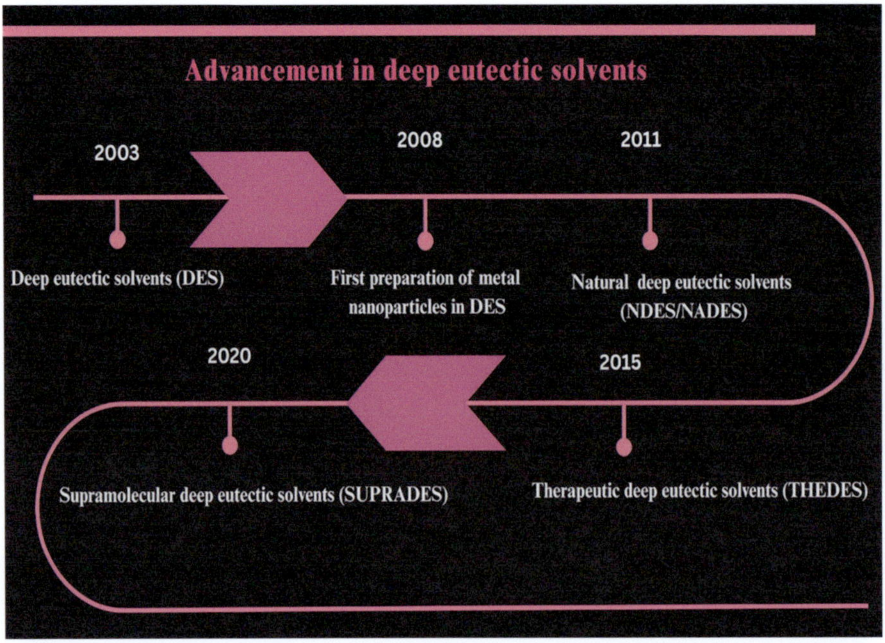

Fig. 8.1 Advancements in DES

8.2 Properties of DES

Certain inorganic substances may be used as components of DES, and this is especially applicable to inorganic molecules. The choice of these elements is the primary determinant of the physical and chemical characteristics of DES, hence influencing their possible uses. Despite differing chemical features, DES and standard IL have similar physical qualities, especially in their solvent tunability for certain chemical processes. The toxicity of DES constituent parts is usually well-defined, but little is known about the eutectic solvents' overall toxic behavior, suggesting that more research is required.

DES have several characteristics that arise from their strong intermolecular forces. They are generally polar in nature, enabling them to dissolve a broad variety of compounds, and are less toxic than traditional organic solvents, hence more environmentally friendly. DES also have a wide range of liquefaction, existing in the liquid state under different conditions, which makes them more usable in different synthesis methods. In addition to their other useful properties, they usually possess low electrical conductivity, which may affect their use in electrochemical processes. DES are formulated from two solid constituents that operate through hydrogen bonding when combined into a novel liquid phase. The resultant state often exhibits a lower freezing point as compared with individual components. Density constitutes a crucial physical feature of solvents. DES density is often assessed with a specific gravity meter. The comparison of measured and calculated densities showed an average error of 1.9%. The experimental determination of DES densities at varying temperatures can be difficult. Consequently, much work has been dedicated to the creation of reliable value estimation systems. The DES density estimates using the Spencer and Danner-modified Rackett equation exhibited an inaccuracy of roughly 1.9%. The viscosity of DES is a critical characteristic that necessitates meticulous attention. DES were synthesized utilizing diminutive cations and fluorinated H-bond donors, which exhibited characteristics of diminished viscosity and enhanced conductivity, with variations reaching up to four orders of magnitude in comparison to traditional ionic liquids. Their elevated viscosity is mostly attributed to extensive hydrogen bonding among the components, which impedes the movement of free entities within the medium. The viscosity of eutectic mixtures is significantly influenced by the chemical properties of the components. Viscosity can vary significantly with temperature, so examining its behavior under different situations is crucial. The conductivities of DES often increase significantly at elevated temperatures, mostly owing to a decrease in viscosity. DES can be synthesized through the reaction of quaternary ammonium salts in conjunction with R-COOH. The physicochemical characteristics of the resultant DES are contingent upon the structural configuration of the R-COOH utilized. Surface tension is anticipated to follow a trajectory akin to viscosity, as it is directly correlated with the intensity of intermolecular forces that lead to the formation of DES [4–7]. Figure 8.2 highlights the general properties of DES.

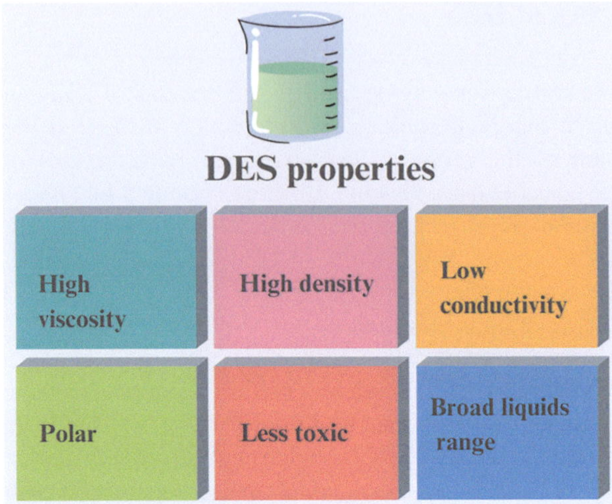

Fig. 8.2 General properties of DES

8.3 Applications of DES

DES acts as a desired solvent for effective cellulose dissolution. A novel one-pot method was developed to create 3D-printable cellulose nanofiber/poly (deep eutectic solvent) (CNF/PDES) conductive elastomers. Adding 5 wt% CNF improved mechanical properties, increasing stress from 59 to 330 KPa and strain from 135% to about 300%. Cellulose nanocrystals were prepared by using DES pretreatment of wood cellulose fibers using choline chloride (ChCl) combined with organic acids at 60–100 °C. Figure 8.3 shows the process of preparation of cellulose nanocrystals by using DES.

DES function as solvents and catalysts in the synthesis of pharmaceuticals, serve as environmentally friendly solvents in the realm of pharmaceutical analysis, and are increasingly being utilized in the formulation and delivery of drugs as solubilizers, stabilizers, therapeutic agents, and mediums for the synthesis of biomaterials. They facilitate the controlled release of ferulic acid within liposomal formulations, significantly enhancing bioavailability and sustainability in comparison to ethanol, exhibiting solubility that is over twenty times greater in DES07 and DES08. Furthermore, DES contribute to the processing of lignocellulosic biomass; for instance, ChCl-pTSA and ChCl-OxAc effectively convert fructose, with ChCl-pTSA attaining yields of 59.2% ethyl levulinate and 76.3% methyl levulinate within a time frame of 20 minutes [8–14]. DES are synthesized through the conjunction of a hydrogen bond donor (HBD) with a hydrogen bond acceptor (HBA); the raw materials of HBA and HBD can be procured from renewable sources. HBA and HBD represent the fundamental components of the DES utilized as catalytic agents in the synthesis of active pharmaceutical ingredients (APIs) and their precursors, as illustrated in Fig. 8.4 [15].

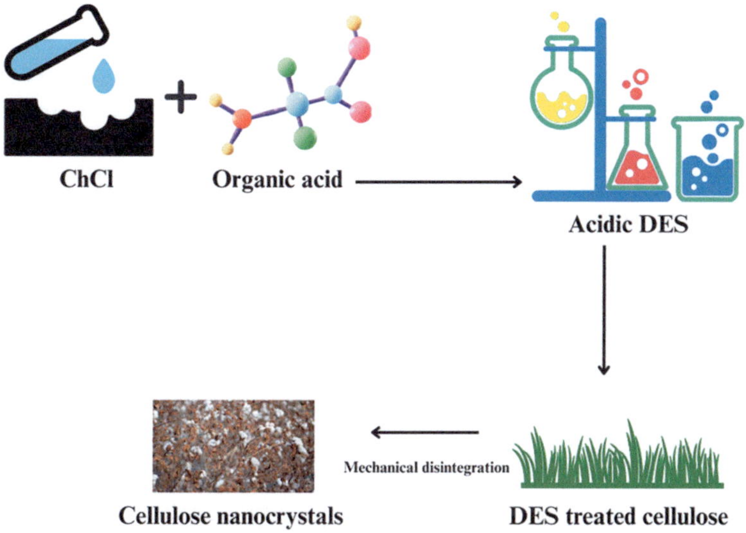

Fig. 8.3 Preparation method of cellulose nanocrystals by using DES

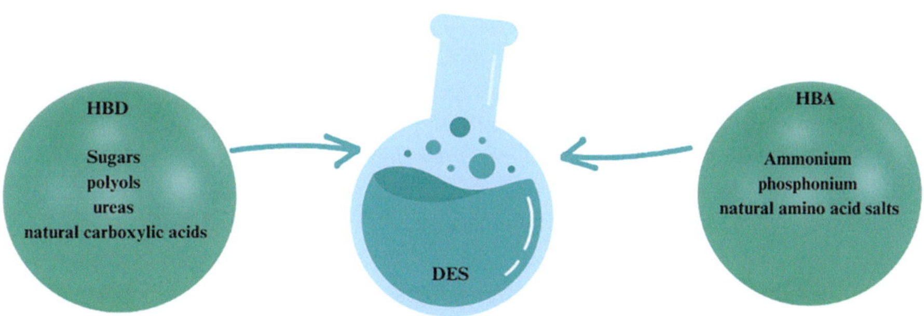

Fig. 8.4 HBD and HBA components of DES work as catalysts in the synthesis of API

Although the majority of DES applications in environmental sample analysis focus on organic contaminants and analysis of inorganic species [16]. DES have unique features such as low volatility, nonflammability, superior thermal stability, and tunable solubility. These features render them extremely versatile and useful in various industries. Some applications, such as wastewater analysis, green synthesis, recyclability, energy conversion, medical application, and, most importantly, inorganic synthesis, are explained in this section. In addition to extraction, DES have been documented for pesticide and insecticide analysis, detection of toxic and harmful substances, decontamination of heavy metals, identification of contaminants in the milk sample, elimination of antibiotics, and the production of sustainable packaging films. These various applications underscore the adaptability and increasing role of DES in food security, ecological sustainability, and drug discovery [17]. DES have been used with 3D printing, a cutting-edge additive manufacturing method transforming industries globally. From nanoscale accuracy to mass production, the technology continues to push boundaries. The incorporation of DES in 3D printing has created new opportunities, especially in the advancement of biomineralization-based materials, and can be used in green material development [18].

8.3.1 DES' in the Preparation of Inorganic Synthesis

Their remarkable chemical reactivity may be regulated and utilized as a synthesis tool for the manufacture of inorganic molecules that are challenging or unfeasible with conventional methods. DES have gained greater recognition as an intrinsically sustainable and resource-efficient medium for the synthesis of inorganic chemicals. The extremely efficient fabrication of Pt NPs with hollow-opened morphology served as an efficient electrocatalyst for the methanol oxidation reaction through an easy template-free solvothermal synthesis in DES [19]. Natural DES were used as eco-friendly solvents in ultrasound- and microwave-assisted extraction methods to prepare biological tissue and plant samples for heavy metal analysis using inductively coupled plasma mass spectrometry (ICP-MS) [20]. The effective extraction of stevioside and rebaudioside A from *Stevia rebaudiana* with the combination of ultrasound and DES was developed [21]. Silver NPs were prepared by using halide-free DES using a wet chemical reduction route with microwave heating in the presence of oleylamine as a surfactant and reducing agent. NPs production was very quick and took place in 30 sec. of microwave heating at 100 °C [22]. Ferroelectric barium titanate was prepared by utilizing a ChCl-malonic acid DES [23]. Hydrophobic DES had excellent extraction performance for nonpolar analytes. Hence, they are used as viable extraction media to substitute harmful chemical solvents and hydrophobic IL [24]. Pyrite could be solubilized by both electrochemical oxidation and reduction in DES Ethaline, which is a mixture of ChCl and ethylene glycol [25]. DES work as a medium to carry out Ru (IV)-catalyzed redox isomerization of allylic alcohols. This reaction was then combined with the selective addition of organolithium or organomagnesium reagents to form ketones during the process, resulting in the production of 3^0 alcohols [26]. DES were applied to enhance the biocatalytic asymmetric reduction of 3-chloropropiophenone into (S)-3-chloro-1-phenylpropanol, using whole cells of *Acetobacter* sp. CCTCC M209061 as the catalyst [27]. DES are employed in the analysis of several inorganic compounds, including NH_3, metal oxides, chlorides, acetates, phosphates, sulfates, and carbonates [28]. A DES-based method with annealing produced a Fe_2O_3-FeS hybrid catalyst for N_2 fixation, achieving an NH_3 yield of 34.31 µg/h·mg/cat, and 18.06% was the Faradaic efficiency at −0.25 V vs. RHE [29]. DES have also been found to be very useful in ionothermal synthesis because they possess outstanding features such as solvating capacity, recyclability, vapor pressure, and thermal stability. The ionothermal method allows the alteration in the properties of the DES solvent.

ChCl-based DES systems have become dominant in the synthesis of nanostructured ceria [30]. The important application of DES is the synthesis of ZnO NPs [31]. Ammonium cations derived from hydrogen bond acceptors or resulting from the breakdown of urea work as effective templating agents. This feature gives a green and effective route for the preparation of 3D supramolecular inorganic frameworks [32]. Fe_3O_4 magnetic NPs were effectively produced utilizing a ChCl-urea DES, with 3-iodopropyltrimethoxysilane serving as a linker, illustrating the efficacy of DES for surface functionalization and improved material properties [33]. Chitin nanocrystals were synthesized utilizing a ChCl-$ZnCl_2$ DES, resulting in a notable yield of 62%, underscoring the solvent's efficacy in biomass valorization [34]. Ferrite NPs, i.e., MFe_2O_4 (M = Mg, Zn, Co, Ni), were manufactured using DES comprising ChCl and maleic acid, markedly decreasing energy consumption during the synthesis process [35]. γ-CoV_2O_6 NPs were produced with a 1:1 ChCl-malonic acid DES with octahedral geometry, demonstrating the impact of DES composition on NPs morphology and structure [36]. Mn_3O_4 NPs synthesized in ethylene glycol-ChCl DES exhibited a consistent spherical morphology with an average diameter of 25 nm, highlighting the control of DES on particle size [37]. A DES-based polymer incorporating ChCl and acrylic acid exhibited antibacterial efficacy against *E. coli, P. fluorescens, S. aureus*, and *B. subtilis*, thereby affirming the medicinal potential of DES-derived materials [38]. Titanium dioxide nanospheres produced using a ChCl/p-toluene sulfonic acid DES exhibited remarkable photocatalytic efficacy of 98% after 3 h, underscoring DES as a viable medium for environment-friendly photocatalysis [39]. Graphite was efficiently dispersed by using DES, including ChCl and tris(hydroxymethyl)propane, demonstrating DES as an excellent, nontoxic dispersant in the fabrication of nanocomposites [40]. Fe_3O_4 integrated with graphene oxide in a ChCl-urea DES produced nanohybrids that served as effective adsorbents for methylene blue (MB) and lead (II), demonstrating their potential for wastewater remediation [41]. Iron oxide NPs produced using a ChCl-citric acid DES attained a photodegradation rate of up to 94% for Rhodamine B under ultraviolet light, highlighting the efficacy of DES in environmental photocatalysis [42]. $Ni_3V_2O_8$/N--doped reduced graphene oxide, produced using ChCl and oxalic acid, had significant degradation efficiencies of 94.6% and 96.7% for methyl orange and crystal violet, respectively, highlighting the efficacy of DES in dye elimination [43]. Magnetic graphene NPs synthesized using a DES of ChCl, itaconic acid, and 3-mercaptopropionic acid exhibited a significant adsorption capacity of 215.1 mg g^{-1} for Hg^{2+}, highlighting the efficacy of DES-assisted synthesis in heavy metal remediation [44]. Nitrogen-doped ceria NPs, produced via DES comprising cetyltrimethylammonium bromide and acetic acid, had a 96% removal effectiveness for sulfamethoxazole [45]. A ZnO/GO nanosheet was produced in a ChCl-ethylene glycol DES (1:2 ratio), demonstrating efficient removal of the pharmaceutical contaminant cefixime trihydrate from water [46]. Molecularly imprinted polymers of Fe_3O_4 were synthesized utilizing a choline-imidazole DES, which preferentially extracted naproxen, illustrating the specificity attainable by DES-based imprinting [47]. Cobalt ferrite NPs produced in a DES of ChCl and p-aminophenol were efficiently employed for the extraction of heavy metals, demonstrating their efficacy in multi-metal ion removal [48]. Magnetic poly(acrylic acid-menthol DES) hydrogels were created for the detection of pesticides in water [49]. The application of ChCl-based DES has demonstrated considerable potential in the synthesis of diverse nanomaterials with improved characteristics. Lignin NPs synthesized employing combinations such as ChCl with ethanolamine, ethylene glycol, or lactic acid showed exceptional dispersibility, and yields reached up to 90% [50]. Lignin nanospheres produced using formic acid and ChCl attained a remarkable 95.0% delignification with significant selectivity [51]. CeO_2 NPs were effectively produced utilizing ChCl and urea, achieving a complete yield of 100% within 100 seconds at 160 °C, highlighting the swift and high-yield potential of this method [52]. Nanosheets of Fe_3O_4 synthesized using ChCl: oxalic acid at a 1:1 ratio attained an impressive ammonia yield rate of 12.09 μg h^{-1} mg^{-1}_cat, indicating their potential for catalytic uses [53]. Finally, levofloxacin-imprinted NPs synthesized with ChCl: acrylic acid exhibited a removal efficiency of 98.3%, signifying their appropriateness for restoration applications [54].

8.3.2 *Inorganic Nanoparticle (NPs) Synthesis by Using DES*

8.3.2.1 Metal NPs

DES (1:4 silver triflate:acetamide) enabled bulk synthesis of silver nanocrystals. Using oleylamine as cosolvent, reductant, and stabilizer, the method achieved high yields of monodispersed nanocrystals via microwave reduction in just 2 minutes at 200 °C [55]. The zwitterionic natural DES are composed of betaine and several carboxylic acids as precursors to synthesize gold nanoparticles. Such NPs had been successfully fabricated with ascorbic acid or sodium borohydride as reducing agents without the use of surfactants [56, 57].

8.3.2.2 Metal Oxide NPs and Inorganic Salt-Based NPs

Exfoliated manganese oxides synthesized with various kinds of DES displayed a large water flux of 5736 Lm^{-2} h^{-1} [58]. Cu_2O nanoclusters supported over TiO_2 NPs from a ChCl and urea (1:2) that displayed an immense charge density of 16.3×10^{19} cm^{-3} [59]. Cu_2O NPs synthesized in an ethylene glycol/ChCl DES demonstrated a remarkable photothermal conversion efficiency of 83.74% with a mere 0.1 wt% Cu_2O [60]. Ag_2O NPs were synthesized in a DES of malonic acid and $ZnCl_2$, using hydrazine hydrate as a reducing agent [61]. Mercuric sulfide, zirconium oxide, manganese oxide, and CuO NPs were synthesized from a choline chloride-urea DES [62]. Corncob biomass was treated with a ChCl and oxalic acid DES in which the solvent was reused and lignin was recovered [63]. $FeMnO_3$ NPs were synthesized using natural DES, which had a detection limit of 5.3 nM for nimesulide [64]. $CuCo_2O_4$ nanocomposites prepared using ChCl and urea (1:2) provided a high specific capacity of 421 mAh g^{-1} at 10 mV s^{-1} [65]. $ZnCo_2O_4$ nanosheets were used for colorimetric detection of pyrophosphate and pyrophosphatase with detection limits of 0.01 mM and 0.004 U/mL, respectively [66]. $MgFe_2O_4$ NPs were synthesized using different DES and found with the linear response range up to 342.6 μM [67]. Selenium NPs had particle sizes of 279 nm and 96.0 nm, which were prepared from propolis extract and dissolved quercetin, respectively, in natural DES. These NPs were rich in antioxidant activity [68]. Table 8.1 tabulates the key features of different inorganic NP materials prepared using DES.

Table 8.1 Various inorganic NPs materials prepared using DES with their key features

Material	DES	Key features	References
Fe_3O_4 magnetic NPs	ChCl-urea	3-iodopropyltrimethoxysilane works as a linker	[33]
Chitin nanocrystals	ChCl-$ZnCl_2$ DES	Yield 62%	[34]
Ferrite NPs MFe_2O_4 (M = Mg, Zn, co, Ni).	ChCl and maleic acid	Less energy required for the synthesis of nanoparticles	[35]
γ-CoV_2O_6	1:1ChCl-malonic acid	Octahedral geometry	[36]
Mn_3O_4	Ethylene glycol-ChCl	Mean particle size was 25 nm and morphology was spherical	[37]
DES-based polymer	ChCl-acrylic acid	Antibacterial activity against *E. coli, P. fluorescens, S. aureus,* and *B. subtilis*	[38]
TiO_2 nanospheres	ChCl/p-toluene sulphonic acid	Photocatalytic efficiency was 98% within 3 h	[39]
Fe_3O_4	ChCl and urea	DES/GO Fe_3O_4 nanohybrids adsorbed MB and Pb(II)	[41]
Iron oxide NPs	ChCl and citric acid	Rhodamine B was degraded up to 94% under UV light	[42]
$Ni_3V_2O_8$/N--doped reduced graphene oxide	ChCl and oxalic acid	Degradation efficiency of 94.6% and 96.7% for methyl orange and crystal violet	[43]
Magnetic graphene NPs	ChCl/itaconic acid/3-mercaptopropionic acid 2:1:1	Adsorption capacity for Hg^{2+} was 215.1 mg/g	[44]
Nitrogen-doped ceria NPs	Cetyltrimethylammoniumbromide and acetic acid	96% removal of sulfamethoxazole	[45]
ZnO/GO nanosheet	ChCl and ethylene glycol at 1:2	Removal of cefixime trihydrate	[46]
Fe_3O_4 molecularly imprinted polymer	Choline-imidazole	Removal of naproxen	[47]

(continued)

Table 8.1 (continued)

Material	DES	Key features	References
Cobalt ferrite NPs	ChCl and p-aminophenol	Zn(II), Ni(II), Cu(II), Pb(II), and Hg(II) were removed	[48]
Magnetic poly(acrylic acid-menthol) DES hydrogel	Acrylic acid-menthol	Detection of pesticides in water	[49]
Lignin NPs	ChCl and ethanolamine, ChCl and ethylene glycol, ChCl and lactic acid	Showed excellent dispersibility and 90% yield	[50]
Lignin nanospheres	Formic acid and ChCl	95.0% delignification with high selectivity	[51]
CeO_2 NPs	Ch and urea	100% yields within 100 sec. at 160 °C	[52]
Fe_3O_4 nanosheets	ChCl and oxalic acid 1:1	12.09 μg h^{-1} mg^{-1}_{cat} NH_3 was produced	[53]
Levofloxacin-imprinted NPs	ChCl: Acrylic acid	Decontamination efficiency for levo-floxacin was 98.3%	[54]

8.3.3 Other Applications

DES have been recognized as an environmentally friendly substitute for traditional solvents, especially in wastewater treatment, owing to their sustainability and biodegradability [69]. DES are extensively used for the determination of heavy metals like Cu, Pb, Cd, Co, Ni, etc. Trace elements can also be detected in wastewater by using DES [70–72]. The increasing demand for DES in wastewater treatment is due to their multitasking ability, which makes them useful in methods like membrane processes, microextraction, and extractive distillation. Furthermore, an innovative water purification method would also utilize DES, which have proven to be efficient and sustainable [73]. Different DES systems, which were prepared from the combination of ChCl, urea, citric acid, ethylene glycol, and phenol, and then applied for the trace determination of niclosamide in pharmaceutical and wastewater samples, employing liquid-liquid microextraction [74]. Supramolecular gels of L-amino acids in the DES (ChCl and phenylacetic acid) have demonstrated great reusability, retaining efficiency for a minimum of 9 cycles in the treatment of running water. This type of gel exhibited an adsorption capability of 1930 mg/g for rhodamine B [75]. A DES of L-menthol and 10-camphorsulfonic acid (0.2 wt%) was used to modify polyethersulfone membranes, achieving 89.1% COD, 100% turbidity, and 51.6% TDS removal from pharmaceutical waste [76]. DES have also been used with solid-phase extraction, which, combined with DES-based dispersive liquid-liquid microextraction, is used for the detection of estrogenic compounds in wastewater before analysis via gas chromatography-mass spectrometry. This method was able to attain limits of detection (LODs) ranging from 0.01 to 0.05 ng/L [77]. In the recovery of metals, hydrophobic DES had been used to recover Ir, Pt, Ru, and Rh from wastewater and achieved the extraction efficiency of 97.57%, 94.26%, and 99.55% for Ir, Ru, and Pt, respectively [78]. DES prepared from dl-menthol and fatty acids were utilized successfully for phenol extraction from actual wastewater samples [79]. Another usage of the DES system consists of ChCl and methacrylic acid (MAA) and is used for the removal of *levofloxacin* from wastewater via adsorption [54]. A new extraction process with dispersive liquid-liquid microextraction and DES solidification has also been utilized for heavy metals analysis in soil and vegetables [80]. In wastewater treatment, cationic cellulose derivatives synthesized with DES have been demonstrated to be a high-performance bio-based alternative, greatly improving flocculation and sludge dewatering. Real wastewater sample tests showed a turbidity reduction rate of up to 93.2%, which was higher than commercial polyacrylamides [81]. DES-based hollow fiber microextraction with HPLC-UV detected raloxifene and ethinylestradiol in pharmaceutical wastewater, with LODs of 5.0 ng/mL and 10 ng/mL, respectively [82]. DES were added with ethanol (50% w/w DES content) and used as an electrolyte medium in DSCs, highlighting good photovoltaic efficiency [83]. Microwave-assisted DES pretreatment was highly effective in attaining elevated sugar yields (99% glucose, 85% xylose yield) in the pretreated wheat straw fibers. The highest amount of total sugar extraction was 619 mg/g for pretreated wheat straw fibers, which was double that of traditional DES pretreatment, occurring at a 1:3 mol ratio, 360 W microwave power, and an 8-minute retention time. 81.5% was the ethanol yield by utilizing *Escherichia coli KO11* [84]. Seven DES were synthesized and utilized for the processing of corn stover. ChCl:formic acid exhibited optimal efficiency with 17.0 g/L glucose in corn stover hydrolysate. It was effectively utilized in butanol fermentation [85]. A novel technique was developed for utilizing DES to extract lignin from woody biomass, achieving elevated yield and purity. These DES treatments were found to selectively extract substantial quantities of lignin from wood, yielding 78% from poplar and 58% from D. fir. The extracted lignin possesses 95% purity [86]. Apple residues, potato peels, coffee silverskin, and brewer's wasted grains were subjected to pretreatment using two DES, namely ChCl-glycerol and ChCl-ethylene glycol, to facilitate the generation of fermentable sugars. The total sugar production was approximately 217 kilotonnes/year [87]. With the growing importance of green chemistry, there is a need for studies to unlock the potential of DES in material engineering, sustainable utilization, and biotechnology. One of the most important uses of DES is their use as a solvent in the controlled synthesis production of metal nanoparticles. Such properties are invaluable in the electrocatalysis process, where nanoparticle size and shape play a critical role in enhancing catalytic efficacy. DES can improve energy conversion, fuel cells, and various catalytic advancements due to alterations in the structure and size of nanoparticles. DES have been utilized for the extraction of bioactive compounds from fruits, vegetables, and milk. One of the primary applications of DES is to solubilize metal ions in solution, which is useful for metal deposition, dissolution, and processing [88]. DES served as a solvent in the preparation of crassula perforata-like TiO_2 at 160 °C for 4–8 h. The photocatalytic water splitting H_2 generation rate was 13.13 mmol·h^{-1}·g^{-1}, which was 18 times higher than commercial P25 TiO_2 [89]. DES was used to synthesize heterophase-junction nano-TiO_2 at 180 °C for 18 h. H_2 generation rate reached 14.51 mmol/h·g, 20 times higher than commercial P25 TiO_2 [90]. Mn_3O_4 nanoparticles were manufactured by using ethaline (ethylene glycol-ChCl) DES. The as-produced Mn_3O_4 nanoparticles were semiconductor in nature due to a band gap of 3.4 eV [91]. CZTS nanoparticles were produced in a ChCl and urea DES. The synthesized CZTS exhibited crystallization in the kesterite phase, with a crystal diameter of 6.5 nm [92]. $CuCo_2O_4$ nanomaterials were prepared utilizing the DES technique for supercapacitor applications. The improved $CuCo_2O_4$ electrode exhibited an exceptional specific capacity of 421 mAh g^{-1} at 10 mV s^{-1} [93]. 1,2-disubstituted or 2-substituted benzimidazoles utilized o-phenylenediamine in conjunction with various aldehydes. The utilization of DES as both a reaction medium and reagent, without the incorporation of an external solvent, offers benefits regarding yields [37]. DES could enhance the biogenic synthesis of silver nanoparticles [94]. A one-step hydrothermal process was utilized to prepare N,Cl-doped chiral carbon dots in which vine tea was used as the precursor, and a natural DES works as a solvent, doping agent, and chiral source [65]. ChCl: MA DES catalyst was prepared to produce 5-hydroxymethylfurfural (5-HMF). The DMSO/H_2O system was utilized for converting biomass into 5-HMF. The 5-HMF yield reached 85.6% at 150 °C after 30 min. The production of 5-HMF had an activation energy of 28.53 kJ/mol. The catalyst was reusable for 5 cycles [95]. $CsPbBr_3$ PNCs were prepared using the ligand-assisted reprecipitation technique, employing natural DES as both solvents and surface ligands. The synthesized $CsPbBr_3$ PNCs demonstrated a high photoluminescence quantum yield around 96.8% [96–99].

8.4 Future and Challenges

DES are gaining significant interest across different fields due to their environmentally friendly nature and exceptional physicochemical properties. They are not only multifunctional as solvent systems but are also regarded as catalysts, adsorbents, and electrolytes. DES have demonstrated efficacy as nanocarriers for the encapsulation and delivery of anticancer drugs, hence enhancing their use in pharmaceuticals and biotechnology [100]. DES have emerged as a viable eco-friendly option due to their minimal toxic effect, biodegradability, and applicability in several fields, including medicines, metal extraction, and energy storage. DES have high viscosity, reduced thermal stability, environmental hazards, and pose challenges for large-scale production. In wastewater treatment, DES may require a high initial investment in a large-scale process; moreover, reusing the solvent also requires additional cost. There is a need to alter the formulations of DES, which helps in reducing the manufacturing cost, improving efficiency and stability while maintaining their environmentally friendly characteristics. The life cycle assessment of DES is needed to understand their environmental impact. Through diligent study and collaboration, DES will work as a standard solvent for green chemistry.

8.5 Conclusion

DES are a novel category of green solvents with significant importance across many scientific and industrial applications. Their capacity to function as reaction media, solubilizing agents, and eco-friendly alternatives to conventional solvents has rendered them highly valuable in fields such as nanotechnology, electrochemistry, medicine, and materials science. Additionally, DES are utilized in the synthesis of inorganic nanoparticles, including metals (e.g., gold, silver, and copper), metal oxides of Fe and Ti, and other nanomaterials such as cellulose and chitin. Despite its numerous advantages, DES nonetheless possesses limitations like elevated viscosity, thermal sensitivity, and difficulties in large-scale production. Addressing these problems via joint research and innovation is essential for achieving their full potential. The advancement of novel DES formulations, including supramolecular and hydrophobic DES, is already broadening their uses and enhancing their efficacy. Further study indicates that DES might significantly contribute to breakthroughs in green chemistry, sustainable manufacturing, and eco-friendly industrial processes. Through continuous research and technical progress, DES possess the capacity to revolutionize several sectors, fostering a more efficient and sustainable future.

Declaration of Generative AI and AI-Assisted Technologies in the Writing Process While preparing this work, the authors utilized the ChatGPT tool to enhance the language and readability of several statements in the manuscript. After using this tool/service, the authors reviewed and edited the content as needed and take full responsibility for the publication content.

Conflict of Interest The authors declare no conflict of interest.

References

1. El Achkar T, Greige-Gerges H, Fourmentin S (2021) Basics and properties of deep eutectic solvents: a review. Environ Chem Lett 19:3397–3408
2. El Achkar T, Moura L, Moufawad T, Ruellan S, Panda S, Longuemart S, Legrand FX, Gomes MC, Landy D, Greige-Gerges H (2020) New generation of supramolecular mixtures: characterization and solubilization studies. Int J Pharm 584:119443
3. Długosz O (2023) Natural deep eutectic solvents in the synthesis of inorganic nanoparticles. Materials 16(2):627
4. Hansen BB, Spittle S, Chen B, Poe D, Zhang Y, Klein JM, Horton A, Adhikari L, Zelovich T, Doherty BW (2020) Deep eutectic solvents: A review of fundamentals and applications. Chem Rev 121(3):1232–1285
5. Davis SJ (2016) Deep eutectic solvents derived from inorganic salts. University of Leicester
6. Yeow AT, Hayyan A, Hayyan M, Junaidi MU, Saleh J, Basirun WJ, Nor MR, Al Abdulmonem W, Salleh MZ, Zuki FM, Hamid MD (2024) A comprehensive review on the physicochemical properties of deep eutectic solvents. Results Chem 7:101378. https://doi.org/10.1016/J.RECHEM.2024.101378

7. Wang H, Kang X, Han B (2024) Electrocatalysis in deep eutectic solvents: from fundamental properties to applications. Chem Sci 15(26):9949–9976. https://doi.org/10.1039/D4SC02318H
8. Zhu J, Shao C, Hao S, Zhang J, Ren W, Wang B, Xiao L, Wang C, Shao L (2025) Recent progress on the dissolution of cellulose in deep eutectic solvents. Ind Crop Prod 228:120844. https://doi.org/10.1016/J.INDCROP.2025.120844
9. Sirvio JA, Visanko M, Liimatainen H (2016) Acidic deep eutectic solvents as hydrolytic media for cellulose nanocrystal production. Biomacromolecules 17(9):3025–3032. https://doi.org/10.1021/ACS.BIOMAC.6B00910/SUPPL_FILE/BM6B00910_SI_001.PDF
10. Zhu G, Liu M, Weng S, Zhang G, Hu Y, Kou Z, Bo C, Hu L, Wu S, Zhou Y (2023) Dye-free and reprintable multi-color DLP 3D printing using $ZnCl_2$-based polymerizable deep eutectic solvents and type I photoinitiators. Chem Eng J 472:144987. https://doi.org/10.1016/J.CEJ.2023.144987
11. Ganorkar SB, Hadole PM, Patil MR, Pardeshi CV, Bobade PS, Shirkhedkar AA, Vander HY (2025) Deep eutectic solvents in analysis, delivery and chemistry of pharmaceuticals. Int J Pharm 672:125278. https://doi.org/10.1016/J.IJPHARM.2025.125278
12. Bennaadja S, Soltani EK, Ghernaout D, Albrahim M, Benguerba Y (2025) Sustainable drug delivery innovations: enhancing ferulic acid release using deep eutectic solvents in dimyristoyl phosphatidylcholine (DMPC) systems. J Mol Liq 429:127572. https://doi.org/10.1016/J.MOLLIQ.2025.127572
13. Bao Y, WangY YC, Xue Z (2025) Deep eutectic solvents for fractionation and valorization of lignocellulose. Green Chem Eng 6(1):21–35. https://doi.org/10.1016/J.GCE.2024.04.001
14. Aslanlı I, Sert M (2025) One pot conversion of fructose to alkyl levulinates catalyzed by deep eutectic solvents in microwave reactor. Renew Energy 244:122698. https://doi.org/10.1016/J.RENENE.2025.122698
15. Falcini C, de Gonzalo G (2024) Deep eutectic solvents as catalysts in the synthesis of active pharmaceutical ingredients and precursors. Catalysts 14(2):120. https://doi.org/10.3390/CATAL14020120
16. Rodríguez-Ramos R, Santana-Mayor Á, Socas-Rodríguez B, Rodríguez-Delgado MÁ (2021) Recent applications of deep eutectic solvents in environmental analysis. Appl Sci 11(11):4779
17. Werner J (2020) Ligandless, deep eutectic solvent-based ultrasound-assisted dispersive liquid-liquid microextraction with solidification of the aqueous phase for preconcentration of lead, cadmium, cobalt and nickel in water samples. J Sep Sci 43(7):1297–1305
18. Singh S, Negi T, Sagar NA, Kumar Y, Tarafdar A, Sirohi R, Sindhu R, Pandey A (2022) Sustainable processes for treatment and management of seafood solid waste. Sci Total Environ 817:152951
19. Wang X, Sun M, Xiang S, Waqas M, Fan Y, Zhong J, Huang K, Chen W, Liu L, Yang J (2020) Template-free synthesis of platinum hollow-opened structures in deep-eutectic solvents and their enhanced performance for methanol electrooxidation. Electrochim Acta 337:135742
20. Santana AP, Andrade DF, Guimarães TG, Amaral CD, Oliveira A, Gonzalez MH (2020) Synthesis of natural deep eutectic solvents using a mixture design for extraction of animal and plant samples prior to ICP-MS analysis. Talanta 216:120956. https://doi.org/10.1016/J.TALANTA.2020.120956
21. Milani G, Vian M, Cavalluzzi MM, Franchini C, Corbo F, Lentini G, Chemat F (2020) Ultrasound and deep eutectic solvents: an efficient combination to tune the mechanism of steviol glycosides extraction. Ultrason Sonochem 69:105255. https://doi.org/10.1016/J.ULTSONCH.2020.105255
22. Adhikari L, Larm NE, Bhawawet N, Baker GA (2018) Rapid microwave-assisted synthesis of silver nanoparticles in a halide-free deep eutectic solvent. ACS Sustain Chem Eng 6(5):5725–5731. https://doi.org/10.1021/ACSSUSCHEMENG.8B00050/SUPPL_FILE/SC8B00050_SI_001.PDF
23. Boston R, Foeller PY, Sinclair DC, Reaney IM (2017) Synthesis of barium titanate using deep eutectic solvents. Inorg Chem 56(1):542–547. https://doi.org/10.1021/ACS.INORGCHEM.6B02432/ASSET/IMAGES/MEDIUM/IC-2016-02432P_0005.GIF

24. Lee J, Jung D, Park K (2019) Hydrophobic deep eutectic solvents for the extraction of organic and inorganic analytes from aqueous environments. Trends Anal Chem 118:853–868. https://doi.org/10.1016/J.TRAC.2019.07.008
25. Abbott AP, Al-Bassam AZM, Goddard A, Harris RC, Jenkin GRT, Nisbet FJ, Wieland M (2017) Dissolution of pyrite and other Fe–S–As minerals using deep eutectic solvents. Green Chem 19(9):2225–2233. https://doi.org/10.1039/C7GC00334J
26. Cicco L, Rodríguez-Álvarez MJ, Perna FM, García-Álvarez J, Capriati V (2017) One-pot sustainable synthesis of tertiary alcohols by combining ruthenium-catalysed isomerisation of allylic alcohols and chemoselective addition of polar organometallic reagents in deep eutectic solvents. Green Chem 19(13):3069–3077. https://doi.org/10.1039/C7GC00458C
27. Xu P, Xu Y, Li XF, Zhao BY, Zong MH, Lou WY (2015) Enhancing asymmetric reduction of 3-chloropropiophenone with immobilized acetobacter sp. CCTCC M209061 cells by using deep eutectic solvents as cosolvents. ACS Sustain Chem Eng 3(4):718–724. https://doi.org/10.1021/ACSSUSCHEMENG.5B00025/SUPPL_FILE/SC5B00025_SI_001.PDF
28. Söldner A, Zach J, König B (2019) Deep eutectic solvents as extraction media for metal salts and oxides exemplarily shown for phosphates from incinerated sewage sludge ash. Green Chem 21(2):321–328
29. Chen T, Ying H, Zhang C, Bi J, Li Z, Hao J (2021) Engineering an Fe_2O_3/FeS hybrid catalyst from a deep eutectic solvent for highly efficient electrocatalytic N_2 fixation. Chem Commun 57(54):6688–6691. https://doi.org/10.1039/D1CC02072B
30. Hammond OS, Edler KJ, Bowron DT, Torrente-Murciano L (2017) Deep eutectic-solvothermal synthesis of nanostructured ceria. Nat Commun 8(1):14150
31. Gontrani L, Donia DT, Maria Bauer E, Tagliatesta P, Carbone M (2023) Novel synthesis of zinc oxide nanoparticles from type IV deep eutectic solvents. Inorg Chim Acta 545:121268. https://doi.org/10.1016/J.ICA.2022.121268
32. Wang Q, Ren H, Zhao Y, Xia X, Huang F, Cui G, Dong B, Tang B (2019) Facile and mild preparation of brookite-rutile heterophase-junction TiO_2 with high photocatalytic activity based on a deep eutectic solvent (DES). J Mater Chem 7(24):14613–14619. https://doi.org/10.1039/C9TA02500F
33. Tavakol H, Keshavarzipour F (2017) Preparation of choline chloride–urea deep eutectic solvent-modified magnetic nanoparticles for synthesis of various 2-amino-4H-Pyran derivatives in water solution. Appl Organomet Chem 31(11):e3811. https://doi.org/10.1002/AOC.3811
34. Hong S, Yuan Y, Yang Q, Chen L, Deng J, Chen W, Lian H, Mota-Morales JD, Liimatainen H (2019) Choline chloride-zinc chloride deep eutectic solvent mediated preparation of partial O-acetylation of chitin nanocrystal in one step reaction. Carbohydr Polym 220:211–218. https://doi.org/10.1016/J.CARBPOL.2019.05.075
35. Söldner A, Zach J, Iwanow M, Gärtner T, Schlosser M, Pfitzner A, König B (2016) Preparation of magnesium, cobalt and nickel ferrite nanoparticles from metal oxides using deep eutectic solvents. Chem A Euro J 22(37):13108–13113. https://doi.org/10.1002/CHEM.201602821
36. Thorat GM, Jadhav HS, Roy A, Chung WJ, Seo JG (2018) Dual role of deep eutectic solvent as a solvent and template for the synthesis of octahedral cobalt vanadate for an oxygen evolution reaction. ACS Sustain Chem Eng 6(12):16255–16266. https://doi.org/10.1021/ACSSUSCHEMENG.8B03119/SUPPL_FILE/SC8B03119_SI_001.PDF
37. Karimi M, Eshraghi MJ (2017) One-pot and green synthesis of Mn_3O_4 nanoparticles using an all-in-one system (solvent, reactant and template) based on ethaline deep eutectic solvent. J Alloys Compd 696:171–176. https://doi.org/10.1016/J.JALLCOM.2016.11.259
38. Fu N, Li L, Liu K, Kim CK, Li J, Zhu T, Li J, Tang B (2019) A choline chloride-acrylic acid deep eutectic solvent polymer based on Fe_3O_4 particles and MoS_2 sheets (Poly(ChCl-AA DES)@Fe_3O_4@MoS_2) with specific recognition and good antibacterial properties for β-Lactoglobulin in milk. Talanta 197:567–577. https://doi.org/10.1016/J.TALANTA.2019.01.072
39. Shahi SK, Kaur N, Sofia Sandhu JS, Shahi, Singh V (2017) Influences of a new templating agent on the synthesis of coral-like TiO_2 nanoparticles and their photocatalytic activity. J Sci Adv Mater Dev 2(3):347–353. https://doi.org/10.1016/J.JSAMD.2017.07.006

40. Honorata Mąka, Spychaj T, Kowalczyk K (2014) Imidazolium and deep eutectic ionic liquids as epoxy resin crosslinkers and graphite nanoplatelets dispersants. J Appl Polym Sci 131(12):1. https://doi.org/10.1002/APP.40401
41. Mehrabi N, Haq UFA, Toufiq Reza M, Aich N (2020) Application of deep eutectic solvent for conjugation of magnetic nanoparticles onto graphene oxide for lead(II) and methylene blue removal. J Environ Chem Eng 8(5):104222. https://doi.org/10.1016/J.JECE.2020.104222
42. Sri S, Pon S, Taj J, George M (2020) Facile synthesis of magnetite nanocubes using deep eutectic solvent: an insight to anticancer and photo-Fenton efficacy. Surf Interfaces 20:100609. https://doi.org/10.1016/J.SURFIN.2020.100609
43. Alharthi FA, Ababtain AS, Aldubeikl HK, Alanazi HS, Hasan I (2023) Deep eutectic solvent-mediated synthesis of $Ni_3V_2O_8$/N-doped RGO for visible-light-driven H_2 evolution and simultaneous degradation of dyes. Inorganics 11(2):67. https://doi.org/10.3390/INORGANICS11020067/S1
44. Chen J, Wang Y, Wei X, Panli X, Wei X, Ni R, Meng J (2018) Magnetic solid-phase extraction for the removal of mercury from water with ternary hydrosulphonyl-based deep eutectic solvent modified magnetic graphene oxide. Talanta 188:454–462. https://doi.org/10.1016/J.TALANTA.2018.06.016
45. Iqbal J, Shah NS, Sayed M, Khan JA, Muhammad N, Khan ZUH, Saif-ur-Rehman. (2020) Synthesis of nitrogen-doped ceria nanoparticles in deep eutectic solvent for the degradation of sulfamethaxazole under solar irradiation and additional antibacterial activities. Chem Eng J 394:124869. https://doi.org/10.1016/J.CEJ.2020.124869
46. Ciğeroğlu Z, Şahin S, Kazan ES (2022) One-pot green preparation of deep eutectic solvent-assisted ZnO/GO nanocomposite for cefixime trihydrate photocatalytic degradation under UV-A irradiation. Biomass Convers Biorefinery 12(1):73–86. https://doi.org/10.1007/S13399-021-01734-0/TABLES/5
47. Husin NA, Muhamad M, Yahaya N, Miskam M, Kamal NNSNM, Asman S, Raoov M, Zain NNM (2021) Application of a new choline-imidazole based deep eutectic solvents in hybrid magnetic molecularly imprinted polymer for efficient and selective removal of naproxen from aqueous samples. Mater Chem Phys 261:124228. https://doi.org/10.1016/J.MATCHEMPHYS.2021.124228
48. Baghaei AM, Pariya MR, Mogaddam A, Farajzadeh MA, Mohebbi A, Sorouraddin SM (2023) Application of deep eutectic solvent functionalized cobalt ferrite nanoparticles in dispersive micro solid phase extraction of some heavy metals from aqueous samples prior to ICP-OES. J Food Compos Anal 117:105125. https://doi.org/10.1016/J.JFCA.2022.105125
49. Jamshidi F, Nouri N, Sereshti H, Aliabadi MHS (2020) Synthesis of magnetic poly (acrylic acid-menthol deep eutectic solvent) hydrogel: application for extraction of pesticides. J Mol Liq 318:114073. https://doi.org/10.1016/J.MOLLIQ.2020.114073
50. Luo T, Wang C, Ji X, Yang G, Chen J, Janaswamy S, Lyu G (2021) Preparation and characterization of size-controlled lignin nanoparticles with deep eutectic solvents by nanoprecipitation. Molecules 26(1):218. https://doi.org/10.3390/MOLECULES26010218
51. Guo Y, Lu X, Shen F, Jinguang H, Huang M, He J, Zhang Y, Deng S, Li Q, Tian D (2022) Insights into lignocellulosic waste fractionation for lignin nanospheres fabrication using acidic/alkaline deep eutectic solvents. Chemosphere 286:131798. https://doi.org/10.1016/J.CHEMOSPHERE.2021.131798
52. Exposito AJ, Barrie PJ, Torrente-Murciano L (2020) Fast synthesis of CeO_2 nanoparticles in a continuous microreactor using deep eutectic reline as solvent. ACS Sustain Chem Eng 8(49):18297–18302. https://doi.org/10.1021/ACSSUSCHEMENG.0C06949/ASSET/IMAGES/MEDIUM/SC0C06949_M008.GIF
53. Ying H, Chen T, Zhang C, Bi J, Li Z, Hao J (2021) Regeneration of porous Fe_3O_4 nanosheets from deep eutectic solvent for high-performance electrocatalytic nitrogen reduction. J Colloid Interface Sci 2021(602):64–72. https://doi.org/10.1016/J.JCIS.2021.05.185
54. Li X, Row KH (2020) Preparation of levofloxacin-imprinted nanoparticles using designed deep eutectic solvents for the selective removal of levofloxacin pollutants from environmental waste water. Analyst 145(8):2958–2965. https://doi.org/10.1039/C9AN02556A

55. Adhikari L, Larm NE, Baker GA (2019) Argentous deep eutectic solvent approach for scaling up the production of colloidal silver nanocrystals. ACS Sustain Chem Eng 7(13):11036–11043. https://doi.org/10.1021/ACSSUSCHEMENG.9B01777/SUPPL_FILE/SC9B01777_SI_001.PDF
56. Yu D, Jiang D, Xue Z, Mu T (2024) Deep eutectic solvents as green solvents for materials preparation. Green Chem 26(13):7478–7507. https://doi.org/10.1039/D4GC00136B
57. Di Crescenzo A, Tiecco M, Zappacosta R, Boncompagni S, Di Profio P, Ettorre V, Fontana A, Germani R, Siani G (2018) Novel zwitterionic natural deep eutectic solvents as environmentally friendly media for spontaneous self-assembly of gold nanoparticles. J Mol Liq 268:371–375. https://doi.org/10.1016/J.MOLLIQ.2018.07.060
58. Aruchamy K, Mahto A, Nagaraj R, Kalpana D, Ghosh D, Mondal D, Nataraj SK (2020) Ultrafast synthesis of exfoliated manganese oxides in deep eutectic solvents for water purification and energy storage. Chem Eng J 379:122327. https://doi.org/10.1016/J.CEJ.2019.122327
59. Mohite SV, Kim S, Lee C, Bae J, Kim Y (2022) Z-scheme heterojunction photocatalyst: deep eutectic solvents-assisted synthesis of Cu_2O nanocluster improved hydrogen production of TiO_2. J Alloys Compd 928:167168. https://doi.org/10.1016/J.JALLCOM.2022.167168
60. Liu C, Yan Y, Sun W, Shi X, Shi N, Huo Y, Zhao J, Said Z (2022) Preparation and thermophysical study on a super stable copper oxide/deep eutectic solvent nanofluid. J Mol Liq 356:119020. https://doi.org/10.1016/J.MOLLIQ.2022.119020
61. Sarjuna K, Ilangeswaran D (2022) Preparation and physico-chemical studies of Ag_2O nanoparticles using newly formed malonic acid and $ZnCl_2$-based deep eutectic solvents. Mater Today Proc 49:2943–2948. https://doi.org/10.1016/J.MATPR.2021.11.355
62. Balaji R, Ilangeswaran D (2022) Choline chloride—urea deep eutectic solvent an efficient media for the preparation of metal nanoparticles. J Indian Chem Soc 99(5):100446. https://doi.org/10.1016/J.JICS.2022.100446
63. Yan G, Zhou Y, Zhao L, Wang W, Yang Y, Zhao X, Chen Y, Yao X (2022) Recycling of deep eutectic solvent for sustainable and efficient pretreatment of corncob. Ind Crop Prod 183:115005. https://doi.org/10.1016/J.INDCROP.2022.115005
64. Joseph XB, Wang SF, Baby JN, George M (2022) Natural deep eutectic solvent assisted synthesis of $FeMnO_3$ entrapped functionalized carbon nanofiber composite: an electrochemical detection of nimesulide. J Mol Liq 367:120421. https://doi.org/10.1016/J.MOLLIQ.2022.120421
65. Shinde SK, Karade SS, Maile NC, Yadav HM, Ghodake GS, Jagadale AD, Kim DY (2021) Green synthesis of novel $CuCo_2O_4$ nanocomposite for stable hybrid supercapacitors by deep eutectic solvents. J Mol Liq 334:116390. https://doi.org/10.1016/J.MOLLIQ.2021.116390
66. Chen J, Wei X, Tang H, Munyemana JC, Guan M, Zhang S, Qiu H (2021) Deep eutectic solvents-assisted synthesis of $ZnCo_2O_4$ nanosheets as peroxidase-like nanozyme and its application in colorimetric logic gate. Talanta 222:121680. https://doi.org/10.1016/J.TALANTA.2020.121680
67. Baby JN, Sriram B, Wang SF, George M (2020) Effect of various deep eutectic solvents on the sustainable synthesis of $MgFe_2O_4$ nanoparticles for simultaneous electrochemical determination of nitrofurantoin and 4-Nitrophenol. ACS Sustain Chem Eng 8(3):1479–1486. https://doi.org/10.1021/ACSSUSCHEMENG.9B05755/SUPPL_FILE/SC9B05755_SI_001.PDF
68. Długosz O, Chmielowiec-Korzeniowska A, Drabik A, Tymczyna L, Banach M (2023) Bioactive selenium nanoparticles synthesized from propolis extract and quercetin based on natural deep eutectic solvents (NDES). J Clust Sci 34(3):1401–1412. https://doi.org/10.1007/S10876-022-02306-6/FIGURES/6
69. Chabib CM, Ali JK, Abi Jaoude M, Alhseinat E, Adeyemi IA, Al Nashef IM (2022) Application of deep eutectic solvents in water treatment processes: a review. J Water Process Eng 47:102663

70. Matong JM, Nyaba L, Nomngongo PN (2017) Determination of As, Cr, Mo, Sb, Se and V in agricultural soil samples by inductively coupled plasma optical emission spectrometry after simple and rapid solvent extraction using choline chloride-oxalic acid deep eutectic solvent. Ecotoxicol Environ Saf 135:152–157
71. Panhwar AH, Tuzen M, Kazi TG (2017) Ultrasonic assisted dispersive liquid-liquid microextraction method based on deep eutectic solvent for speciation, preconcentration and determination of selenium species (IV) and (VI) in water and food samples. Talanta 175:352–358
72. Zounr RA, Tuzen M, Khuhawar MY (2017) Ultrasound assisted deep eutectic solvent based on dispersive liquid liquid microextraction of arsenic speciation in water and environmental samples by electrothermal atomic absorption spectrometry. J Mol Liq 242:441–446
73. Chabib CM, Ali JK, Abi Jaoude M, Alhseinat E, Adeyemi IA, Al Nashef IM (2022) Application of deep eutectic solvents in water treatment processes: a review. J Water Proc Eng 47:102663
74. Faraz N, Haq HU, Balal Arain M, Castro-Muñoz R, Boczkaj G, Khan A (2021) Deep eutectic solvent based method for analysis of Niclosamide in pharmaceutical and wastewater samples—a green analytical chemistry approach. J Mol Liq 335:116142. https://doi.org/10.1016/J.MOLLIQ.2021.116142
75. Matong JM, Nyaba L, Nomngongo PN (2020) Environmentally friendly eutectogels comprising l-amino acids and deep eutectic solvents: efficient materials for wastewater treatment. ChemPlusChem 85(2):301–311. https://doi.org/10.1002/CPLU.202000017
76. Moradi G, Rahimi M, Zinadini S, Shamsipur M, Babajani N (2022) Natural deep eutectic solvent modified nanofiltration membranes with superior antifouling properties for pharmaceutical wastewater treatment. Chem Eng J.448:137704. https://doi.org/10.1016/J.CEJ.2022.137704
77. Davoodi R, Nodehi RN, Rastkari N, Zinatizadeh AA, Mahvi AH, Fattahi. (2020) Solid-phase extraction followed by deep eutectic solvent based dispersive liquid–liquid microextraction and GC-MS detection of the estrogenic compounds in wastewater samples. New J Chem N 44(23):9844–9851. https://doi.org/10.1039/D0NJ00911C
78. Wang Y, Song Y, Li Y, Li J, Bao J, Zhang X, Li B (2023) Study on the extraction and separation of precious metals from wastewater using a hydrophobic deep eutectic solvent. J Environ Chem Eng 11(5):111102. https://doi.org/10.1016/J.JECE.2023.111102
79. Cheng H, Huang Y, Lv H, Li L, Meng Q, Yuan M, Liang Y, Jin M (2022) Insights into the liquid extraction mechanism of actual high-strength phenolic wastewater by hydrophobic deep eutectic solvents. J Mol Liq 368:120609. https://doi.org/10.1016/J.MOLLIQ.2022.120609
80. Habibollahi MH, Karimyan K, Arfaeinia H, Mirzaei N, Safari Y, Akramipour R, Sharafi H, Fattahi N (2019) Extraction and determination of heavy metals in soil and vegetables irrigated with treated municipal wastewater using new mode of dispersive liquid–liquid microextraction based on the solidified deep eutectic solvent followed by GFAAS. J Sci Food Agric 99(2):656–665. https://doi.org/10.1002/JSFA.9230
81. Vuoti S, Narasimha K, Reinikainen K (2018) Green wastewater treatment flocculants and fixatives prepared from cellulose using high-consistency processing and deep eutectic solvents. J Water Process Eng 26:83–91. https://doi.org/10.1016/J.JWPE.2018.09.003
82. Seidi S, Alavi L, Jabbari A, Shanehsaz M (2019) Three-phase carrier-mediated hollow fiber microextraction based on deep eutectic solvent followed by HPLC–UV for determination of raloxifene and ethinylestradiol in pharmaceutical wastewater treatment plants. J Iran Chem Soc 16(5):1007–1018. https://doi.org/10.1007/S13738-018-01572-4/TABLES/4
83. Sharma A, Sharma R, Thakur RC, Singh L (2023) An overview of deep eutectic solvents: alternative for organic electrolytes, aqueous systems and ionic liquids for electrochemical energy storage. J Energy Chem 82:592–626
84. Nguyen PT, Nguyen TDT, Nguyen VS, Dang DTX, Le HM, Wei TC, Tran PH (2019) Application of deep eutectic solvent from phenol and choline chloride in electrolyte to improve stability performance in dye-sensitized solar cells. J Mol Liq 277:157–162. https://doi.org/10.1016/J.MOLLIQ.2018.12.114

85. Nguyen TD, Nguyen TDT, Nguyen PT, Tran PH (2018) Dye-sensitized solar cells using deep eutectic solvents mixed with ethanol as an effective electrolyte medium. Science and Technology Development Journal 21(1):15–23. https://doi.org/10.32508/stdj.v21i1.424
86. Nguyen D, Van Huynh T, Nguyen VS, Doan CPL, Nguyen HT, Wei TC, Tran PH, Nguyen PT (2021) Choline chloride-based deep eutectic solvents as effective electrolytes for dye-sensitized solar cells. RSC Advances 11(35):21560–21566. https://doi.org/10.1039/D1RA03273A
87. Isci A, Erdem GM, Bagder Elmaci S, Sakiyan O, Lamp A, Kaltschmitt M (2020) Effect of microwave-assisted deep eutectic solvent pretreatment on lignocellulosic structure and bioconversion of wheat straw. Cellulose 27(15):8949–8962. https://doi.org/10.1007/S10570-020-03371-8/FIGURES/4
88. Xu GC, Ding JC, Han RZ, Dong JJ, Ni Y (2016) Enhancing cellulose accessibility of corn Stover by deep eutectic solvent pretreatment for butanol fermentation. Bioresour Technol 203:364–369. https://doi.org/10.1016/J.BIORTECH.2015.11.002
89. Alvarez-Vasco C, Ma R, Quintero M, Guo M, Geleynse S, Ramasamy KK, Wolcott M, Zhang X (2016) Unique low-molecular-weight lignin with high purity extracted from wood by deep eutectic solvents (DES): a source of lignin for valorization. Green Chem 18(19):5133–5141. https://doi.org/10.1039/C6GC01007E
90. Procentese A, Raganati F, Olivieri G, Russo ME, Rehmann L, Marzocchella A (2018) Deep eutectic solvents pretreatment of agro-industrial food waste. Biotechnol Biofuels Bioprod 11(1):1–12. https://doi.org/10.1186/S13068-018-1034-Y/TABLES/4
91. Mota-Morales JD, Sánchez-Leija RJ, Carranza A, Pojman JA, del Monte F, Luna-Bárcenas G (2018) Free-radical polymerizations of and in deep eutectic solvents: green synthesis of functional materials. Prog Polym Sci 78:139–153
92. Wang Q (2018) Controllable green synthesis of crassula peforata-like TiO_2 with high photocatalytic activity based on deep eutectic solvent (DES). J Chem Eng 348:811–819. https://doi.org/10.1016/J.CEJ.2018.05.020
93. Wang KY, Liu HW, Zhang S, Ding D, Cheng L, Wang C (2019) Selenidostannates and a silver selenidostannate synthesized in deep eutectic solvents: crystal structures and thermochromic study. Inorg Chem 58(5):2942–2953
94. Karimi M, Eshraghi MJ, Jahangir V (2016) A facile and green synthetic approach based on deep eutectic solvents toward synthesis of CZTS nanoparticles. Mater Lett 171:100–103. https://doi.org/10.1016/J.MATLET.2016.02.065
95. Di GML, Cassano R, Costanzo P, Cano NH, Maiuolo L, Nardi M, Nicoletta FP, Oliverio M, Procopio A (2019) Green synthesis of privileged benzimidazole scaffolds using active deep eutectic solvent. Molecules 24(16):2885. https://doi.org/10.3390/MOLECULES24162885
96. Liu Y, Kang S, Li K, Chen J, Bae B, Hwang I, Ahn EY, Park Y, Chun KH, Lee J (2022) Ecofriendly and enhanced biogenic synthesis of silver nanoparticles using deep eutectic solvent-based green tea extracts. J Clean Prod 379:134655. https://doi.org/10.1016/J.JCLEPRO.2022.134655
97. Wang M, Li C, Zhou M, Xia Z, Huang Y (2022) Natural deep eutectic solvent assisted synthesis and applications of chiral carbon dots. Green Chem 24(17):6696–6706. https://doi.org/10.1039/D2GC01949C
98. Amesho KTT, Chen SC, Wu TY, Ponnusamy VK, Lin YC (2023) Green synthesis of 5-hydroxymethylfurfural from biomass-derived carbohydrates using deep eutectic solvents as environmentally benign catalyst. Environ Technol Innov 29:102982. https://doi.org/10.1016/J.ETI.2022.102982
99. Lu H, Tan X, Huang G, Wu S, Zhou Y, Zhang J, Zheng Q, Chen T, Li F, Cai Z, Zeng J, Zhang M (2022) Green synthesis of highly stable $CsPbBr_3$ perovskite nanocrystals using natural deep eutectic solvents as solvents and surface ligands. Nanoscale 14(46):17222–17229. https://doi.org/10.1039/D2NR04173A
100. González-Campos JB, Pérez-Nava A, Valle-Sánchez M, Delgado-Rangel LH (2024) Deep eutectic solvents applications aligned to 2030. United Nations Agenda for Sustainable Development. Chem Eng Process Process Intensif 199:109751

Chapter 9
Extraction of Phenolic Compounds from Plant Matrices Using Eutectic Solvents

Lucas dos Santos Silva, Cristiane Nunes da Silva, Welison Oliveira Santos, Filipe Smith Buarque, and Bernardo Dias Ribeiro

9.1 Introduction

Phenolic compounds are among the primary bioactive constituents present in plant matrices. These secondary metabolites feature aromatic rings containing at least one hydroxyl group and are biosynthesized through the shikimic acid and phenylpropanoid pathway [1]. They may occur in free form or be bound to other components within the plant matrix, such as sugars. Based on their structure, phenolic acids, flavonoids, anthocyanins, stilbenes, and polymeric phenolic compounds such as lignin and tannins [2, 3]. Consumption of phenolic compounds from plant-derived foods has been extensively linked to various health benefits, including antioxidant, antihypertensive, anti-inflammatory, and hypoglycemic properties [3]. Moreover, they exhibit critical technological properties, acting as natural colorants, preservatives, and functional additives in the food, cosmetic, and pharmaceutical industries [4, 5].

From a structural standpoint, phenolic acids are the most straightforward group. They are typically categorized into hydroxybenzoic acids such as gallic and protocatechuic acid and hydroxycinnamic acids including caffeic, ferulic, and p-coumaric acid which are commonly found in fruits, vegetables, and whole grains.

L. dos Santos Silva · F. S. Buarque
School of Chemistry, Federal University of Rio de Janeiro, Rio de Janeiro, Brazil

C. N. da Silva · W. O. Santos
Chemistry Institute, Federal University of Rio de Janeiro, Rio de Janeiro, Brazil

B. D. Ribeiro (✉)
School of Chemistry, Federal University of Rio de Janeiro, Rio de Janeiro, Brazil

Chemistry Institute, Federal University of Rio de Janeiro, Rio de Janeiro, Brazil
e-mail: bernardo@eq.ufrj.br

K. Kuperkar et al. (eds.), *Deep Eutectic Solvents*, Environmental Footprints and Eco-design of Products and Processes,
https://doi.org/10.1007/978-3-032-03801-2_9

Flavonoids, the largest class of phenolic compounds, consist of flavones, flavanols, flavanols, flavanones, isoflavones, and anthocyanins, which contribute to the wide range of color and functional properties of fruits and vegetables [6]. Furthermore, they are the main phenolic compounds in the human diet [1]. Anthocyanins, belonging to the flavonoid class, are water-soluble pigments responsible for colors spanning red to blue tones in flowers, fruits, and vegetables and are considered potent natural antioxidants. Stilbenes, such as resveratrol, are the least abundant phenolic compounds found mainly in grapes and wines, with extensive studies demonstrating their anti-inflammatory, antioxidant, cardioprotective, and anticancer properties. Polymeric phenolic compounds include phenolic groups with complex chemical structures, such as lignins, responsible for giving rigidity to plant cell walls, and tannins, known for their astringent properties and their use in the food and pharmaceutical sectors [7].

Effective and selective techniques are crucial for isolating and extracting these bioactive substances. However, developing extraction processes that are simultaneously effective, environmentally sustainable, and economically viable remains a significant challenge [6, 7]. Traditional organic solvents like methanol, ethanol, hexane, and ethyl acetate are commonly used to extract phenolic compounds from plant materials. Nevertheless, these solvents often result in high solvent consumption, the generation of toxic waste, and negative environmental impacts, including air, soil, and water contamination [8, 9].

To address this challenge, eutectic solvents (ES) have emerged as environmentally friendly alternatives for the extraction of bioactive compounds. These solvents are formed by blending hydrogen bond donors (HBD) and hydrogen bond acceptors (HBA), resulting in a system where the entropy of phase transitions is reduced. This interaction leads to a considerable decrease in the melting points of the individual components [10]. A wide variety of HBD and HBA combinations can produce eutectic mixtures, whose physicochemical characteristics can be fine-tuned by adjusting the molecular structures and proportions of the components. Since no chemical reaction occurs between the HBD and HBA, ES can be prepared in a broad range of molar ratios commonly 1:2, 1:1, and 2:1 [11]. This versatility allows the customization of solvents for specific extraction goals, enabling the design of optimized systems for targeting desired compounds [12, 13]. Furthermore, these solvents exhibit low melting points and desirable features like being biodegradable, nontoxic, and thermally stable [14, 15]. Additionally, the interaction between these components (HBA and HBD) establishes a hydrogen bonding interaction, thereby lowering the melting temperature of the resulting eutectic system and providing unique solvation properties [16–18]. Studies indicate that ES can not only enhance extraction efficiency and preserve the bioactivity of extracted compounds but also outperform conventional solvents in extracting bioactive compounds from plants [16, 19–22].

Choi et al. [23] introduced natural ES, consisting entirely of naturally occurring, nontoxic substances like organic acids, amino acids, and sugars. Widely recognized by the scientific community, these solvents are considered ideal green alternatives due to their 100% atomic efficiency and their composition of renewable, biocompatible, biodegradable, and cost-effective substances. Their natural presence in cells links them to health-related applications, particularly within the pharmaceutical, cosmetic, and food sectors. Numerous studies have investigated the extraction of phenolic compounds from coproducts [24, 25], with research exploring the use of natural ES for removing phenolic compounds from sources such as coffee seeds [26], grapefruit peel [27], and mulberry leaves [28].

To enhance phenolic compounds' extraction efficiency, natural eutectic solvents have been integrated with innovative methods like microwave-assisted extraction (MAE) and ultrasound-assisted extraction (UAE). These technologies are more advanced extraction tools that provide higher yields and require fewer solvents, in addition to low temperatures and brief extraction durations that help prevent the breakdown of heat-sensitive compounds, thus helping to improve extraction efficiency. Phenolic compounds extracted with ES undergo different separation and purification steps before their utilization within the food, cosmetics, and pharmaceutical industries, sectors that demand more sustainable and environmentally responsible processes [29, 30]. ES can be separated by adding an antisolvent (water and ethanol) or using microporous resins and reused [31, 32]. However, depending on the composition of the ES, there is no need for separation processes, and the ES can act as an active component together with the target compound [14]. Therefore, this chapter explores the capability of eutectic solvents to extract phenolic compounds from plant matrices. The characteristics of phenolic compounds and their subclasses, their biological and technological importance, traditional extraction methods and their limitations, and the advantages and challenges of using eutectic solvents as an eco-friendly option for extracting these bioactive compounds will be addressed.

9.2 Phenolic Compounds

Phenolic compounds constitute one of the broadest and most diverse secondary metabolites in plant matrices, with more than 8000 structures already identified. These compounds play crucial functional and structural roles, contributing to plant defense mechanisms against environmental stressors and pathogens and as integral components of plant cell walls [33]. Biosynthetically, phenolic compounds originate from multiple metabolic pathways, with the pentose phosphate pathway, the shikimic acid pathway, and the phenylpropanoid pathway being the most significant [34]. These molecules feature at least one aromatic ring attached to hydroxyl groups, and when several hydroxyl groups are present, they are categorized as polyphenols. Structurally, phenolic compounds can range from simple molecules to high molecular weight molecules and polymeric structures [35]. This vast structural diversity allows phenolic compounds to be categorized into several major classes, including flavonoids, phenolic acids, stilbenes, and tannins [1]. In addition, lignin will also be considered a phenolic compound in this chapter, as it aligns with the definition used.

The demand for phenolic compounds for industries and the consumer public has recently increased due to the technological effects on food, the physiological effects they produce on the human body, and their classic nutritional properties. The bioactive compound is often incapable of promoting bioactivity but produces or induces metabolite effects, or effects on the host's gut microbiome, after consumption [36]. They are constantly associated with several beneficial properties for the human body, including antioxidant, antimicrobial, anticancer, anti-inflammatory, cardioprotective, neuroprotective, ultraviolet (UV), and the ability to protect the immune system and reduce damage from cell oxidation and cardiovascular complications [35, 37]. Due to these beneficial properties, phenolic compounds have extensive applications across multiple industrial sectors, including food, cosmetics, and pharmaceuticals [38]. Their natural bioactivity and functional versatility make them valuable ingredients in developing health-promoting products and sustainable solutions for various industries. Due to these beneficial properties, phenolic compounds have extensive applications across multiple industrial sectors, including food, cosmetics, and pharmaceuticals [38]. Their natural bioactivity and functional versatility make them valuable ingredients in developing health-promoting products and sustainable solutions for various industries.

9.2.1 Phenolic Acids

Phenolic acids represent a group of organic acids widely distributed in various plant sources, including seeds, fruit skins, berries, vegetable leaves, and flowers [39, 40]. These compounds belong to the broader group of phenolic compounds and are characterized explicitly by at least one functional group of carboxylic acid (-COOH) within their structure. Based on their chemical backbone, phenolic acids can be classified into two major subgroups: hydroxybenzoic and hydroxycinnamic acid [34].

These compounds are defined by a C6-C1 structure, where a carboxyl group (-COOH) is attached to a benzene ring. This subgroup includes some of the most well-known phenolic acids, such as p-hydroxybenzoic, gallic, protocatechuic, and vanillic acids, which are commonly found in fruits, tea, and certain medicinal plants [41]. On the other hand, hydroxycinnamic acid derivatives are characterized by a C6-C3 structure (C_6H_5-CH=CH-COOH), where the original cinnamic acid backbone may undergo hydroxylation at different positions. The key representatives of this class include p-hydroxycinnamic, p-coumaric, caffeic, and ferulic acids, which are abundant in coffee, cereals, and various fruits and vegetables [34, 40]. Due to their diverse structures and functional properties, phenolic acids contribute significantly to plant-based foods' antioxidant, antimicrobial, and anti-inflammatory potential. Their general structural framework is illustrated in Fig. 9.1, highlighting the differences between hydroxybenzoic and hydroxycinnamic acid derivatives.

Hydroxybenzoic acids	R1	R2	R3	R4
ρ-hydroxybenzoic acid	- OH	- H	- H	- H
Vanillic acid	- OH	- H	$-OCH_3$	- H
Gallic acid	- OH	- OH	- OH	- H
Syringic acid	- OH	$-OCH_3$	$-OCH_3$	- H

Hydroxycinnamic acids	R1	R2	R3	R4	R5
ρ-coumaric acid	- OH	- H	- H	- H	-H
Caffeic acid	- OH	- OH	-H	- H	-H
Sinapic acid	- OH	$- OCH_3$	$- OCH_3$	- H	-H
Ferulic acid	- OH	$-OCH_3$	-H	- H	-H

Fig. 9.1 General chemical structures of phenolic acids (adapted from Córdova and Medina, [42])

As mentioned, phenolic acids are found in a large number of plants, especially in foods and beverages such as coffee and tea [43, 44], medicinal herbs [45, 46], wines [47], and many more, and they can also be used as functional additives in foods, for example for preservation, color retention and microbial growth delayer [41].

9.2.2 *Flavonoids*

Flavonoids are one of the most extensively studied phenolic compounds, widely distributed across various plant species, particularly in fruits and vegetables [48]. These compounds are crucial in mediating plant responses to biotic and abiotic environmental stimuli. For instance, they contribute to the pigmentation and flavor profile of fruits and flowers, participate in the biosynthesis of specific enzymes and vitamins, and function as potent antioxidants, protecting plants from oxidative stress [34, 48]. Owing to their widespread presence in plant-based foods, including fruits, vegetables, legumes, flowers, leaves, and seeds, flavonoids have gained significant research interest, particularly for their antioxidant activity, which is regarded as their primary biological function [49]. This antioxidant capacity is fundamental to plant physiology and human health, as flavonoids reduce oxidative stress, inflammation, and the risk of chronic diseases.

Structurally, all flavonoids share a common C6-C3-C6 backbone, consisting of two aromatic rings (designated as A and B) connected by a heterocyclic ring (C), which contains three carbon atoms and one oxygen [1]. Depending on the oxidation state of the central heterocyclic ring, flavonoids can be classified into seven major subclasses: flavanols, flavones, isoflavones, flavanones, flavanols, anthocyanidins, and chalcones (Fig. 9.2). Their chemical structures frequently undergo modifications such as glycosylation and acylation, primarily at sites containing methyl and hydroxyl groups on the aromatic rings, which influence their solubility, bioavailability, and biological activity [48]. Due to their diverse structures and wide-ranging

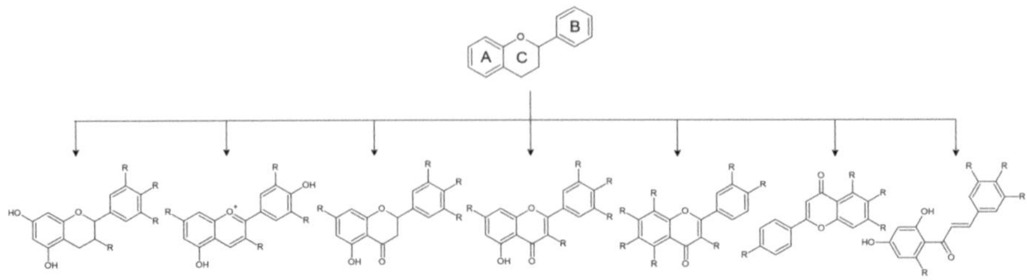

Fig. 9.2 Basic structure of (**a**) flavonoids and (**b**) their subclasses (adapted from Shen et al. [48])

bioactivities, including antioxidant, anti-inflammatory, antimicrobial, and cardioprotective properties, flavonoids have found extensive applications in the pharmaceutical, nutraceutical, cosmetic, and food industries. For this reason, research efforts persist in evaluating their potential benefits in health maintenance and disease prevention [37].

9.2.3 Anthocyanins

Although anthocyanins are a subclass of flavonoids, they will be treated independently in this chapter due to their notable scientific relevance, unique properties, and commercial importance. As one of the most extensively studied phenolic compounds, anthocyanins have gained significant attention for their diverse bioactivities and widespread applications in the food, pharmaceutical, and cosmetic industries [50].

Anthocyanins are water-soluble pigments characterized by a structure based on the flavylium cation, which is hydroxylated at different positions (Fig. 9.3). These compounds exist predominantly in their glycosylated forms, known as anthocyanins, which exhibit greater stability compared to their non-glycosylated counterparts, anthocyanidins. Some of the most common anthocyanins include cyanidin, malvidin, delphinidin, and pelargonidin, which are widely found in nature [50].

These pigments are responsible for the vibrant coloration of many fruits, vegetables, and flowers, with their hues ranging from red and pink to purple and blue, depending on the pH of their environment [1]. Under acidic conditions, anthocyanins retain a positive charge on the oxygen of the C ring, typically appearing as red pigments. At neutral pH, the formation of uncharged quinone structures results in purple coloration. As the pH increases and the environment becomes more alkaline, anthocyanins shift toward bluish hues [51]. However, these compounds become less stable at higher pH levels, leading to degradation and a loss of color intensity [50]. Due to their pH-sensitive color changes and antioxidant activity, anthocyanins are extensively utilized as natural food colorants and show promise for applications in functional foods, nutraceuticals, and pharmaceuticals.

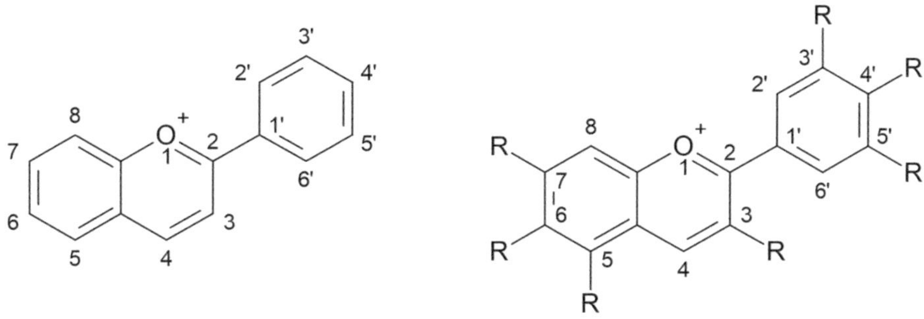

Fig. 9.3 General chemical structure of flavylium cation (on the left) and possible sites for hydroxylation and posterior glycosylation (adapted from Mattioli et al. [50])

Fig. 9.4 Basic structure of the two isomeric forms of stilbenes (adapted from Teka et al. [53])

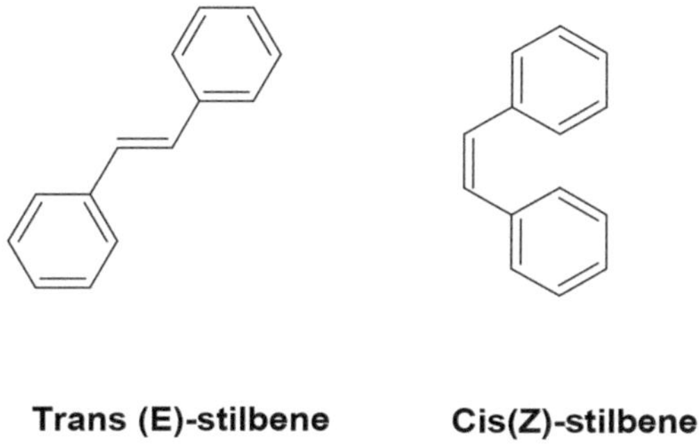

9.2.4 *Stilbenes*

Stilbenes comprise a group of diphenyl compounds defined by a C6-C2-C6 structure, consisting of two aromatic rings connected via an ethylene bridge (Fig. 9.4). These compounds exist in two isomeric forms: E-1,2-diphenylethylene (trans configuration) and Z-1,2-diphenylethylene (cis configuration). Among these, the E (trans) form is the most prevalent and is considered more biologically active due to its enhanced stability and bioavailability [52]. Like other phenolic compounds, plants synthesize stilbenes as defense mechanisms against various environmental stresses, particularly ultraviolet (UV) radiation and pathogenic infections. These protective functions contribute to their presence in various plant families, including grapes, peanuts, and different types of berries, where they serve as natural antioxidants and antimicrobial agents [53].

Among stilbenes, resveratrol is the most well-known and widely studied due to its numerous health-promoting properties, including anti-inflammatory, antioxidant, cardioprotective, and anticancer effects. It has been extensively investigated for its potential role in longevity, metabolic health, and neuroprotection. Although resveratrol is the most prominent stilbene, other compounds within this class, such as pterostilbene and piceatannol, have also been recognized for their biological activity and potential therapeutic benefits [52, 53].

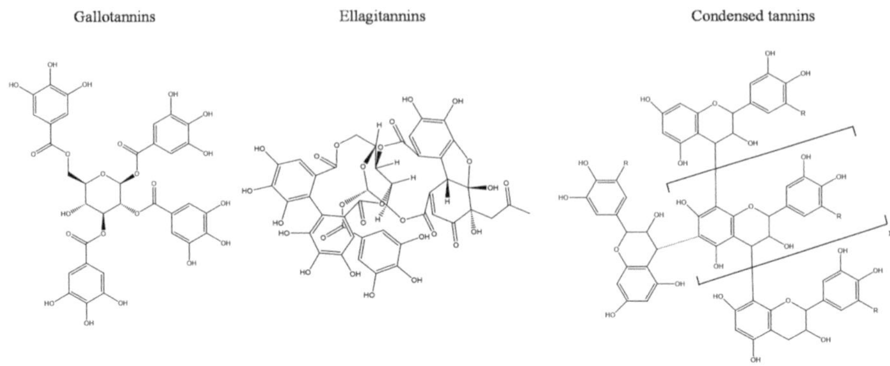

Fig. 9.5 General structure of hydrolyzable tannins (gallotannins and ellagitannins) and condensed tannins (adapted from Besharati et al. [56])

9.2.5 *Tannins*

Unlike the previously mentioned phenolic compounds, tannins are phenolic compounds with a higher molecular weight, typically ranging from 500 to 30,000 Da. Tannins are generally divided into two primary categories: hydrolyzable tannins (also known as pyrogallol tannins) and condensed tannins (also referred to as proanthocyanidins) [54]. Hydrolyzable tannins are esters derived from gallic acid (gallotannins) or ellagic acid (ellagitannins), which are linked to carbohydrates, usually glucose. As the name suggests, these tannins can be hydrolyzed into their monomers through hydrolysis reactions, releasing phenolic acids. On the other hand, condensed tannins are oligomers or polymers composed of flavan-3-ol (catechin) and/or flavan-3,4-diols (leucoanthocyanidins), linked by covalent carbon-carbon bonds. Unlike hydrolyzable tannins, they do not contain a carbohydrate core in their structure [55]. The general structure of tannins can be seen in Fig. 9.5.

Tannins are present in various types of plants, from trees such as maple, oak, acacia, and eucalyptus [54] to other structures such as seeds, roots, leaves, and fruits, especially grapes, persimmons, berries, bananas, and many others, which are recognized as rich in tannins [57]. Their presence in plants is associated with a defensive agent and contributes to various characteristics such as astringency, flavor, aroma, color, etc. [57].

9.2.6 *Lignin*

Lignin is a complex natural polymer formed through the oxidative coupling of 4-hydroxycinnamyl alcohols, also known as monolignols [58]. As the most plentiful renewable source of aromatic compounds [59], it is essential for preserving the structural integrity of plant cell walls, offering mechanical strength and facilitating water conduction in vascular plants [58]. Found in nearly all plant species except

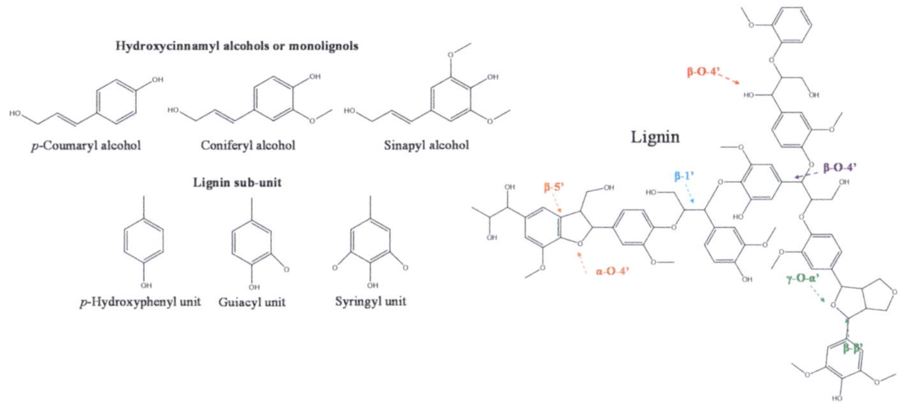

Fig. 9.6 Chemical structure of monolignols and lignin submits (G, S, and H), and a schematic representation of lignin polymeric structure with the main lignin inter-unit linkage motifs, with the occurrence in different types of feedstocks (adapted from Slabon and Rodrigues [62])

bryophytes, its composition varies significantly among hardwoods, softwoods, and grasses, each exhibiting distinct lignin content and structural characteristics [60, 61].

Structurally, lignin comprises three primary p-hydroxycinnamyl alcohols: p-coumaryl alcohol, coniferyl alcohol, and sinapyl alcohol. These monolignols vary in the number of methoxy groups attached to their aromatic rings, and upon polymerization, they form the three primary lignin subunits: p-hydroxyphenyl (H), guaiacyl (G), and syringyl (S) units. A variety of characteristic C interconnects these units, C–C and C–O linkages (e.g., β–O–4′, β–β–5′, β–β) that determine the polymer's complexity and properties [58]. The composition and structural arrangement of lignin vary depending on plant species, tissue type, cell type, and developmental stage (Fig. 9.6). In addition to the three primary monolignols, various other monomeric and dimeric compounds, such as hydroxycinnamic acids, hydroxycinnamaldehydes, cinnamoyl esters, and flavonoids, can also serve as precursors for lignin polymerization, further diversifying its structure and function [58].

In recent years, lignin has emerged as a highly relevant topic in the scientific community. Research institutes and biorefineries are actively exploring its potential for diverse applications, including its use as a carbon fiber precursor [63], a substitute for phenol-formaldehyde resins [64], in nanocomposites and nanohybrid assemblies [65], in the production of fine chemicals [66] and biofuels [67], and even in medicinal nanoparticles for pharmaceutical applications [68].

9.3 Application of Eutectic Solvents in the Extraction of Phenolic Compounds

Table 9.1 and 9.2 shows the optimized conditions of different subclasses of phenolic compounds (phenolic acids, flavonoids, anthocyanins, stilbenes, lignin, and tannins) extracted by eutectic solvents combined or not by emerging technologies (MAE, UAE, and PEF) in different plant matrices.

Table 9.1 Extraction of phenolic compounds from plants using eutectic solvents

Material	Extraction method/ES composition	Optimal extraction conditions	Target compounds	Extraction yield	References
Artemisia argyi leaves	ES-UAE ChCl:malic acid:urea (2:1:2)	SLR: 1:57.5, 54% water, 200 W, 40 kHz, 23.5 min	Phenolic acids	22.80 mg/g	Duan et al. [69]
Artichoke outer petals	ES-UAE ChCl:lactic acid (1:2)	SLR: 1:20, 37 kHz, 200 W, 30 min	Phenolic acids	5.92 mg/g	Ozkan et al. [70]
Rosmarinus officinalis	ES-UAE Ammonium acetate:lactic acid (1:2)	SLR: 1:20, 37 kHz, 320 W, 5 min	Phenolic acids	8.40 mg/g	Pizani et al. [71]
Green coffee seeds	ES-MAE ChCl:urea (1:2)	SLR: 1:10, 60 °C, 25% water, 10 min	Phenolic acids	40.42 mg/g	Herrera-Beltran et al. [72]
Tomato pomace	ES-UAE ChCl:lactic acid (1:2)	SLR: 1:20, 10% water, 30 kHz, 100 W, 60 min	Phenolic acids	73.27 μg/g	Vasyliev et al. [73]
Black mulberry pomace	ES Choline chloride:lactic acid	SLR: 1:70, 38.9% water, 67.9 °C, 24.2 min	Flavonoids	16.8 mg RE/g	Vo et al. [74]
Selaginella moellendorffii leaves	ES-UAE Choline chloride:lactic acid (1:2)	SLR: 1:24, 24% water, 260 W, 43 min	Flavonoids	5.72 mg/g	Lei et al. [75]
Noni pomace	ES-PEF Choline chloride:oxalic acid	SLR: 1:20, 20% water, 60 °C, 38.45 min PEF: 1.60 kV/cm, 85 pulses SLR: 1:20, 20% water, 80 °C, 50 min, PEF: 1.65 kV/cm, 110 pulses	Flavonoids (rutin and quercetin)	16.21 mg/g rutin 19.85 mg/g quercetin	Li et al. [76]
Ribes mandshuricum leaves	ES-MAE Choline chloride:oxalic acid (1:2)	SLR: 1:27, 25% water, 54 °C, 600 W, 10 min	Flavonoids	10.51 mg/g	Wang et al. [77]
Moringa oleifera leaves	ES Choline chloride:lactic acid (1:15)	SLR: 1:61, 66 °C, 67 min	Flavonoids	54.8 mg/g	Wang et al. [78]

Table 9.2 (continued)

Material	Extraction method/ES composition	Optimal extraction conditions	Target compounds	Extraction yield	References
Citrus peel (*Tangerina nobis*)	ES Choline chloride:levulinic acid:N-methylurea (1:1.2:0.8)	SLR: 1:50, 20% water, 50 °C, 25 min	Flavonoids (hesperidin, nobiletin and tangeretin)	65.82 mg/g	Xu et al. [79]
Raspberry powder	ES-UAE Choline chloride:1,4-butanediol	SLR: 1:3, 210 W, 29% water, 32 min, 51 °C	Anthocyanins	1.378 mg/g	Xue et al. [15]
Black rice bran	NAES-UAE Lactic acid:fructose	SLR: 1:18, amplitude 21%, 12% water, 11,25 min	Anthocyanins	32.19 mg/L	Thakur et al. [80]
Purple Perilla leaves	UMAE-ES Choline chloride:ethylene glycol:lactic acid (1:3:0.7)	SLR: 1:10, MAE 410 W, UAE 357 W, 20% water, 26 min, 58 °C	Anthocyanins	619.62 mg/100 g	Han et al. [81]
Chokeberry	NAES-UAE Choline chloride:malic acid (1:1)	SLR: 1:30 g/ml, 40% water, 90 min, 42 °C	Cyanidin-3-O-galactoside, cyanidin-3-O-glucoside, cyanidin-3-O-arabinoside TAC	2.62 mg/g DW 0.36 mg/g DW 1.25 mg/g DW 5.92 mg/g DW	Jovanović et al. [82]
Blackberry	NAES-HSH-CBE Chloride:citric acid: glucose (1:1:1)	SLR: 1:22, 30% water, homogenization time of 60 s, speed 12.000 rpm, negative pressure – 0,08 Mpa, 30 min	Anthocyanins	6.05 mg/g DW	Guo et al. [83]
Blueberry	NAES-UAE Choline chloride:sorbitol (1:1)	SLR: 10 mL/g, 34% water, 37 min, 48 °C	Anthocyanins (cyanidin-3-O-glucoside)	0.27 mg/g DW	Jovanović et al. [84]
Chilean berries	NAES-UAE Lactic acid:glucose (8:1)	SLR: 25 mg/1 mL, 35 KHz, 40% water, 90 min, 42 °C	Anthocyanins	3.3 mg/g DW	Velásquez et al. [85]
Pigeon pea leaves	ES-UAE Thymol:coumarin (2:1)	SLR: 1:50, 200 W, 20 min	Stilbenes	5.51 mg/g	Yue et al. [86]
Polygon cuspidatum root	ES Tetrabutylammonium chloride:ethylene glycol	SLR: 1:50, 85 °C	Resveratrol	12.26 mg/g	Sun et al. [87]

(continued)

Table 9.2 (continued)

Material	Extraction method/ES composition	Optimal extraction conditions	Target compounds	Extraction yield	References
Peanut root	ES-UAE Choline chloride:1,4-butanediol (1/3 g/g)	SLR: 1:30, 400 W, 40% water, 40 min, 55 °C	Resveratrol	38.34 mg/kg	Chen et al. [88]
Seeds of *Cichorium intybus*	ES-MAE Glycerol:sodium acetate (3:1)	SLR: 34 mL, 320 W, 2 min	Tannins	15.48 mg TAE/g	Imtiaz et al. [89]
Medicinal plant *Alchemilla Vulgaris L.*	ES Choline chloride:urea (1:2)	SLR: 50 mg/1Ml, 30% water, 70 °C, 30 min	Hydrolyzable tannins	181.26 μgTAE/mg	Kovac et al. [46]
Medicinal plant *Geum japonicum*	ES-UAE L-proline:lactic acid	SLR: 1:27, 29% water, 48 °C, 300 W, 48 min	Tannins	67.5 mg/g	Wang et al. [90]
Onion peel	ES-UAE Choline chloride:urea (1:1)	SLR: 1:10, duty cycle 10%, 60 °C, 3 h	Tannins	1705.80 μg/g	Sukor et al. [91]
Corncob	ES ChCl:oxalic acid (1:1)	SLR: 90 °C, 24 h	Lignin	98.5%	Zhang et al. [92]
Poplar	ES ChCl:lactic acid (1:2)	SLR: 1:10, 145 °C, 6 h	Lignin	78.5%	Alvarez-Vasco et al. [93]
Rice straw	ES ChCl:Lactic acid (1:5)	SLR: 1:10, 60 °C, 12 h	Lignin	60%	Kumar et al. [94]
Corn Stover	ES ChCl:glycerol (1:2)	SLR: 1:10, 180 °C, 2 h	Lignin	60%	Xu et al. [95]
Rice straw	ES ChCl:glycerol (1:2)	SLR: 1:19, 150 °C, 15 h	Lignin	74%	Hossain et al. [96]

SLR (solid-liquid ratio) expressed in g/mL, *UAE* ultrasound-assisted extraction, *MAE* microwave-assisted extraction, *UMAE* ultrasound microwave-assisted extraction, *HSH-CBE* high-speed homogenization and cavitation-explosion extraction, *ES* eutectic solvent, *NAES* natural eutectic solvent, *ChCl* choline chloride, *DW* dry weight, *TAE* tannic acid equivalent, *RE* Rutin equivalent, *TAC* total anthocyanin content

9.3.1 Phenolic Acids

Although most studies on the extraction of phenolic compounds primarily focus on total phenolics, in many cases, phenolic acids constitute the predominant fraction of the extracts [73, 97, 98]. In recent years, eutectic solvents have been increasingly applied extensively to extract phenolic acids, often combined with enhanced methodologies such as UAE and MAE, as an alternative to conventional solvents.

Gao et al. [99] investigated the efficiency of phenolic acid extraction from *Asparagopsis taxiformis*, comparing a eutectic solvent (betaine:levulinic acid, 1:2 molar ratio) with conventional solvents (water, 70% ethanol, 70% methanol, and 70% acetone) under ultrasound-assisted extraction conditions. Their results demonstrated that the eutectic solvent significantly outperformed traditional solvents, yielding 6.74 and 153.11 times more phenolic acids. In a similar study, Liu et al. [100] tested 16 eutectic solvents to extract phenolic compounds from *Paederia scandens* (Lour.) Merr. using UAE. Among the solvents evaluated, the mixture of choline chloride and ethylene glycol in a 1:2 molar ratio exhibited the highest extraction efficiency. The authors compared the phenolic acid content in extracts obtained with this eutectic solvent to that extracted using conventional solvents, such as ultrapure water, 80% ethanol, and 70% methanol. Their findings revealed that for the primary phenolic acid in the extract (2-hydroxybenzoic acid), extraction with the eutectic solvent yielded similar results to water extraction (42.20 μg/g DW with ES vs. 44.26 μg/g DW with water) and higher yields than methanol and ethanol (38.88 μg/g DW and 39.58 μg/g DW, respectively) [100].

Likewise, Zhu et al. [101] examined the extraction of phenolic acids from *Morinda citrifolia* L. using various eutectic and conventional solvents. Among the eutectic solvents tested, choline chloride:tetramethylene glycol, choline chloride:trimethylene glycol, and proline:ethylene glycol yielded substantially greater levels of six distinct phenolic acids (4-caffeoylquinic acid, chlorogenic acid, protocatechuic acid, p-hydroxycinnamic acid, ferulic acid, and o-coumaric acid) compared to conventional solvents. Only gallic acid showed higher extraction efficiency with conventional solvents. Together, these results emphasize the potential of eutectic solvents as effective alternatives to conventional solvents in the extraction of phenolic acid.

The efficiency of phenolic acid extraction using eutectic solvents is primarily influenced by their physicochemical properties, including polarity, pH, and viscosity. Polarity is closely related to the solvent's ability to form hydrogen bonds with target molecules. Higher polarity generally enhances the extraction of phenolic acids, as observed with organic acid-based eutectic solvents, which exhibit greater polarity than sugar- or polyol-based solvents. For example, Ozkan [70] investigated the extraction of five phenolic acids (gallic, ferulic, syringic, sinapic, and chlorogenic acids) using six different eutectic solvents, ChCl and betaine as HBA, and sucrose, glycerol, and organic acids as HBD. Their results demonstrated that solvent polarity significantly impacted phenolic acid yields. The combinations of ChCl:lactic acid (1:2), betaine:citric acid (1:1), and betaine:lactic acid (1:2) were particularly effective for chlorogenic and ferulic acid extraction. For sinapic acid, ChCl:lactic acid and betaine:citric acid also exhibited the highest extraction yields compared to less polar eutectic solvents.

The pH level of the eutectic solvent is another important factor influencing the extraction of phenolic acid, with some studies suggesting that an acidic or neutral environment enhances extraction efficiency. For instance, Vasyliev et al. [73] conducted a comparative study of the extraction of phenolic compounds from tomato pomace using two eutectic solvents: choline chloride combined with 1,2-propanediol (1:2 molar ratio) and choline chloride paired with lactic acid (1:2 molar ratio). The extract produced using choline chloride:lactic acid showed considerably higher concentrations of gallic, caffeic, chlorogenic, and ferulic acids (14.4, 5.03, 52.33, and 1.51 μg/g, respectively) than ChCl:1,2-propanediol (10.2, 2.03, 37.23, and 0.87 μg/g, respectively). These findings indicate that eutectic solvents with organic acids can enhance phenolic acid extraction due to their ability to maintain a favorable pH environment.

Another critical parameter influencing phenolic acid extraction is the viscosity of eutectic solvents. Researchers often incorporate water into the solvent formulation to decrease viscosity and improve extraction efficiency. Pizani et al. [71] examined how varying water content on rosmarinic acid extraction from *Rosmarinus officinalis* using ammonium acetate:lactic acid with UAE (240 W, 5 min). They found that raising content from 20% to 50% caused a decrease in extraction yield, probably because of the disruption of the hydrogen bonding network within the eutectic solvent. They did not explore water concentrations below 20%, arguing that the solvent viscosity remained too high at lower levels, impeding ultrasonic wave propagation and reducing efficiency. Moreover, Yu et al. [102] observed a decrease in extraction yields when the water content is raised from 20% to 80% in a ternary eutectic solvent (ChCl:1,4-butanediol:ascorbic acid, 1:1:0.2). However, the authors found that adding 20% water improved extraction efficiency compared to using the eutectic solvent without water.

Although the adding of water helps adjust viscosity, it simultaneously changes other physicochemical characteristics like polarity, density, and conductivity, which have a notable impact on extraction efficiency [103]. Nevertheless, too much water can compromise the eutectic solvent's structure, weaken hydrogen bonding, and decrease its ability to interact with phenolic acids [46].

9.3.2 Flavonoids

Flavonoids represent a major subclass of phenolic compounds and are well known for their strong biological properties [31, 48, 104]. ES have gained attention as sustainable and efficient alternatives for flavonoid extraction from plant matrices. ES and flavonoids exhibit comparable polarities and intermolecular forces, including hydrogen bonding, dipole-dipole interactions, and van der Waals forces, and enhance their solubilization capacity [105]. These factors are key contributors to mass transfer and the overall efficiency of extraction. However, various parameters, including pH, polarity, cosolvent (water) concentration, solid-liquid ratio, time of extraction, temperature, and integration with emerging technologies, significantly influence the extraction yield of flavonoids using ES.

The polarity of ES plays a pivotal role in determining the efficiency of flavonoid extraction. Xu et al. [79] explored chloride choline (HBA) ES systems combined with different hydrogen bond donors (HBDs) for extracting flavanones (hesperidin and narirutin) from citrus fruits. The highest yields were achieved using a lower polarity ES system comprising choline chloride and levulinic acid (65.82 mg/g). This indicates optimal flavonoid extraction occurs when the solvent and target compounds have similar polarities. Citrus flavonoids, in particular, exhibit lower polarity due to their multiple methoxyl groups. Conversely, more polar sugar-based ES resulted in significantly lower extraction yields (<4.40 mg/g), likely due to weaker intermolecular interactions with citrus flavonoids. These results emphasize the importance of customizing ES characteristics to suit specific plant matrices, thereby enhancing flavonoid extraction efficiency.

ES are known for their high viscosity, which can hinder extraction efficiency by slowing mass transfer. Water is commonly added to reduce viscosity and improve solvent diffusion within the plant matrix. Vo et al. [74] evaluated how varying water content (10–50%) in a choline chloride:lactic acid (2:1) ES influences flavonoid extraction from mulberry. The study reported a marked improvement in flavonoid yield at 40% water content (>25 mg RE/g), compared to 10% (<14 mg RE/g), due to reduced viscosity and enhanced mass transfer. However, excessive water (50%) disrupted intermolecular interactions between ES and flavonoids, lowering extraction efficiency. The optimal water content was determined to be 38.9%, yielding 16.8 mg RE/g of total flavonoids. Li et al. [76] also found that increasing water content from 20% to 80% in choline chloride:oxalic acid ES decreased flavonoid extraction efficiency. The highest yields were obtained at 20% water content (19.85 mg/g quercetin and 16.21 mg/g rutin). These results emphasize the importance of balancing solvent structural integrity and flavonoid solubility to maximize extraction efficiency. Additionally, while water modulates viscosity, this modification influences other essential properties of ES such as polarity, density, and conductivity that play a direct role in extraction efficiency.

The use of ES in conjunction with innovative extraction techniques, including MAE and UAE, has been extensively explored to improve flavonoid yield while reducing processing time. Lei et al. [75] demonstrated that using ES (choline chloride:lactic acid, 1:2) with UAE resulted in significantly higher flavonoid extraction (5.72 mg/g) compared to conventional methods like maceration (1.90 mg/g) and percolation (1.85 mg/g). The increased efficiency was attributed to acoustic cavitation, which improves solvent infiltration, breaks down plant cell walls, and accelerates solute dissolution. In the same way, Wang et al. [77] combined ES with MAE to extract flavonoids from *Ribes mandshuricum leaves*. Using a choline chloride:lactic acid (1:2) ES under optimal conditions (25% water, solid-liquid ratio of 27 mg/L, 54 °C, and 10 min extraction time), they achieved a flavonoid yield of 10.51 mg/g, significantly higher than thermal reflux extraction (6.83 mg/g in

120 min). The enhanced efficiency of MAE is attributed to electromagnetic radiation (300 MHz to 300 GHz), which increases solvent penetration, raises temperature and pressure, and facilitates cell membrane rupture, releasing greater amounts of extractable flavonoids [106].

9.3.3 Anthocyanins

Various factors govern the extraction efficiency of anthocyanins when using eutectic solvents (ES), such as the ratio between solid and liquid phases, viscosity, water concentration, pH, and polarity characteristics of the ES. These parameters are critical in determining extraction efficiency and the preservation of anthocyanins, as evidenced in various plant matrices [105].

The proportion of solid to liquid significantly impacts mass transfer, enabling deeper solvent penetration into the plant tissues and enhancing the release of anthocyanins. It has been observed in literature that higher solid-to-liquid ratios, up to an optimal point, tend to improve the effectiveness of the extraction process, as observed in raspberry [15], rice bran [80], and perilla leaves [81]. However, excessively high solid-to-liquid ratios can hinder diffusion, reducing extraction efficiency.

The viscosity of ES is another determining factor, as highly viscous solvents limit molecular mobility and impede extraction. Research has demonstrated that ES containing choline chloride combined with organic acids exhibits different viscosities and extraction performances [15, 81, 107]. Introducing water into the eutectic solvent system may reduce its viscosity, facilitating compound extraction. On the other hand, too much water may compromise the solvent's structural organization, leading to reduced extraction performance, as observed in blackberry [83] and perilla leaves [81].

Another key parameter is the pH of the eutectic solvent, as it directly influences the chemical stability of anthocyanins. Acidic formulations for instance, those combining choline chloride with lactic acid tend to provide higher extraction yields, as this pH range favors the preservation of anthocyanins in their stable flavylium cationic form [80, 84, 85]. Conversely, alkaline ES, such as choline chloride-urea, yield lower anthocyanin recovery due to compound degradation.

Furthermore, the polarity of ES significantly influences extraction efficiency, as anthocyanins are highly polar compounds. Studies indicate that highly polar ES enhance extraction, as observed in raspberry [15], black rice bran [80], and blackberry [107]. This enhanced extraction efficiency is likely associated with forming strong hydrogen bonds between anthocyanins and the extraction solvents. Therefore, optimizing extraction parameters is essential to maximize anthocyanin recovery. Careful selection of ES based on polarity, pH, water content, and solid-to-liquid ratio is crucial for ensuring efficient and sustainable extraction of these valuable bioactive compounds.

9.3.4 Stilbenes

The extraction of stilbenes using ES presents an effective and more sustainable alternative than conventional organic solvents. However, as evidenced in the studies presented in Table 9.2, the efficiency of stilbene extraction is influenced by several critical factors, such as the type of SE used (hydrophilic or hydrophobic), the solid/liquid ratio, the water content in the solvent, the temperature, and the extraction time. These parameters affect the physical-chemical properties of the solvents, directly reflecting the extraction yield of the target compounds present in the plant matrices.

Sun et al. [87] conducted a comparative study on the extraction of resveratrol from the root of *Polygonum cuspidatum*, using SE and traditional organic solvents (methanol and ethanol). The results indicated that SE provided a higher resveratrol extraction yield (5.33–9.00 mg/g) when compared to methanol (5.00 mg/g) and ethanol (5.08 mg/g). Among the SE evaluated, the combination of tetrabutylammonium chloride and ethylene glycol (TBAC-EG) presented the highest yield, reaching 9.00 mg/g. This superior performance was associated with the low viscosity of the system (67.81 mPa s), the slightly acidic character (pH = 5.59), and the intense hydrophobic interactions between SE and resveratrol. It is important to highlight that quaternary ammonium salts with longer carbon chains, such as TBAC, intensified these interactions, contributing to greater extraction efficiency. Conversely, ES, composed of L-proline:glycerol, betaine:glycerol, and L-carnitine:glycerol, exhibited lower extraction efficiency due to its higher viscosity. To mitigate this issue, the researchers explored the influence of varying water concentrations (ranging from 30% to 70% v/v) on the efficiency of resveratrol recovery. Adding water reduced resveratrol solubility due to its strong hydrophobicity, with the highest yield achieved at water concentrations below 40% (>11.0 mg/g). This suggests that a higher ES content facilitates stronger intermolecular interactions between the solvent system and the bioactive compound, which tends to enhance extraction yield.

Yue et al. [86] evaluated the recovery of stilbenes from chickpea foliage employing hydrophilic eutectic solvents derived from chlorides and hydrophobic ones formulated with thymol. Their findings indicated that HES outperformed hydrophilic ES due to structural similarities between HES and stilbenes, which enhanced intermolecular interactions. Furthermore, HES demonstrated lower viscosity, likely contributing to improved extraction efficiency. Among the different HES formulations tested, those based on coumarins and vanillin as hydrogen bond donors (HBD) exhibited superior yields compared to those using octanoic acid, decanoic acid, and camphor as HBD. This behavior is likely associated with the pronounced hydrogen bonding capacity and π–π stacking interactions between stilbenes and solvents based on thymol or coumarins.

Wang et al. [63] studied the use of ES to convert polydatin into resveratrol from *Polygonum cuspidatum* root. The 1:1 choline chloride and oxalic acid mixture achieved the best extraction performance, yielding 8.0 ± 0.03 mg/g. The authors attributed this to the high acidity of the eutectic mixture (pH = 1.23), which promoted polydatin hydrolysis, thereby increasing resveratrol yield. However, while ES formed by choline chloride and citric acid (1:1) also exhibited low pH, its high viscosity impeded the extraction process, resulting in a lower yield (2.5 mg/g). Notably, the study did not address the potential degradation of resveratrol at extremely acidic pH, which could compromise compound stability during extraction. The authors examined the impact of different water concentrations (10–50%) on extraction efficiency to further optimize the process. The highest yield (>7.5 mg/g) was observed at 30% water content, whereas higher concentrations (>30%) led to decreased yields (e.g., 50% water resulted in 5.8 mg/g). In addition, Chen et al. [88] reported that 30% water addition maximized resveratrol extraction (28.26 ± 0.77 mg/kg) from peanut roots using ES (choline chloride:1,4-butanediol). It is important to highlight that water concentrations below 30% were insufficient to fully dissolve the compounds due to high system viscosity, which restricted mass transfer. In contrast, excessive water content (>30%) weakened the interactions between ES and resveratrol, reducing extraction efficiency. While water addition lowers viscosity, it also alters solvent polarity, which in turn influences extraction yield.

Sun et al. [87] assessed how varying the temperature range (50–90 °C) affects the resveratrol recovery rate. An increase in temperature to 80 °C enhanced the extraction efficiency, resulting in a resveratrol yield exceeding 9.0 mg/g. However, using a higher temperature (90 °C) caused a reduction in the yield (>8.5 mg/g). In their studies, Wang et al. [102] verified the resveratrol yield employing a eutectic mixture of choline chloride and oxalic acid while adjusting the temperature within the 50–80 °C interval. The best yield, surpassing 9.5 mg/g, was obtained when the process was conducted at 75 °C, with a reduction in the resveratrol content in the extraction at 80 °C (<8.5 mg/g). It should be emphasized that, although higher temperatures help decrease ES viscosity and enhance both mass transfer and compound diffusion in the solvent, however, elevated thermal conditions may lead to resveratrol degradation due to the increased temperature. This can cause a decrease in the pH of ES. At low pH, resveratrol has low stability; that is, it becomes unstable and can be degraded [108]. In addition, further investigations are necessary to clarify how processing time affects the thermal stability of stilbenes extracted using ES.

9.3.5 Polymeric Compounds Phenolic

9.3.5.1 Tannins

Tannins represent one of the major phenolic classes widely found in plants, recognized as polymeric phenolic compounds (hydrolyzable and condensed tannins). These compounds have limited or no solubility in water, showing the importance of testing eutectic solvents, an environmentally friendly and viable option for the recovery of these active substances, as a replacement for conventional organic solvents [46, 55].

ES have gained attention as efficient systems for recovering tannins, standing out for their efficiency in solubilizing these compounds and for being sustainable options when compared with traditional solvents. Research findings confirm that both the composition of the eutectic solvent and the operational parameters are critical in determining the process efficiency, making it necessary to fine-tune the conditions to maximize yields [55].

The composition of ES significantly impacts the extraction of tannins, as polarity and viscosity of the solvent system are mainly governed by the interaction between hydrogen bond donors (HBD) and acceptors (HBA). Wang et al. [90] demonstrated that a natural eutectic system combining L-proline and lactic acid (Pro:La) was highly efficient in extracting total tannins from *Geum japonicum*, outperforming conventional solvents due to its excellent polar compatibility with target compounds. Similarly, Kovac et al. [46] identified that the ES formulated with choline chloride and urea in a 1:2 ratio proved to be the most effective in extracting hydrolyzable tannins (*Alchemilla vulgaris*), highlighting that the choice of HBD can directly influence the extraction yield. Adding water to ES can change its viscosity and extractability. Kovac et al. [46] observed that a water content of 30% significantly improved the extraction of hydrolyzable tannins and phenolic acids due to reduced solvent viscosity and increased mass transfer. Nevertheless, when the water content exceeded the optimal level, when increased to 50%, a reduction in extraction efficiency was observed for some compounds, probably due to the dilution of the interactions between ES and tannins. This behavior was corroborated by According to Wang et al. [90], incorporating 29% water into NAES (Pro:La) represented the optimal condition for extracting total tannins, while higher levels compromised the efficiency of the process. The reduction in solvent viscosity with the moderate water incorporation enhances the migration of compounds from the plant matrix into the extracting medium. Still, excessive amounts can reduce interactions between tannins and ES components.

The use of auxiliary extraction techniques, including ultrasound (UAE), microwave (MAE), and infrared (IR), has demonstrated positive effects on tannin extraction when combined with ES. In the study by Sukor et al. [91], ES-mediated UAE showed a higher yield than methanol in the recovery of tannic acid from onion skins. The 10% duty cycle was ideal for maximizing extraction, while higher intensities contributed to compound degradation. Rajha et al. [109] demonstrated that infrared (IR)-assisted extraction was the most efficient in extracting tannins from pomegranate peel, surpassing solid-liquid extraction and UAE. The use of IR resulted in an excellent breakdown of cellular structures and better facilitated solvent infiltration into plant structures, promoting the release of target molecules. Furthermore, Imtiaz et al. [89] demonstrated the efficiency of MAE in isolating tannins from *Cichorium intybus*, highlighting that the power and extraction time were critical factors in avoiding thermal degradation of the compounds.

A central question in ES research is its effectiveness relative to conventional solvents. Wang et al. [90] evidenced that total tannin extraction from *Geum japonicum* using NAES (Pro:La) resulted in higher yields than ethanol extraction. Kovac et al. [46] reported that the eutectic system composed of choline chloride and urea

(1:2) achieved superior yields of hydrolyzable tannins when compared to aqueous ethanol. Rajha et al. [109] also compared ES with ethanol and methanol for recovering tannins present in pomegranate peel, concluding that selected ES presented comparable or superior efficiency to organic solvents, with the advantage of being biodegradable and less toxic.

The reviewed studies indicate that tannin extraction using ES is a promising approach for industrial applications, especially in replacing toxic solvents and obtaining more selective extracts with more excellent bioactivity. Imtiaz et al. [89] highlighted the possibility of scaling extraction technology based on ES, mainly when associated with microwaves, due to the reduced extraction time and less degradation of the compounds. Furthermore, Wang et al. [90] suggest that the application of NAES could be considered in the formulation of new hypoglycemic drugs since the extracted tannins demonstrated vigorous inhibitory activity against α-glucosidase. This application reinforces the ability of ES to yield bioactive constituents relevant to both pharmaceutical and nutraceutical fields.

The studies analyzed demonstrate that ES have several advantages in the extraction of tannins, including better selectivity, greater compatibility with phenolic compounds, and lower environmental impact than conventional solvents. ES composition, water content, operational parameters, and auxiliary techniques such as ultrasound, microwave, and infrared strongly influence extraction efficiency. Although the results are promising, challenges such as the standardization of ES and assessing extract stability remain a necessary step so that the application of these solvents becomes viable on an industrial scale. Future research should explore the stability of the extracted tannins and the bioactivity and potential application of these compounds in different sectors, such as pharmaceuticals, cosmetics, and food.

9.3.5.2 Lignin

In comparison to other classes of phenolic compounds, lignin extraction has very particular characteristics. It is not an exaggeration to attribute much of the growing interest in eutectic solvents to their interaction with lignocellulosic biomass, mainly due to their ability to solubilize lignin [110]. In recent decades, considering the intense race for developing environmentally sustainable strategies, many studies have focused on establishing green and sustainable methods for lignin extraction. In this sense, using eutectic solvents as an extraction solvent aligns well, considering their eco-friendly properties [110].

Lignin extraction using different eutectic solvents has been demonstrated in several studies with varying matrices of plants, such as corncob [92, 111], wheat straw [112, 113], eucalyptus [114], poplar [93], and a variety of other lignocellulosic biomasses (Table 9.2). Most of the studies use eutectic solvents with choline chloride (ChCl) or amino acids, such as betaine and proline, as HBAs associated with different types of HBDs, commonly carboxylic acids, polyols, and amines [115].

During lignin extraction with eutectic solvents, both the yield and the properties/characteristics of the extracted lignin are affected by many factors, especially the type of eutectic solvent, extraction time and temperature, the HBA/HBD molar ratio, and the addition of catalysts or enhancers [116]. The presence and content of water are key factors that significantly influence the extraction performance. The nature and quantity of functional groups present in the components of these solvents interact with lignin; this impacts both the amount of lignin recovered and its chemical and physical properties [116]. Eutectic solvents using carboxylic acids as HBD generally perform better in lignin extraction than those containing other functional groups. Several studies report lignin removal rates higher than 90%. For example, Zhang et al. [92] achieved 98.5% delignification of corncob using ChCl:oxalic acid as the solvent. Chen et al. [117] removed 98.1% of lignin from poplar using ChCl:lactic acid as the solvent. The same study evaluated the number of functional groups using eutectic solvents with succinic acid, a dicarboxylic compound, and citric acid, classified as tricarboxylic. The authors observed that the monocarboxylic acid's extraction capacity was higher than that of eutectic solvents with more functionalized acids. Tan et al. [118] also demonstrated this trend, observing a decrease in extraction yield using ChCl:lactic acid (33.5%) compared to ChCl:malic acid (22.9%) and ChCl:citric acid (10.7%), all at the same molar ratio. The high delignification capacity of eutectic solvents containing carboxylic acids is a result of the increased availability of proton-donating groups, which promote the cleavage of lignin-carbohydrate complex bonds [116]. Another advantage of this type of solvent is the high purity achieved. For example, Alvarez-Vasco et al. [93] obtained 95% lignin purity from poplar using a eutectic mixture of choline chloride and lactic acid as the solvent.

Compared to eutectic solvents with carboxylic acids as HBD, those containing polyols (typically glycerol and ethylene glycol) or carbohydrates (e.g., glucose, fructose, xylitol) as HBD tend to have lower extraction efficiency. This phenomenon may be attributed to the neutrality of hydroxyl groups compared to carboxyl groups, which makes lignin dissolution more difficult [116]. To achieve high extraction yields, treatment with this type of solvent must be significantly more severe [119]. For example, to achieve 59% lignin removal from corncob, Procentese et al. [112] treated the biomass with ChCl:glycerol for 15 hours at 150 °C. Xu et al. [95] achieved similar delignification for corn stover using the same eutectic solvent for 2 hours at 180 °C.

Despite the high extraction efficiency and purity of the obtained lignin, treatment with eutectic solvents containing carboxylic acids causes greater chemical modification in the lignin structure, especially in the β-O-4′ bond content, leading to a higher degree of condensation [120]. On the other hand, despite having a lower yield, the use of polyol-based eutectic solvents has a protective effect. It tends to produce extracted lignins with greater retention of their original characteristics, especially β-O-4′ bonds [115].

Extraction time and temperature have similar effects on lignin extraction. In general, longer times and higher temperatures favor greater extraction [121]. Higher temperatures tend to decrease eutectic solvents' viscosity and surface tension, improving their diffusivity and, consequently, their interaction with biomass [121]. Mankar et al. [122] investigated the influence of temperature on lignin extraction from coconut husk using ChCl:lactic acid eutectic solvents, observing an increase from 10.9% at 90 °C to 73.9% at 150 °C. The same study evaluated extraction time between 5 and 40 minutes, with yields increasing from 35.1% (5 min) to 73.9% (20 min). However, the improvement in extraction with increasing time and temperature has limits. Excessively high temperatures or prolonged extraction processes can lead to greater lignin condensation/recondensation, significantly altering the properties of the extracted products. Moreover, increasing these parameters is associated with higher costs and reduced economic feasibility.

The addition of water as a catalyst significantly enhances the efficiency of lignin extraction. For example, Kumar et al. [94] investigated how the addition of water affects lignin extraction and observed that incorporation 5 wt% water into the ChCl:lactic acid solvent enhanced lignin extraction from rice straw by 22%. The presence of water can notably decrease the viscosity of certain eutectic solvents, which facilitates mass transfer and thereby improves the delignification process [116].

9.4 Conclusion and Future Perspectives

Applying eutectic solvents (ES) to extract phenolic compounds presents a sustainable and efficient alternative to conventional organic solvents. The studies analyzed in this chapter demonstrate that ES can significantly improve extraction efficiency, selectivity, and compound stability while minimizing environmental impact. However, several challenges remain, including the standardization of ES formulations, optimization of extraction conditions, and economic feasibility for large-scale applications.

Further investigations should aim at developing novel ES with enhanced selectivity for specific phenolic subclasses and investigating their recyclability to improve process sustainability. Additionally, integrating ES with advanced extraction technologies such as eutectic solvent-modified solid-phase extraction and enzyme-assisted extraction could further enhance yields and bioavailability. Furthermore, a deeper understanding of ES interactions with plant matrices and bioactive compounds will be crucial for optimizing industrial applications. Expanding the use of ES in pharmaceutical and nutraceutical industries, where purity and bioactivity retention are critical, is another promising avenue. By addressing these challenges, ES-based extraction methods can be refined to support the transition toward greener, more sustainable extraction technologies in multiple industries.

AI Declaration While preparing this work, the authors utilized the AI tools to enhance the language and readability of the manuscript text. After using this tool/service, the authors reviewed and edited the content as needed and take full responsibility for the publication content.

Conflict of Interest The authors declare no conflict of interest.

References

1. Alara OR, Abdurahman NH, Ukaegbu CI (2021) Extraction of phenolic compounds: a review. Curr Res Food Sci 4:200–214. https://doi.org/10.1016/j.crfs.2021.03.011
2. Alu'datt MH, Rababah T, Alhamad MN, Al-Mahasneh MA, Almajwal A, Gammoh S, Ereifej K, Johargy A, Alli I (2017) A review of phenolic compounds in oil-bearing plants: distribution, identification and occurrence of phenolic compounds. Food Chem 218:99–106. https://doi.org/10.1016/j.foodchem.2016.09.057
3. Singh B, Singh JP, Kaur A, Singh N (2017) Phenolic composition and antioxidant potential of grain legume seeds: a review. Food Res Int 101:1–16. https://doi.org/10.1016/j.foodres.2017.09.026
4. Lima TSP, Buarque FS, Pinto REP, Vasconcelos ALDS, Soares CMF, Souza RL, Lima ÁS (2024) Water-free two-phase systems: phase equilibria and partitioning of vitamins from tomatoes (Solanum lycopersicum). J Chem Technol Biotechnol 99:437–445. https://doi.org/10.1002/jctb.7545
5. Inocêncio ES, Buarque FS, Ferreira LF, Soares CM, Lima ÁS, Souza RLD (2025) Exploring the potential of Licuri (Syagrus coronata) using sustainable techniques and solvents for extracting bioactive compounds. Sustainability 17:1507–1520. https://doi.org/10.3390/su17041507
6. Lianza M, Marincich L, Antognoni F (2022) The greening of anthocyanins: eco-friendly techniques for their recovery from Agri-food by-products. Antioxidants 11:2169. https://doi.org/10.3390/antiox11112169
7. Palos-Hernández A, Gutiérrez Fernández MY, Escuadra Burrieza J, Pérez-Iglesias JL, Gonzalez-Paramas AM (2022) Obtaining green extracts rich in phenolic compounds from underexploited food by-products using natural deep eutectic solvents: opportunities and challenges. Sustain Chem Pharm 29:100773. https://doi.org/10.1016/j.scp.2022.100773
8. Bashir I, Dar AH, Dash KK, Pandey VK, Fayaz U, Shams R, Srivastava S, Singh R (2023) Deep eutectic solvents for extraction of functional components from plant-based products: a promising approach. Sustain Chem Pharm 33:101102. https://doi.org/10.1016/j.scp.2023.101102
9. Bidwell C, Indra S, Luke C, Kakoma MK (2023) A review of modern and conventional extraction techniques and their applications for extracting phytochemicals from plants. Scientific African 19:e01585. https://doi.org/10.1016/j.sciaf.2023.e01585
10. Buarque FS, Gautério GV, Coelho MAZ, Lemes AC, Ribeiro BD (2022) Aqueous two-phase systems based on ionic liquids and deep eutectic solvents as a tool for the recovery of non-protein bioactive compounds—a review. PRO 11:31–64. https://doi.org/10.3390/pr11010031
11. Souza IN, Rodrigues LC, Soares CM, Buarque FS, Souza RL, Lima ÁS (2024) Deep eutectic solvent-based aqueous two-phase systems and their application in partitioning of phenol compounds. Molecules 29:4383. https://doi.org/10.3390/molecules29184383
12. Santos JPJ, Borges MM, Lima TSP, Camêlo LCA, Costa F, Soares CMF, Lima ÁS (2024) Integrative process of extraction and purification of tagitinin C using ethanolic two-phase system based on protic ionic liquids. J Mol Liq 414:126007. https://doi.org/10.1016/j.molliq.2024.126007
13. Buarque FS, Ribeiro BD, Freire MG, Coelho MA, Pereira MM (2025) Assessing the role of deep eutectic solvents in Yarrowia lipolytica inhibition. J Biotechnol 398:1–10. https://doi.org/10.1016/j.jbiotec.2024.11.016
14. Cannavacciuolo C, Pagliari S, Frigerio J, Giustra CM, Labra M, Campone L (2023) Natural deep eutectic solvents (NADESs) combined with sustainable extraction techniques: a review of the green chemistry approach in food analysis. Foods 12:56. https://doi.org/10.3390/foods12010056
15. Xue H, Tan J, Li Q, Tang J, Cai X (2020) Optimization ultrasound-assisted deep eutectic solvent extraction of anthocyanins from raspberry using response surface methodology coupled with genetic algorithm. Foods 9(10):1409. https://doi.org/10.3390/foods9101409

16. Socas-Rodríguez B, Santana-Mayor Á, Herrera-Herrera AV, Rodríguez-Delgado MÁ (2020) Deep eutectic solvents. In: Green sustainable process for chemical and environmental engineering and science. Elsevier, pp 123–177. https://doi.org/10.1016/B978-0-12-817386-2.00005-6
17. Cvjetko-Bubalo M, Vidović S, Radojčić Redovniković I, Jokić S (2018) New perspective in extraction of plant biologically active compounds by green solvents. Food Bioprod Process 109:52–73. https://doi.org/10.1016/j.fbp.2018.03.001
18. Smith EL, Abbott AP, Ryder KS (2014) Deep eutectic solvents (DESs) and their applications. Chem Rev 114:11060–11082. dx.doi.org. https://doi.org/10.1021/cr300162p
19. Vázquez-González M, Fernández-Prior Á, Bermúdez Oria A, Rodríguez-Juan EM, Pérez-Rubio AG, Fernández-Bolaños J, Rodríguez-Gutiérrez G (2020) Utilization of strawberry and raspberry waste for the extraction of bioactive compounds by deep eutectic solvents. LWT 130:109645. https://doi.org/10.1016/j.lwt.2020.109645
20. Aziz A, Noreen S, Khalid W, Mubarik F, Niazi MK, Koraqi H (2022) Extraction of bioactive compounds from different vegetable sprouts and their potential role in the formulation of functional foods against various disorders: a literature-based review. Molecules (Basel, Switzerland) 27:7320. https://doi.org/10.3390/molecules27217320
21. Koraqi H, Qazimi B, Çesko C, Trajkovska Petkoska A (2022) Environmentally friendly extraction of bioactive compounds from Rosa canina L. fruits using deep eutectic solvent (DES) as green extraction media. Acta Chim Slov 69:3. https://doi.org/10.17344/acsi.2022.7559
22. Ling JKU, Hadinoto K (2022) Deep eutectic solvent as green solvent in extraction of biological macromolecules: a review. Int J Mol Sci 23(6):3381. https://doi.org/10.3390/ijms23063381
23. Choi YH, Verpoorte R (2019) Green solvents for the extraction of bioactive compounds from natural products using ionic liquids and deep eutectic solvents. Curr Opin Food Sci 26:87–93. https://doi.org/10.1016/j.cofs.2019.04.003
24. Buarque FS, Silva SAME, Ribeiro BD (2023) Choline chloride-based deep eutectic solvent as an inhibitor of metalloproteases (collagenase and elastase) in cosmetic formulation. 3 Biotech 13:219. https://doi.org/10.1007/s13205-023-03602-2
25. Vanda H, Dai Y, Wilson EG, Verpoorte R, Choi YH (2018) Green solvents from ionic liquids and deep eutectic solvents to natural deep eutectic solvents. C R Chim 21:628–638. https://doi.org/10.1016/j.crci.2018.04.002
26. Ahmad I, Pertiwi AS, Kembaren YH, Rahman A, Munâ A (2018) Application of natural deep eutectic solvent-based ultrasonic assisted extraction of total polyphenolic and caffeine content from Coffe beans (Coffea beans L.) for instant food products. J Appl Pharm Sci 8:138–143. https://doi.org/10.7324/JAPS.2018.8819
27. El Kantar S, Rajha HN, Boussetta N, Vorobiev E, Maroun RG, Louka N (2019) Green extraction of polyphenols from grapefruit peels using high voltage electrical discharges, deep eutectic solvents and aqueous glycerol. Food Chem 295:165–171. https://doi.org/10.1016/j.foodchem.2019.05.111
28. Gao C, Cai C, Liu J, Wang Y, Chen Y, Wang L, Tan Z (2020) Extraction and preliminary purification of polysaccharides from Camellia oleifera Abel. Seed cake using a thermoseparating aqueous two-phase system based on EOPO copolymer and deep eutectic solvents. Food Chem 313:126164. https://doi.org/10.1016/j.foodchem.2020.126164
29. Cvetanović A (2019) Extractions without organic solvents: advantages and disadvantages. Chem Afr 2:343–349. https://doi.org/10.1007/s42250-019-00070-1
30. Fernández MA, Boiteux J, Espino M, Gomez FJV, Silva MF (2018) Natural deep eutectic solvents-mediated extractions: the way forward for sustainable analytical developments. Anal Chim Acta 1038:1–10. https://doi.org/10.1016/j.aca.2018.07.059
31. Liu W, Cui X, Zhong Y, Ma R, Liu B, Xia Y (2023) Phenolic metabolites as therapeutic in inflammation and neoplasms: molecular pathways explaining their efficacy. Pharmacol Res 193:106812. https://doi.org/10.1016/j.phrs.2023.106812
32. Isci A, Kaltschmitt M (2022) Recovery and recycling of deep eutectic solvents in biomass conversions: a review. Biomass Convers Biorefinery 12:197–226. https://doi.org/10.1007/s13399-021-01860-9

33. Rocchetti G, Gregorio RP, Lorenzo JM, Barba FJ, Oliveira PG, Prieto MA, Simal-Gandara J, Mosele JI, Motilva M, Tomas M, Patrone V, Capanoglu E, Lucini L (2022) Functional implications of bound phenolic compounds and phenolics–food interaction: a review. Comp Rev Food Sci Food Safe 21:811–842. https://doi.org/10.1111/1541-4337.12921
34. De Araújo FF, De Paulo FD, Neri-Numa IA, Pastore GM (2021) Polyphenols and their applications: an approach in food chemistry and innovation potential. Food Chem 338:127535. https://doi.org/10.1016/j.foodchem.2020.127535
35. Gil-Martín E, Forbes-Hernández T, Romero A, Cianciosi D, Giampieri F, Battino M (2022) Influence of the extraction method on the recovery of bioactive phenolic compounds from food industry by-products. Food Chem 378:131918. https://doi.org/10.1016/j.foodchem.2021.131918
36. Santos A, Gomes PGC, Buarque FS, Soares CMF, Bjerk TR, Lima ÁS (2023) Novel strategy for extraction and partitioning of phenol compounds from industrial residue of seriguela (Spondia purpurea L.) using aqueous two-phase systems. Food Bioprod Process 141:219–229. https://doi.org/10.1016/j.fbp.2023.08.011
37. Buarque FS, Guimarães DE, Soares CM, Souza RL, Pereira MM, Lima ÁS (2021) Ethanolic two-phase system formed by polypropylene glycol, ethylene glycol and/or ionic liquid (phase-forming or adjuvant) as a platform to phase separation and partitioning study. J Mol Liq 344:117702. https://doi.org/10.1016/j.molliq.2021.117702
38. Albuquerque BR, Heleno SA, Oliveira MBPP, Barros L, Ferreira ICFR (2021) Phenolic compounds: current industrial applications, limitations and future challenges. Food Funct 12:14–29. https://doi.org/10.1039/D0FO02324H
39. Ratnavathi CV (2019) Grain structure, quality, and nutrition. In: Breeding Sorghum for diverse end uses. Elsevier, pp 193–207. https://doi.org/10.1016/B978-0-08-101879-8.00012-7
40. Kumar N, Goel N (2019) Phenolic acids: natural versatile molecules with promising therapeutic applications. Biotechnol Rep 24:e00370. https://doi.org/10.1016/j.btre.2019.e00370
41. Rashmi HB, Negi PS (2020) Phenolic acids from vegetables: a review on processing stability and health benefits. Food Res Int 136:109298. https://doi.org/10.1016/j.foodres.2020.109298
42. Córdova MLF, Medina AR (2014) Analytical methods for determination of polyphenols in beer. In: Processing and impact on antioxidants in beverages. Elsevier, pp 289–299
43. Angeloni S, Nzekoue FK, Navarini L, Sagratini G, Torregiani E, Vittori S, Caprioli G (2020) An analytical method for the simultaneous quantification of 30 bioactive compounds in spent coffee ground by HPLC-MS/MS. J Mass Spectrom 55:e4519. https://doi.org/10.1002/jms.4519
44. Liu Y, Liu H-Y, Yang X, Zhu F, Wu D-T, Li H-B, Gan R-Y (2022) Green extraction, chemical composition, and in vitro antioxidant activity of theabrownins from Kangzhuan dark tea. Curr Res Food Sci 5:1944–1954. https://doi.org/10.1016/j.crfs.2022.10.019
45. Ivanović M, Albreht A, Krajnc P, Vovk I, Razboršek MI (2021) Sustainable ultrasound-assisted extraction of valuable phenolics from inflorescences of Helichrysum arenarium L. using natural deep eutectic solvents. Ind Crop Prod 160:113102. https://doi.org/10.1016/j.indcrop.2020.113102
46. Kovač MJ, Jokić S, Jerković I, Molnar M (2022) Optimization of deep eutectic solvent extraction of phenolic acids and tannins from Alchemilla vulgaris L. Plants 11:474. https://doi.org/10.3390/plants11040474
47. Mitrović D, Sredović Ignjatović I, Kozarski M, Popović- Đorđević J (2024) Wine is more than just a beverage: chemical diversity, health benefits, and immunomodulating potential of wine polyphenols. Food Safety Health 2:196–212. https://doi.org/10.1002/fsh3.12036
48. Shen N, Wang T, Gan Q, Liu S, Wang L, Jin B (2022) Plant flavonoids: classification, distribution, biosynthesis, and antioxidant activity. Food Chem 383:132531. https://doi.org/10.1016/j.foodchem.2022.132531
49. Williamson G, Kay CD, Crozier A (2018) The bioavailability, transport, and bioactivity of dietary flavonoids: a review from a historical perspective. Comp Rev Food Sci Food Safe 17:1054–1112. https://doi.org/10.1111/1541-4337.12351

50. Mattioli R, Francioso A, Mosca L, Silva P (2020) Anthocyanins: a comprehensive review of their chemical properties and health effects on cardiovascular and neurodegenerative diseases. Molecules 25:3809. https://doi.org/10.3390/molecules25173809
51. Khoo HE, Azlan A, Tang ST, Lim SM (2017) Anthocyanidins and anthocyanins: colored pigments as food, pharmaceutical ingredients, and the potential health benefits. Food Nutr Res 61:1361779. https://doi.org/10.1080/16546628.2017.1361779
52. Pecyna P, Wargula J, Murias M, Kucinska M (2020) More than resveratrol: new insights into stilbene-based compounds. Biomolecules 10:1111. https://doi.org/10.3390/biom10081111
53. Teka T, Zhang L, Ge X, Li Y, Han L, Yan X (2022) Stilbenes: source plants, chemistry, biosynthesis, pharmacology, application and problems related to their clinical application-A comprehensive review. Phytochemistry 197:113128. https://doi.org/10.1016/j.phytochem.2022.113128
54. Farha AK, Yang Q-Q, Kim G, Li H-B, Zhu F, Liu H-Y, Gan R-Y, Corke H (2020) Tannins as an alternative to antibiotics. Food Biosci 38:100751. https://doi.org/10.1016/j.fbio.2020.100751
55. Das AK, Islam MN, Faruk MO, Ashaduzzaman M, Dungani R (2020) Review on tannins: extraction processes, applications and possibilities. S Afr J Bot 135:58–70. https://doi.org/10.1016/j.sajb.2020.08.008
56. Besharati M, Maggiolino A, Palangi V, Kaya A, Jabbar M, Eseceli H, De Palo P, Lorenzo JM (2022) Tannin in ruminant nutrition: review. Molecules 27:8273. https://doi.org/10.3390/molecules27238273
57. Rauf A, Imran M, Abu-Izneid T, Iahtisham-Ul-Haq PS, Pan X, Naz S, Sanches Silva A, Saeed F, Rasul Suleria HA (2019) Proanthocyanidins: a comprehensive review. Biomed Pharmacother 116:108999. https://doi.org/10.1016/j.biopha.2019.108999
58. Chen F (2019b) Lignin biosynthesis and control of lignin composition. In: Lignin: biosynthesis, functions and economic significance. Biochemistry Research Trends
59. Singh M, Lee SC, Won K (2024) Lignin phenolation by graft copolymerization to boost its reactivity. Int J Biol Macromol 266:131258. https://doi.org/10.1016/j.ijbiomac.2024.131258
60. Calvo-Flores FG, Dobado JA, Isac-Gárcia J, Martín-Martínez FJ (2015) Lignin and lignans as renewable raw materials: chemistry, technology and applications. Wiley, Chichester
61. Weng J, Chapple C (2010) The origin and evolution of lignin biosynthesis. New Phytol 187:273–285. https://doi.org/10.1111/j.1469-8137.2010.03327.x
62. Slabon A, Rodrigues BVM (2025) To break, or not to break: is selective depolymerization of lignin a *Riemann hypothesis* rather than a solution? Green Chem 27:2178–2183. https://doi.org/10.1039/D4GC05439C
63. Wang J-D, Fu L-N, Wang L-T, Cai Z-H, Wang Y-Q, Yang Q, Fu Y-J (2021) Simultaneous transformation and extraction of resveratrol from Polygonum cuspidatum using acidic natural deep eutectic solvent. Ind Crop Prod 173:114140. https://doi.org/10.1016/j.indcrop.2021.114140
64. Xu C, Ferdosian F (2017) Conversion of lignin into bio-based chemicals and materials. Springer, Berlin Heidelberg, Berlin, Heidelberg
65. Oliveira Santos RP, Ramos LA, Frollini E (2019) Cellulose and/or lignin in fiber-aligned electrospun PET mats: the influence on materials end-properties. Cellulose 26:617–630. https://doi.org/10.1007/s10570-018-02234-7
66. Wendisch VF, Kim Y, Lee J-H (2018) Chemicals from lignin: recent depolymerization techniques and upgrading extended pathways. Curr Opin Green Sustain Chem 14:33–39. https://doi.org/10.1016/j.cogsc.2018.05.006
67. Amezcua-Allieri MA, Aburto J (2018) Conversion of lignin to heat and power, chemicals or fuels into the transition energy strategy. In: Poletto M (ed) Lignin–trends and applications. InTech. https://doi.org/10.5772/intechopen.71211
68. Kumar R, Butreddy A, Kommineni N, Reddy PG, Bunekar N, Sarkar C, Dutt S, Mishra VK, Aadil KR, Mishra YK, Oupicky D, Kaushik A (2021) Lignin: drug/gene delivery and tissue engineering applications. IJN 16:2419–2441. https://doi.org/10.2147/IJN.S303462
69. Duan L, Zhang C, Zhang C, Xue Z, Zheng Y, Guo L (2019) Green extraction of phenolic acids from Artemisia argyi leaves by tailor-made ternary deep eutectic solvents. Molecules 24:2842. https://doi.org/10.3390/molecules24152842

70. Ozkan G (2024) Valorization of artichoke outer petals by using ultrasound-assisted extraction and natural deep eutectic solvents (NADES) for the recovery of phenolic compounds. J Sci Food Agric 104:2744–2749. https://doi.org/10.1002/jsfa.13158
71. Pizani RS, Viganó J, Contieri LS, Strieder MM, Kamikawashi RK, Vilegas W, De Souza Mesquita LM, Rostagno MA (2024) New selective and sustainable ultrasound-assisted extraction procedure to recover carnosic and rosmarinic acids from Rosmarinus officinalis by sequential use of bio-based solvents. Food Chem 435:137540. https://doi.org/10.1016/j.foodchem.2023.137540
72. Herrera-Beltrán IV, Méndez-Torruco M, Matus MH, Domínguez Z (2023) Study of the chlorogenic acid extraction with choline chloride-based deep eutectic solvents and its non-covalent interactions analysis. J Physical Organic Chem 36:e4560. https://doi.org/10.1002/poc.4560
73. Vasyliev G, Lyudmyla K, Hladun K, Skiba M, Vorobyova V (2022) Valorization of tomato pomace: extraction of value-added components by deep eutectic solvents and their application in the formulation of cosmetic emulsions. Biomass Conv Bioref 12:95–111. https://doi.org/10.1007/s13399-022-02337-z
74. Vo TP, Pham TV, Weina K, Tran TNH, Vo LTV, Nguyen PT, Bui TLH, Phan TH, Nguyen DQ (2023) Green extraction of phenolics and flavonoids from black mulberry fruit using natural deep eutectic solvents: optimization and surface morphology. BMC Chem 17:119. https://doi.org/10.1186/s13065-023-01041-x
75. Lei J, Wang Y, Li W, Fu S, Zhou J, Lu D, Wang C, Sheng X, Zhang M, Xiao S, Sun C, Wang G (2022) Natural green deep eutectic solvents-based eco-friendly and efficient extraction of flavonoids from Selaginella moellendorffii: process optimization, composition identification and biological activity. Sep Purif Technol 283:120203. https://doi.org/10.1016/j.seppur.2021.120203
76. Li J, Chen W, Niu D, Wang R, Xu F-Y, Chen B-R, Lin J-W, Tang Z-S, Zeng X-A (2022) Efficient and green strategy based on pulsed electric field coupled with deep eutectic solvents for recovering flavonoids and preparing flavonoid aglycones from noni-processing wastes. J Clean Prod 368:133019. https://doi.org/10.1016/j.jclepro.2022.133019
77. Wang W, Xiao S-Q, Li L-Y, Gai Q-Y (2023) Deep eutectic solvent-based microwave-assisted extraction for the extraction of seven Main flavonoids from Ribes mandshuricum (maxim.). Kom Leaves Separations 10:191. https://doi.org/10.3390/separations10030191
78. Wang Y-Y, Peng C, Zhang Y, Wang Z-R, Chen Y-M, Dong J-F, Xiao M-L, Li D-L, Li W, Zou Q-J, Zhang K, Wei P (2022) Optimization, identification and bioactivity of flavonoids extracted from Moringa oleifera leaves by deep eutectic solvent. Food Biosci 47:101687. https://doi.org/10.1016/j.fbio.2022.101687
79. Xu M, Ran L, Chen N, Fan X, Ren D, Yi L (2019) Polarity-dependent extraction of flavonoids from citrus peel waste using a tailor-made deep eutectic solvent. Food Chem 297:124970. https://doi.org/10.1016/j.foodchem.2019.124970
80. Thakur R, Gupta V, Dhar P, Deka SC, Das AB (2022) Ultrasound-assisted extraction of anthocyanin from black rice bran using natural deep eutectic solvents: optimization, diffusivity, and stability. J Food Process Preserv 46(3):16309. https://doi.org/10.1111/jfpp.16309
81. Han X, Huang J, Cai J, Yu C, Jiang S, Ye Q, Yang S, Pan W, Zhang Q, Wang Y (2023) Microwave-ultrasonic technique development coupled with natural deep eutectic solvents in anthocyanin extraction from perilla leaves (Perilla frutescens var. acuta). J Sci Food Agric 103, 6:3006–3016. https://doi.org/10.1002/jsfa.12343
82. Jovanovic MS, Krgovic N, Radan M, Cujic-Nikolic N, Mudric J, Lazarevic Z, Savikin K (2023) Natural deep eutectic solvents combined with cyclodextrins: a novel strategy for chokeberry anthocyanins extraction. Food Chem 405:134816. https://doi.org/10.1016/j.foodchem.2022.134816
83. Guo N, Jiang Y-W, Wang L-T, Niu L-J, Liu Z-M, Fu Y-J (2019) Natural deep eutectic solvents couple with integrative extraction technique as an effective approach for mulberry anthocyanin extraction. Food Chem 296:78–85. https://doi.org/10.1016/j.foodchem.2019.05.196

84. Jovanovic MS, Krgovic N, Zivkovic J, Stevic T, Zdunic G, Bigovic D, Savikin K (2022) Ultrasound-assisted natural deep eutectic solvents extraction of bilberry anthocyanins: optimization, bioactivities, and storage stability. Plants 11(20):2680. https://doi.org/10.3390/plants11202680
85. Velásquez P, Bustos D, Montenegro G, Giordano A (2021) Ultrasound-assisted extraction of anthocyanins using natural deep eutectic solvents and their incorporation in edible films. Molecules 26(4):984. https://doi.org/10.3390/molecules26040984
86. Yue G-L, Chang Y-H, Yu L, Xu J, Wan N, Duan C-J, Zhang Y, Fu Y-J (2023) Extraction and separation characteristic stilbene compounds from pigeon pea leaves using V-type thymol-based natural deep eutectic solvent systems. Chem Eng Res Des 193:600–612. https://doi.org/10.1016/j.cherd.2023.04.004
87. Sun B, Zheng Y-L, Yang S-K, Zhang J-R, Cheng X-Y, Ghiladi R, Ma Z, Wang J, Deng W-W (2021) One-pot method based on deep eutectic solvent for extraction and conversion of polydatin to resveratrol from Polygonum cuspidatum. Food Chem 343:128498. https://doi.org/10.1016/j.foodchem.2020.128498
88. Chen J, Jiang X, Yang G, Bi Y, Liu W (2018) Green and efficient extraction of resveratrol from peanut roots using deep eutectic solvents. J Chem 9:1. https://doi.org/10.1155/2018/4091930
89. Imtiaz F, Saif Z, Sajid A, Nazir A, Manzoor Q, Saleem A, Nisa M, Farooq A, Al-Mijalli SH, Iqbal M (2024) Role of glycerol-based deep eutectic solvents for extraction of phytochemicals from Cichorium intybus seeds: optimization by response surface methodology. Microchem J 199:110083. https://doi.org/10.1016/j.microc.2024.110083
90. Wang Z, Wang S, Zhang Y, Bi W (2024) Switching from deep eutectic solvents to deep eutectic systems for natural product extraction. Green Chem Eng 6:36. https://doi.org/10.1016/j.gce.2024.05.002
91. Sukor NF, Panner Selvam V, Jusoh R, Kamarudin NS, Abd Rahim S (2021) Intensified DES mediated ultrasound extraction of tannic acid from onion peel. J Food Eng 296:110437. https://doi.org/10.1016/j.jfoodeng.2020.110437
92. Zhang C-W, Xia S-Q, Ma P-S (2016) Facile pretreatment of lignocellulosic biomass using deep eutectic solvents. Bioresour Technol 219:1–5. https://doi.org/10.1016/j.biortech.2016.07.026
93. Alvarez-Vasco C, Ma R, Quintero M, Guo M, Geleynse S, Ramasamy KK, Wolcott M, Zhang X (2016) Unique low-molecular-weight lignin with high purity extracted from wood by deep eutectic solvents (DES): a source of lignin for valorization. Green Chem 18:5133–5141. https://doi.org/10.1039/C6GC01007E
94. Kumar AK, Parikh BS, Pravakar M (2016) Natural deep eutectic solvent mediated pretreatment of rice straw: bioanalytical characterization of lignin extract and enzymatic hydrolysis of pretreated biomass residue. Environ Sci Pollut Res 23:9265–9275. https://doi.org/10.1007/s11356-015-4780-4
95. Xu F, Sun J, Wehrs M, Kim KH, Rau SS, Chan AM, Simmons BA, Mukhopadhyay A, Singh S (2018) Biocompatible choline-based deep eutectic solvents enable one-pot production of cellulosic ethanol. ACS Sustain Chem Eng 6:8914–8919. https://doi.org/10.1021/acssuschemeng.8b01271
96. Hossain MA, Rahaman MS, Yelle D, Shang H, Sun Z, Renneckar S, Dong J, Tulaphol S, Sathitsuksanoh N (2021) Effects of polyol-based deep eutectic solvents on the efficiency of rice straw enzymatic hydrolysis. Ind Crop Prod 167:113480. https://doi.org/10.1016/j.indcrop.2021.113480
97. Alrugaibah M, Washington TL, Yagiz Y, Gu L (2021) Ultrasound-assisted extraction of phenolic acids, flavonols, and flavan-3-ols from muscadine grape skins and seeds using natural deep eutectic solvents and predictive modelling by artificial neural networking. Ultrason Sonochem 79:105773. https://doi.org/10.1016/j.ultsonch.2021.105773
98. Mansinhos I, Gonçalves S, Rodríguez-Solana R, Ordóñez-Díaz JL, Moreno-Rojas JM, Romano A (2021) Ultrasonic-assisted extraction and natural deep eutectic solvents combination: a green strategy to improve the recovery of phenolic compounds from Lavandula pedunculata subsp. lusitanica (Chaytor) Franco. Antioxidants 10:582. https://doi.org/10.3390/antiox10040582

99. Gao H, Wang Y, Guo Z, Liu Y, Wu Q, Xiao J (2022) Optimization of ultrasound-assisted extraction of phenolics from Asparagopsis taxiformis with deep eutectic solvent and their characterization by ultra-high-performance liquid chromatography-mass spectrometry. Front Nutr 9:1036436. https://doi.org/10.3389/fnut.2022.1036436
100. Liu Y, Zhe W, Zhang R, Peng Z, Wang Y, Gao H, Guo Z, Xiao J (2022) Ultrasonic-assisted extraction of polyphenolic compounds from Paederia scandens (Lour.) Merr. Using deep eutectic solvent: optimization, identification, and comparison with traditional methods. Ultrason Sonochem 86:106005. https://doi.org/10.1016/j.ultsonch.2022.106005
101. Zhu H, Zhang J, Li C, Liu S, Wang L (2020) Morinda citrifolia L. leaves extracts obtained by traditional and eco-friendly extraction solvents: relation between phenolic compositions and biological properties by multivariate analysis. Ind Crop Prod 153:112586. https://doi.org/10.1016/j.indcrop.2020.112586
102. Yu L, Cao L, Chang Y-H, Duan C-J, Liu C, Zhao X-L, Yue G-L, Wang X-Q, Fu Y-J (2021) Enhanced extraction performance of iridoids, phenolic acids from Eucommia ulmoides leaves by tailor-made ternary deep eutectic solvent. Microchem J 161:105788. https://doi.org/10.1016/j.microc.2020.105788
103. Dai W, Dai L, Chu D, Pang R, Deng J, Wang S, Li J, Chen H, Zheng X (2024) Deep eutectic solvents as efficient extractants of caffeoylquinic acids from Blumea aromatica: a comparative analysis of content and antioxidant potential. Arab J Chem 17:105596. https://doi.org/10.1016/j.arabjc.2023.105596
104. Roy A, Khan A, Ahmad I, Alghamdi S, Rajab BS, Babalghith AO, Alshahrani MY, Islam S, Islam MR (2022) Flavonoids a bioactive compound from medicinal plants and its therapeutic applications. Biomed Res Int 2022:5445291. https://doi.org/10.1155/2022/5445291
105. Aktas H, Kurek MA (2024) Deep eutectic solvents for the extraction of polyphenols from food plants. Food Chem 444:138629. https://doi.org/10.1016/j.foodchem.2024.138629
106. Bondam AF, Silveira DD, Pozzada SJ, Hoffman JF (2022) Phenolic compounds from coffee by-products: extraction and application in the food and pharmaceutical industries. TrAC Trends Anal Chem 123:172–186. https://doi.org/10.1016/j.tifs.2022.03.013
107. Zannou O, Koca I (2022) Greener extraction of anthocyanins and antioxidant activity from blackberry (rubus spp) using natural deep eutectic solvents. LWT 158:113184. https://doi.org/10.1016/j.lwt.2022.113184
108. Nejrotti S, Iannicelli M, Jamil SS, Arnodo D, Blangetti M, Prandi C (2020) Natural deep eutectic solvents as an efficient and reusable active system for the Nazarov cyclization. Green Chem 22:110–117. https://doi.org/10.1039/C9GC03465J
109. Rajha HN, Mhanna T, El Kantar S, El Khoury A, Louka N, Maroun RG (2019) Innovative process of polyphenol recovery from pomegranate peels by combining green deep eutectic solvents and a new infrared technology. LWT 111:138–146
110. Zhou M, Fakayode OA, Ren M, Li H, Liang J, Zhou C (2024) Green and sustainable extraction of lignin by deep eutectic solvent, its antioxidant activity, and applications in the food industry. Crit Rev Food Sci Nutr 64:7201–7219. https://doi.org/10.1080/10408398.2023.2181762
111. Procentese A, Johnson E, Orr V, Garruto Campanile A, Wood JA, Marzocchella A, Rehmann L (2015) Deep eutectic solvent pretreatment and subsequent saccharification of corncob. Bioresour Technol 192:31–36. https://doi.org/10.1016/j.biortech.2015.05.053
112. Zhao Z, Chen X, Ali MF, Abdeltawab AA, Yakout SM, Yu G (2018) Pretreatment of wheat straw using basic ethanolamine-based deep eutectic solvents for improving enzymatic hydrolysis. Bioresour Technol 263:325–333. https://doi.org/10.1016/j.biortech.2018.05.016
113. Liu Y, Zheng J, Xiao J, He X, Zhang K, Yuan S, Peng Z, Chen Z, Lin X (2019) Enhanced enzymatic hydrolysis and lignin extraction of wheat straw by Triethylbenzyl ammonium chloride/lactic acid-based deep eutectic solvent pretreatment. ACS Omega 4:19829–19839. https://doi.org/10.1021/acsomega.9b02709
114. Shen X-J, Wen J-L, Mei Q-Q, Chen X, Sun D, Yuan T-Q, Sun R-C (2019) Facile fractionation of lignocelluloses by biomass-derived deep eutectic solvent (DES) pretreatment for cellulose enzymatic hydrolysis and lignin valorization. Green Chem 21:275–283. https://doi.org/10.1039/C8GC03064B

115. Chen Z, Ragauskas A, Wan C (2020) Lignin extraction and upgrading using deep eutectic solvents. Ind Crop Prod 147:112241. https://doi.org/10.1016/j.indcrop.2020.112241
116. Hong S, Shen X-J, Xue Z, Sun Z, Yuan T-Q (2020) Structure–function relationships of deep eutectic solvents for lignin extraction and chemical transformation. Green Chem 22:7219–7232. https://doi.org/10.1039/D0GC02439B
117. Chen Y, Zhang L, Yu J, Lu Y, Jiang B, Fan Y, Wang Z (2019) High-purity lignin isolated from poplar wood meal through dissolving treatment with deep eutectic solvents. R Soc Open Sci 6:181757. https://doi.org/10.1098/rsos.181757
118. Tan YT, Ngoh GC, Chua ASM (2019) Effect of functional groups in acid constituent of deep eutectic solvent for extraction of reactive lignin. Bioresour Technol 281:359–366. https://doi.org/10.1016/j.biortech.2019.02.010
119. Zhao X, Lyu G, Meng X, Liu Y, Wang Z, Yoo CG (2024) Novel ternary deep eutectic solvent fractionation for effective utilization of willow. Bioresour Technol 407:131148. https://doi.org/10.1016/j.biortech.2024.131148
120. Li P, Li T, Wu S (2024) Process parameters and product characterization for efficient extraction of lignin with deep eutectic solvents: a review. Int J Biol Macromol 280:136053. https://doi.org/10.1016/j.ijbiomac.2024.136053
121. Zafar S, Cherakkara A, Misnon II, Yang CC, Jose R (2024) Graphene-like materials from biomass using deep eutectic solvents: a review. ACS Sustain Resour Manag 1(9):2047–2073
122. Mankar AR, Pandey A, Pant KK (2022) Microwave-assisted extraction of lignin from coconut coir using deep eutectic solvents and its valorization to aromatics. Bioresour Technol 345:126528. https://doi.org/10.1016/j.biortech.2021.126528

Chapter 10
Deep Eutectic Solvents as Sustainable Tool in Extraction and Microextraction Techniques

Sofia Aquino Monteiro, Natália Stram Bierhals, and Josias Merib

10.1 Introduction

Despite significant developments, innovations, and improvements in analytical instrumentation observed in recent years, sample preparation remains a critical stage of chemical analysis. Efforts devoted to designing high-performance separation instruments, high-resolution optical devices, as well as mass spectrometers capable of providing accurate results have emphasized a tremendous evolution in the field of analytical instrumentation. However, numerous analytical methodologies that employ these sophisticated instruments still require a previous sample preparation step [1, 2].

Sample preparation is generally necessary due to issues associated with matrix complexity, as well as to enhance the compatibility of samples with analytical instruments [3]. In addition, especially for trace analysis, some strategies that permit preconcentration of analytes can be adopted to improve method detectability. In general, complex matrices need to be subjected to a sample preparation step. This stage is typically the most critical of the experimental workflow. Therefore, the success of the analysis directly depends on properly performed sample preparation [4].

S. A. Monteiro · N. S. Bierhals
Programa de Pós-Graduação em Biociências, Universidade Federal de Ciências da Saúde de Porto Alegre, Porto Alegre, RS, Brazil
e-mail: sofiaam@ufcspa.edu.br; nataliabi@ufcspa.edu.br

J. Merib (✉)
Programa de Pós-Graduação em Biociências, Universidade Federal de Ciências da Saúde de Porto Alegre, Porto Alegre, RS, Brazil

Departamento de Farmacociências, Universidade Federal de Ciências da Saúde de Porto Alegre, Porto Alegre, RS, Brazil
e-mail: josias@ufcspa.edu.br

K. Kuperkar et al. (eds.), *Deep Eutectic Solvents*, Environmental Footprints and Eco-design of Products and Processes,
https://doi.org/10.1007/978-3-032-03801-2_10

Traditional sample preparation methodologies include liquid–liquid extraction (LLE) and solid-phase extraction (SPE), which have been widely employed for the determination of various analytes. These techniques demonstrate satisfactory performance with high extraction efficiency, being included in numerous official protocols worldwide. On the other hand, a large consumption of toxic solvents is a significant limitation associated with risks to the environment and analysts' health [5]. In this regard, strategies that employ lower volumes of solvents have gathered attention in the last decades, particularly with miniaturized approaches, including various configurations of solid- and liquid-based microextractions [6, 7].

Liquid-phase microextraction (LPME) involves numerous configurations in which a solvent (typically an organic solvent) is the extraction phase. The three main configurations of this sample preparation approach are single drop microextraction (SDME) [7], hollow-fiber liquid-phase microextraction (HF-LPME) [8], and dispersive liquid–liquid microextraction (DLLME) [9]. All configurations previously mentioned require significantly lower amounts of solvents and samples compared to traditional approaches.

Briefly, in SDME, a small solvent droplet (typically 1–3 μL) is suspended either directly in the sample or in its headspace to extract the analytes, effectively concentrating them for analysis. In this process, the droplet is usually held at the tip of a microsyringe during the microextraction step. After the extraction procedure, the enriched droplet is then retracted and injected into an analytical system, such as a gas chromatograph [7].

Introduced in 1999 [8], the HF-LPME technique utilizes a porous hollow fiber that is impregnated with an organic solvent, forming a supported liquid membrane (SLM). Depending on the setup, analytes can be transferred from the sample to an aqueous acceptor phase inside the fiber (three-phase) or directly to the organic phase (two-phase). This design enhances selectivity and reduces solvent waste [10]. Afterward, the organic solvent or aqueous solution enriched with the analytes is subjected to instrumental analysis. There are many alternative membrane-based configurations in sample preparation, most of which employ a solvent immobilized in the porous membrane for the analyte extraction.

Originally introduced in 2006 by Rezaee et al. [9], DLLME involves a three-component system: an aqueous sample, a dispersive solvent, and an extraction solvent. In practice, the dispersive and extraction solvents are injected into the aqueous phase, producing a turbid mixture with a large surface area that enhances mass transfer and accelerates the extraction process. Following extraction, a centrifugation step is generally required to achieve a suitable phase separation. Then, the enriched extraction solvent is recovered and subjected to instrumental analysis. A scheme of the three previously mentioned microextraction techniques (SDME, HF-LPME, and DLLME) is shown in Fig. 10.1.

In parallel with the use of lower quantities of toxic solvents through microextraction approaches, recent trends have focused on developing protocols that employ alternative and greener materials. Modern concepts based on sustainability are driving the development of novel protocols in which more sustainable materials, solvents, and procedures are adopted. In this context, sustainable strategies have

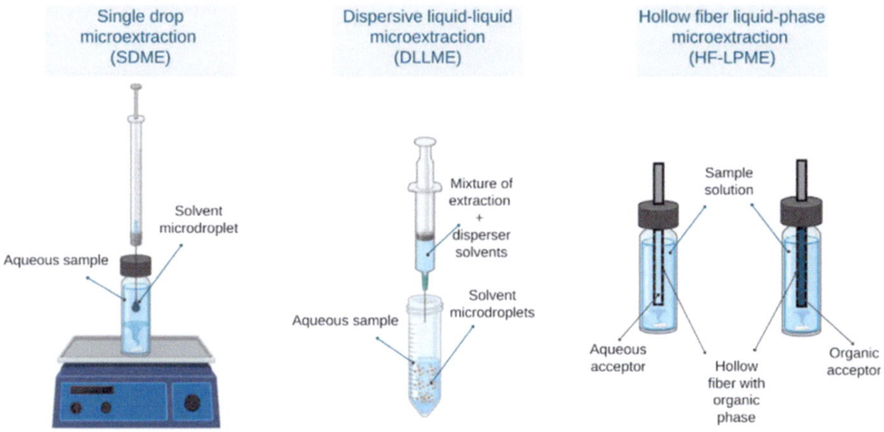

Fig. 10.1 Scheme of the main configurations of LPME

been reported exploiting the use of supramolecular solvents (SUPRAS) [11, 12] and deep eutectic solvents (DESs) as extraction phases [13].

10.2 Deep Eutectic Solvents in Sample Preparation Approaches

Deep eutectic solvents were first described by Abbot et al. in 2003 during investigations involving choline chloride and urea mixtures, which exhibited a melting point significantly lower than that of their individual components [14]. The marked depression can be associated with strong hydrogen bonding interactions between hydrogen bond donors (HBDs) and acceptors (HBAs), leading to a highly stable eutectic mixture [15].

The production of DESs is a straightforward process without requiring time-consuming or laborious procedures. Basically, it consists of mixing specific amounts of the components under agitation, heating, vortexing, or sonication until a homogeneous liquid is formed [16]. Following production, DESs can be easily characterized by Fourier-transform infrared spectroscopy (FTIR), proton nuclear magnetic resonance (H^1NMR), and differential scanning calorimetry (DSC), among others. Additionally, the measurement of physical properties like density and viscosity provides complementary information that is also useful in solvent-based sample preparation approaches.

The remarkable evolution of microextraction strategies has also been based on exploiting alternative and efficient solvents and sorbents. In this context, DESs stand out as a valuable alternative to toxic organic solvents due to their physicochemical properties, such as low vapor pressure, biodegradability, tunability, and varied viscosities and densities. These compounds can also be obtained from natural sources, resulting in natural deep eutectic solvents (NADES), which are considered as the solvents for the twenty-first century [17].

The applicability of DESs and NADES has increased notably. The use of these solvents is not limited to sample preparation but also involves other fields such as catalysis [18], organic synthesis [19], gas adsorption [20, 21], sensors [22], and separation science [23]. These studies emphasize their importance and efficiency, as well as the green aspects associated with these revolutionary solvents.

Initially, most DESs consisted of hydrophilic compounds, generally composed of choline chloride and urea, which limited their use in aqueous sample extractions due to poor phase separation. However, more recent developments have introduced hydrophobic DESs (HDESs), and the application of these compounds to analyze water samples has significantly increased. HDESs have shown improved phase separation when extracting analytes from aqueous matrices, addressing earlier limitations of hydrophilic systems [24]. Particularly, the production is performed using constituents with limited water solubility, such as tetraalkyl quaternary ammonium/phosphorus salts, monoterpenes, phenols, glycols, and carboxylic acids with medium/long alkyl chains [25]. Some physicochemical properties of these solvents can be checked elsewhere [26]. Figure 10.2 displays commonly used hydrogen bond donors (HBDs) and acceptors (HBDs) components that are typically employed in DES formation for analytical purposes.

DESs exhibit enormous potential to be exploited in sample preparation workflows. In this regard, several review articles focusing on the solvents employed in analytical sample preparation have been reported in the previous years [26–29]. Importantly, analytes with varied physicochemical properties have been successfully determined in varied matrices. The bar graph of Fig. 10.3 was obtained using the *Scopus* database employing the descriptors "deep eutectic solvents" and

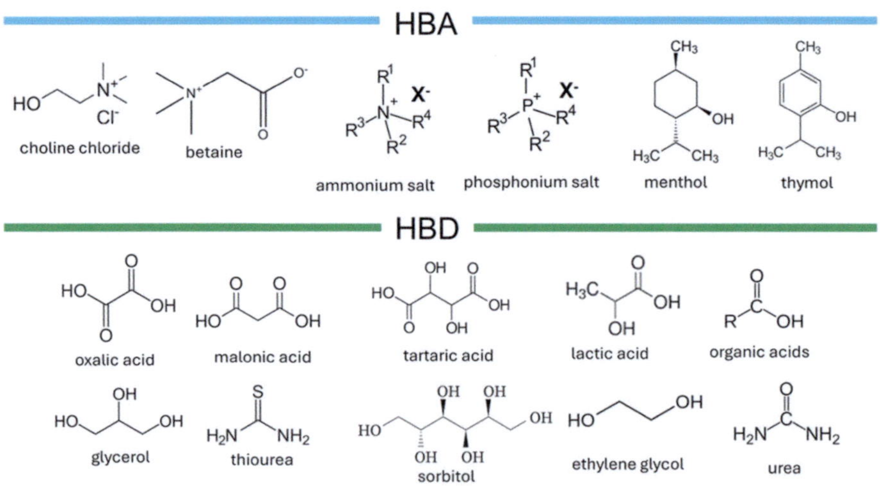

Fig. 10.2 Examples of HBA and HBD generally used for analytical extractions. Reproduced with permission from ref. [26]

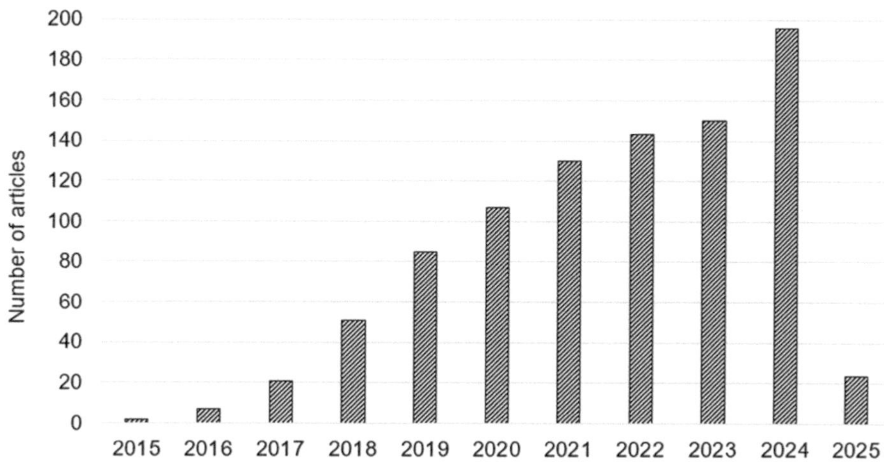

Fig. 10.3 Bar graph representing the annual production of research papers involving deep eutectic solvents in microextraction approaches. Data obtained from the *Scopus* database in February 2025

"microextraction" from 2015 to 2025 (up to February 2025). A total of 916 research papers were published in this period, with the production constantly increasing year by year.

In the following sections, some important contributions involving the application of DESs in sample preparation approaches are highlighted. Specifically, studies focused on environmental and food samples, extraction of bioactive compounds, and analysis of biological matrices that have been developed in the last 5 years are discussed and evaluated. Additionally, perspectives and trends in the application of DESs in sample preparation are pointed out, with emphasis on the automation of analytical workflows and recent developments that hold high potential to be exploited in the future.

10.3 Applications of DESs in Environmental Samples

The quantification of environmental pollutants is a critical subject in chemical analysis. Numerous compounds can reach diverse matrices, such as water, air, plants, and soil, resulting in several environmental and health issues [30]. Importantly, some environmental pollutants are capable of accumulating in the environment and organisms, causing a severe impact [31]. Numerous organic compounds are harmful to human beings and animals; therefore, monitoring their presence and concentration is a considerable analytical challenge.

Some compounds are regulated by legislations that define their maximum allowable concentration (MAC). In this case, typical contaminants such as most organochlorine pesticides (OPs), polycyclic aromatic hydrocarbons (PAHs), and polychlorinated biphenyls (PCBs) exhibit strict regulations in different matrices.

Additionally, some emerging contaminants are not yet regulated by specific legislation controlling their concentrations. Some of these toxic compounds include pharmaceuticals, illicit drugs, artificial sweeteners, pesticides, and microplastics with the potential to be harmful to several ecosystems [30].

As previously mentioned, DESs have been exploited as a sustainable alternative to classical methodologies. These solvents are also gaining attention for the analysis of environmental contaminants, as demonstrated by Quintanilla et al., who recently proposed the incorporation of DES into a polymeric film for the determination of organophosphorus pesticides [32]. A DES based on dodecanoic acid and lidocaine was immobilized in a cellulosic film, which was employed for the extractions. Then, a desorption step was performed with ethyl acetate, and the extract was analyzed by gas chromatography coupled to mass spectrometry (GC-MS).

Another sustainable approach involves the use of NADES for the determination of organochlorine and organophosphorus pesticides from drinking water. In a recent study, Hristozova et al. proposed a vortex-assisted DLLME procedure using a mixture of DL-menthol and decanoic acid as the extraction phase, followed by GC-MS/MS [33]. In this approach, 10 mL of the sample and only 45 μL of the NADES were required. The method featured high efficiency and satisfactory sensitivity for all analytes. Importantly, authors investigated the so-called "matrix-induced response enchantment effect" that demonstrated a favorable result of the NADES during the GC-MS/MS analysis by increasing sensitivity for all analytes.

In parallel, Morelli et al. developed a NADES-based workflow combined with hollow-fiber microporous membrane liquid–liquid microextraction (HF-MMLLE) to determine emerging contaminants [34]. A thymol:camphor DES was immobilized within porous polypropylene fibers, enabling selective extraction through a supported liquid membrane setup. This approach achieved high-throughput processing while maintaining sensitivity for emerging contaminants using high-performance liquid chromatography coupled to diode array detector (HPLC-DAD).

Other strategies have emerged as potentially useful for expanding the applications of DES in analytical extractions. In this context, homogeneous liquid–liquid microextraction with switchable deep eutectic solvents (SDES-HLLME) has been evaluated for the determination of triazole fungicides [35]. This approach required 200 μL of a temperature-switchable deep eutectic solvent and 4 mL of the water sample, which were heated above the upper critical solution temperature (UCST), resulting in a homogeneous solution. After sonication and centrifugation, the mixture was allowed to settle, and phase separation was achieved. Then, the extraction phase was collected, diluted, and analyzed by high-performance liquid chromatography coupled to ultraviolet detector (HPLC-UV). The DES composed of N-methyldiethanolamine/4-bromophenol exhibited the best extraction performance and permitted limits of quantification (LOQs) from 7.8 to 8.7 μg L^{-1} using HPLC-UV. Interestingly, with this sample preparation approach, the extraction solvent was capable of reversibly switching between hydrophilic and hydrophobic states based on temperature changes.

A recent trend in analytical sample preparation is the use of magneto-active solvents as extraction phases [36–38]. These approaches also benefit from the extraction efficiency and sustainable aspects of DES; in addition, they exhibit paramagnetic properties that enable easier phase separation using small magnets. In this context, a ferrofluid-based magnetic LPME was recently developed to determine metronidazole [39]. In this study, a deep eutectic solvent composed of DL-carnitine and octanoic acid was combined with surface-modified Fe_3O_4 nanoparticles to form a ferrofluidic phase. This hybrid system, stabilized by oleic acid coatings on the magnetic particles, enabled magnetically assisted microextraction when applied to aqueous samples, offering a simplified separation and effective analyte recovery. The final extract was analyzed by HPLC-UV, achieving an LOQ of 0.387 ng mL^{-1}.

Despite being a relatively emerging field, the applications of magnetic DES (MDES) are gaining attention. Makos-Chełstowska et al. reported an investigation on the capacity of these compounds to extract furfural and 5-hydroxymethylfurfural [40]. Interestingly, a theoretical approach based on the conductor-like screening model for realistic solvents (COSMO-RS) was employed for an initial screening. A liquid–liquid extraction was employed, and the magnetic DES composed of menthol, thymol, and $FeCl_3$ demonstrated the best performance.

Another critical topic in analytical sample preparation is the analysis of solid samples. These samples may provide additional challenges due to their high complexity; therefore, proposing strategies to overcome these challenges is not trivial. In this context, hydrophobic DES have been evaluated, followed by GC-MS for the analysis of sludge samples [41]. Satisfactory results were obtained through the combination of a DLLME procedure with sample introduction via pyrolyzer. Emerging contaminants were successfully determined using 300 μL of DL-menthol:acetic acid (1:1) and 350 μL of acetonitrile (ACN). This mixture was added to 4 mL of filtered sludge sample for 20 s, centrifuged at 8000 rpm, and then cooled to 25 °C for 20 min. Importantly, sample introduction via pyrolyzer can overcome some problems caused by the direct injection of a DES-rich extract into the GC injector.

10.4 Applications of DESs in Food Analysis

The matrix complexity and the low concentrations at which pollutants are typically found in food matrices are another significant analytical challenge. In general, these substances are undesirable and exhibit toxic properties that can lead to health issues for the consumer, depending on the concentration and level of exposure. Therefore, sample preparation is critical in food analysis, and the application of DESs has increased in recent years.

Approaches involving both classical extractions and microextraction-based techniques have also been adapted to enhance sustainability. For example, Zhao et al. proposed a solid–liquid extraction (SLE) followed by an LPME workflow using hydrophilic and hydrophobic DES for the determination of parathion in cereals [42]. This strategy involved the first extraction performed with 0.1 g of cereal using 200 μL of hydrophilic DES (choline chloride and propylene glycol). After centrifugation, the supernatant was collected, and 100 μL of NaOH was employed. Afterward, 100 μL of hydrophobic DES composed of tetrabutylammonium bromide and terpineol was injected to extract a yellow product formed. Integrated digital image colorimetry using a smartphone permitted an LOQ of 0.01 mg/kg.

Methodologies based on QuEChERS (Quick, Easy, Cheap, Rugged, and Safe) have been widely used to analyze food samples. Briefly, this approach consists of extraction, partition, and cleaning. The extraction and partition steps are performed with acetonitrile, followed by salting-out (the first configuration employed MgSO4 and NaCl). Finally, a dispersive SPE (d-SPE) employing a mixture of cleaning sorbents (anhydrous $MgSO_4$ and primary secondary amine (PSA)) is employed to remove matrix interferences [43, 44]. Notably, some aspects of these methodologies have been modified to increase the sustainable aspects of their steps. Recently, Soltani and Sereshti proposed a QuEChERS methodology based on DES to determine pesticides in tea [45]. In their study, the partition step was performed with choline chloride: polyethylene glycol, followed by an in-syringe dispersive SPE cleanup using magnetic three-dimensional graphene aerogel (3DGA-Fe_3O_4). Using this approach combined with GC-MS, LOQs from 0.70 to 1.90 μg kg^{-1} were obtained, and the analysis of green and black tea samples was successfully performed.

Due to the physicochemical properties, the analysis of acrylamide in food matrices is also challenging. Posta et al. proposed a DES-assisted matrix solid-phase dispersion (MSPD) to determine acrylamide in bread [46]. This procedure was based on homogenizing the sample with a solid dispersion material (Florisil®). Then, a choline chloride-glycerol DES (50 μL) was added during the homogenization to improve analyte extraction, thus reducing the amount of organic solvent. Finally, 2 mL of methanol was applied for the elution step, followed by HPLC-MS analysis that provided a limit of quantification of 35 μg/kg for acrylamide.

The evolution of DES-based techniques for analyzing food samples has been evident recently. Hao et al. proposed an approach using a water-based deep eutectic solvent (WDES) for the determination of bisphenol A in edible oil [47]. In this case, water not only plays a role in the production of WDES but also has an influence on the viscosity and improves solvent diffusion and mass transfer. Among the evaluated compositions, tetrabutylammonium chloride (TBAC) and H_2O were chosen. In the extraction step, 0.5 g of edible oil and 3 mL of n-hexane were used. Following, 100 μL of WDES was employed, and the solution was subjected to vortexing for 3 min. After centrifuging, the WDES (90 ± 2 μL) was collected and solubilized in 200 μL of methanol, followed by HPLC analysis.

Other studies have also exploited different configurations of LPME in food analysis. Recently, Mu et al. performed the determination of pesticides in beeswax employing DESs composed of proline and oxalic acid [48]. Importantly, they also investigated the extraction mechanisms through molecular dynamics (MD) simulations. Soylak et al. comprehensively examined DES with varied properties (hydrophilic, hydrophobic, ionic, and nonionic) to extract analytes of diverse chemical natures (anionic, cationic, and nonionic). In this study, HF-LPME was employed with analysis performed using UV-Vis spectrophotometry [49]. Another recent approach was developed by Wu et al. for the determination of flavonoids and saponins from functional foods. In this study, a DES-assisted mechanochemical extraction based on K_2CO_3:glycerol (1:7) was proposed, followed by UHPLC-Q-TOF-MS [50]. Furthermore, DLLME based on magnetic DES has been successfully exploited to determine succinate dehydrogenase inhibitors in juice, wine, water, and vinegar using 150 μL of MDES [51]. In this study, the MDES composed of heptyltriphenylphosphonium bromide, ferric chloride, and hexanoic acid exhibited the best performance, and HPLC-DAD analysis provided LOQs from 3.8 to 9.8 $\mu g\ L^{-1}$.

10.5 DESs Applied to Extract Bioactive Compounds

Bioactive compounds are natural molecules with potential medicinal applications. These compounds can directly influence human metabolism and physiology, promoting the treatment and prevention of diseases because of their anti-inflammatory, antioxidant, anti-lipidemic, and anti-depressive properties. The main sources of these compounds include plants, roots, herbs, spices, fruits, nuts, vegetables, cereals, meat, and seafood [52]. In particular, the development of alternative analytical methodologies for extracting bioactive compounds focuses on reducing the consumption of toxic solvents. Additionally, studies have investigated the use of alternative types of DES, such as magnetic deep eutectic solvents [53] and switchable deep eutectic solvents [54].

Switchable-DES has the capacity of "switching" between different forms by varying pH, temperature, or pressure. Yang et al. recently exploited the use of a temperature-induced SDES to determine flavonoids in tea [54]. The combination of benzamide and malic acid (2:1 molar ratio) provided the best response. The simple and straightforward extraction step consisted of mixing 6 mL of SDES and 0.1 g of sweet tea powder at 70 °C, followed by centrifuging for 5 min. After collecting the enriched SDES, the final extract was analyzed by HPLC. A scheme of this approach is shown in Fig. 10.4.

A subclass of compounds termed switchable supramolecular DES (SS-DES) has been effectively applied to isolate flavonoids from *Scutellariae Radix*, followed by HPLC-UV analysis [55]. This experimental workflow consisted of 0.1 g of *Scutellariae Radix* powder mixed with a certain amount of SS-DES, followed by adding 1 mL of H_2SO_4, which produced a cloudy solution. After centrifugation, phase separation, and filtration, the extract was diluted with acetic acid and subjected to instrumental analysis. N, N-dimethylbenzylamine:octanoic acid SDES achieved the highest efficiency for flavonoid extraction.

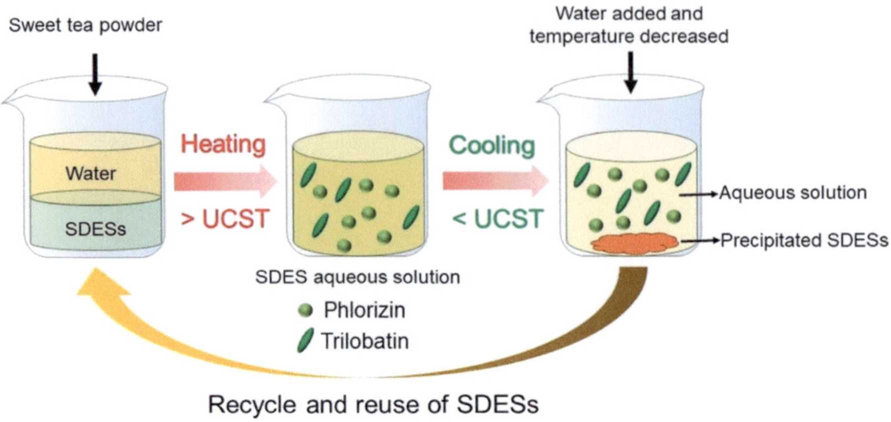

Fig. 10.4 Scheme of the experimental workflow proposed by Yang et al. Reproduced with permission from ref. [54]

Jovanović et al. proposed an ultrasound-based methodology for extracting gentiopicroside, isogentisin, and isovitexin from *Gentiana asclepiadea L.* biowaste [56]. In this study, choline chloride and 1,2-propanediol proved to be the best extraction solvents. Additionally, molecular dynamics simulations were carried out to examine the binding interactions between the analytes and solvents. These interactions were mostly attributed to an extensive network of hydrogen bonds and electrostatic forces that facilitate stable binding with the analytes. Another recent study focused on the isolation of carotenoids in persimmon peels using hydrophobic-NADES [57]. The NADES composed of thymol and menthol was chosen due to its high efficiency associated with its strong antioxidant capacity.

Another important approach in the sample preparation of natural products consists of solid eutectic systems (SES). This strategy was examined to extract phenolic compounds from *Lactuca sativa* using matrix solid-phase dispersion (MSPD). SES, composed of thymol and citric acid, combined with NADES consisting of lactic acid, glucose, and water, was employed [58]. Notably, this approach achieved high scores on green metrics used to evaluate the sustainability of the procedure.

The use of DES for determining bioactive compounds offers a promising alternative to conventional organic solvents due to their versatility, sustainability, and ability to extract varied bioactive substances from challenging matrices. The application of DES also features significant advantages over conventional organic solvents, including greater extraction efficiency, lower toxicity, and higher selectivity for specific compounds. In addition, these solvents can be easily tailored to promote the extraction of bioactive compounds, including flavonoids, alkaloids, and phenolic acids. Therefore, DES not only serve as sustainable alternatives but also broaden the possibilities for analyzing and applying bioactive compounds in fields such as pharmacology, nutrition, and cosmetology.

10.6 Applications of DESs for the Analysis of Biological Samples

The complexity of biological samples also requires proper sample preparation. Biological matrices typically consist of cells, proteins, electrolytes, minerals, hormones, biotransformation products, and xenobiotics that can significantly interfere with instrumental analysis. The number of studies exploiting DES as the extraction solvent in biological samples has been increasing. These studies include the analysis of urine [59, 60], whole blood [61], plasma [62], serum [63], and saliva [64]. In this context, NADES have been applied in an LPME-based strategy for determining pharmaceuticals from urine, followed by LC-MS/MS [60]. In this study, L-menthol and octanoic acid DES provided the highest efficiency for analyzing 20 mL of urine. Following the extraction procedure, a refrigeration step was important to facilitate the removal of the extraction solvent.

In addition, Andrade et al. combined molecular dynamics simulations with experimental data to assess the type and composition of NADES for determining steroidal hormones [65]. Specifically, molecular dynamics simulations were employed to assess the extraction capacity of a menthol:decanoic acid NADES, demonstrating that van der Waals interactions exhibited intense influence on the extraction of the analyzed compounds. An ultrasound-assisted LPME experimental approach permitted satisfactory performance when combined with HPLC-DAD.

Less conventional biological matrices, such as exhaled air and saliva, have also been explored. Norouzi et al. reported an analytical methodology to determine oxymorphone and morphine in exhaled breath condensate samples [66]. The sample preparation involved microwave-enhanced air-assisted liquid–liquid microextraction (AALLME), followed by GC-MS analysis. A ternary deep eutectic composed of choline chloride-menthol-phenylacetic acid was employed as the extraction phase, offering satisfactory results in terms of analytical development with LOQs of 7.2 and 5.2 $\mu g\ L^{-1}$ for morphine and oxymorphone, respectively. In this technique, AALLME combined with microwave irradiation was an alternative to dispersive solvents typically used, which can exhibit certain toxicity compared to green solvents. Although exhaled air remains relatively unexplored, this biological matrix has many advantages, such as simple, noninvasive, and painless collection, and does not require a highly trained professional. On the other hand, the need for a derivatization step can lead to a more labor-intensive procedure and an increased possibility of analyte loss.

High-throughput methodologies using DES have gained increasing attention in recent years. In this context, membrane-based approaches can be highlighted since the solvent can be immobilized in the porous polymeric membranes, allowing for the analysis of multiple samples. A membrane-based procedure called parallel artificial liquid membrane extraction (PALME) was proposed to determine nerve agent biomarkers in human urine, followed by LC-MS/MS [59]. In this case, the

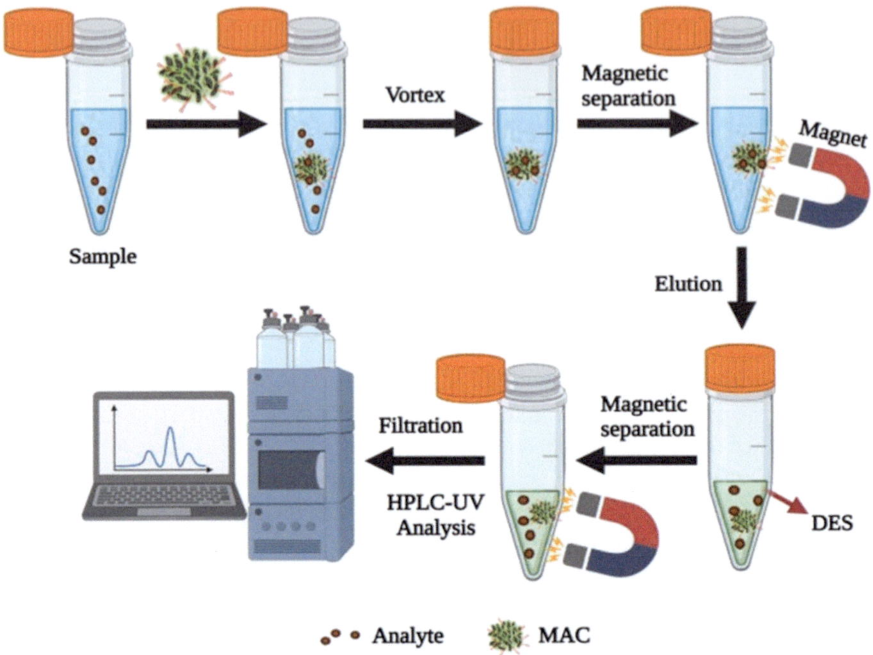

Fig. 10.5 Illustration of the experimental workflow developed by Pasupuleti et al. Reproduced with permission from ref. [67]

coumarin:thymol deep eutectic solvent was immobilized in the porous flat membrane, which provided a high-throughput configuration with LOQs from 0.04 to 5.35 μg L^{-1}. Importantly, this methodology allowed for processing 192 samples within 120 min, which is particularly important for routine applications.

Magneto-assisted approaches have also been evaluated in DES-based sample preparation of biological matrices. Figure 10.5 illustrates a magnetically assisted solid-phase extraction protocol applied for paraben analysis, adapted from the work by Pasupuleti et al. [67], in which DESs and functionalized carbon-based sorbents were utilized (reproduced with permission). This approach required only 10 mg of sorbent and 200 μL of DES, permitting a sensitive methodology with LOQs ranging from 0.3 to 0.5 μg L^{-1}.

The number of studies involving DES in extractions and microextraction approaches has grown significantly over the past 5 years. Therefore, this chapter does not provide an exhaustive literature overview. A more detailed description of the analytical features of selected studies that exploited DESs in sample preparation is provided in Table 10.1.

Table 10.1 Studies involving the use of DES in extraction and microextraction techniques reported within the last 5 years

Sample preparation technique	Analytes	Sample matrix	Solvent amount	Type of DES	Sample amount	Extraction time	Instrumentation	LOQ	Ref.
Environmental contaminants									
S-TFME and (PT)-TFME	Pesticides	Water	–	Dodecanoic acid:lidocaine	10 mL 20 mL	60 min	GC-MS	1.3–4.3 μg L^{-1}	[32]
Vortex-assisted DLLME	Pesticides	Drinking water	45 μL	DL-menthol and decanoic acid	10 mL	3 min	GC-MS/MS	0.2–78 ng L^{-1}	[33]
HF-MMLLE	Emerging contaminants	Water	–	Thymol: camphor	1.5 mL	50 min	HPLC-DAD	1–20 μg L^{-1}	[34]
SDES-HLLME	Triazole pesticides	Water	200 μL	N-ethyldiethanolamine: 4-bromophenol	4 mL	< 7 min	HPLC-UV	7.8–8.7 μg L^{-1}	[35]
VA-LPME	Metronidazole	Water	160 μL	Octanoic acid and DL-carnitine	10 mL	8 min	HPLC-UV	0.387 μg L^{-1}	[39]
DLLME	Bisphenol A and parabens	Sludge	300 μL	DL-menthol:acetic acid	4 mL	20 s	Py-GC-MS	83–166 μg L^{-1}	[41]
FF-MDLLME	Chiral pesticides	River water	50 μl	L-menthol:thymol	5 mL	15 min	HPLC-MS/MS	0.24–1.20 μg L^{-1}	[77]
Analysis of food samples									
Solid–liquid extraction and liquid–liquid microextraction	Parathion	Cereals	200 μL	Choline chloride:propylene glycol	0.1 g	–	Colorimetry	0.01 mg kg^{-1}	[42]
Matrix solid-phase dispersion (MSPD) extraction	Acrylamide	Bread	50 μL + 2 mL of methanol	Choline chloride-glycerol	0.2 g	–	HPLC-MS	35 μg kg^{-1}	[46]
LLME	Bisphenol A	Edible oils	100 μL	TBAC:H_2O	0.5 g	3 min	HPLC	0.6 μg L^{-1}	[47]
HF(3)-LPME	Amaranth, quercetin methylene blue	Fruits	–	Menthol/TBAB-As SLM, malic acid/citric acid/water	25 mL	30 min	UV-vis spectrometer	20–50 μg L^{-1}	[49]

(continued)

Table 10.1 (continued)

Sample preparation technique	Analytes	Sample matrix	Solvent amount	Type of DES	Sample amount	Extraction time	Instrumentation	LOQ	Ref.
Assisted mechanochemical extraction method	Flavonoids and saponins	Functional food	80 μL	K_2CO_3:glycerol	0.5 g	5.3 min	UHPLC-Q-TOF/MS	17.61–69.44 μg L^{-1}	[50]
LPME	Thiabendazole	Fruit juice	–	Formic acid:L-menthol	–	5 min	LC-MS/MS	0.45–1.42 μg L^{-1}	[74]
Liquid–liquid microextraction (TRDES HLLME)	Bisphenols	Mineral water and soft drinks	280 mg	Lidocaine: heptanoic acid	3.4 g	6 min	HPLC-FLD	0.03–0.1 μg L^{-1}	[78]
DLLME	Curcumin	Water and food	2 mL	Octylamine (OA) succinic acid (SA)	40 mL	-	UV-vis spectrophotometry	35 μg L^{-1}	[79]
Extraction of bioactive compounds									
DLLME	Succinate dehydrogenase inhibitors	Juice, vinegar, wine, and water	150 μL	Quaternary phosphorus salt, metal chloride, and fatty acid	5 mL	30 s	HPLC-DAD	0.0038–0.0098 mg L^{-1}	[51]
VALLME	Patulin	Fruits and fruit juices	560 μL	Trihexyltetradecyl phosphonium chloride:$MnCl_2$: octanoic acid	10 mL	300 s	UV-Vis	0.2–1.2 μg L^{-1}	[53]
Temperature-induced SDESs	Flavonoids	Sweet tea	6 mL	Benzamide:malic acid	0.1 g	15 min	HPLC	–	[54]
LPME	Flavonoids	*Scutellariae radix*	6 mL	DMBA-octanoic acid	0.1 g	3.3 min	HPLC-UV	–	[55]
UAE	Isogentisin gentiopicrosid and isovitexin	*Gentiana asclepiadea L.*	10 mL	Choline chloride: 1,2-propanediol	–	30 s	HPLC-DAD	–	[56]

UAE	Carotenoids	Persimmon peels	2 mL	Thymol:menthol	100 mg	13.8 min	UHPLC-DAD	–	[57]
Solid-phase dispersion	Phenolic compounds	Lettuce samples	1430 μL	Lactic acid:glucose:water	100 mg	≅6 min	HRMS and HPLC-UV	5.1–6.7 μg g^{-1}	[58]
UAE	Osthole	*Cnidium monnieri* (L.) cuss	1 mL NaDES +200 μL SHS	Xylose: levulinic acid	0.067 g	30 min	HPLC-DAD	0.1 mg L^{-1}	[80]
UAE	Astaxanthin	Black tiger shrimp (*Penaeus monodon*) shells	15 mL	Choline chloride and lactic acid	1 g	33 min	HPLC	–	[81]
MSPD	Lindenenyl acetate, isolinderalactone, norisoboldine, linderalactone, linderene, and linderane	*Linderae radix*	500 μL	DL-menthol:acetic acid	0.1 g	1.5 min	UHPLC	33.66–279.33 ng mL^{-1}	[82]
Analysis of biological samples									
PALME	Alkyl methyl phosphonic acids	Urine	4 μL	Coumarin:thymol	1.95 mL	120 min	LC-MS/MS	0.04–5.35 ng mL^{-1}	[59]
DLLME	Bicalutamide	Human plasma	150 μL	Borneol:capric acid	5 mL	1 min	Spectrophotometric	0.066 and 0.146 μg mL^{-1}	[62]
D-μ-SPE	Dopamine, epinephrine, and norepinephrine	Serum and urine	15 μL	Choline chloride and urea	10 mL	–	HPLC-UV	0.78–1.20 μg L^{-1}	[63]
UALLME	Cortisol and cortisone	Human saliva	10 μL	Trioctylmethylammonium chloride: pentafluorophenol	0.5 mL	≅ 25 min	LC-UV	6.9 and 6.0 pmol mL^{-1}	[64]

(continued)

Table 10.1 (continued)

Sample preparation technique	Analytes	Sample matrix	Solvent amount	Type of DES	Sample amount	Extraction time	Instrumentation	LOQ	Ref.
LPME	Pharmaceuticals	Urine	300 μL	Menthol-octanoic acid	20 mL	5 min	LC-MS/MS	15–208 ng L^{-1}	[60]
UALLME	Steroidal hormones	Urine	150 μL	Menthol:decanoic acid	3 mL	3 min	HPLC-DAD	10 and 25 μg L^{-1}	[65]
ALLME	Oxymorphone and morphine	Exhaled breath condensate	80 μL	Choline chloride:menthol: phenylacetic acid	5 mL	≅ 6 min	GC	7.2 and 5.2 ng mL^{-1}	[66]
MSPE	Parabens	Water and urine	200 μL	DL-menthol and acetic acid	3 mL	≅ 11 min	HPLC-UV	0.3–0.5 ng mL^{-1}	[67]
LPME	Psychoactive substances	Urine and oral fluid	300 μL	L-menthol:octanoic acid	5 mL	5 min	LC-MS/MS	0.0006–0.05 μg L^{-1}	[83]
ALLME	Warfarin	Urine and plasma	60 μL	Borneol:decanoic acid	10 mL	11 min	HPLC-UV	1.65–8.91 μg L^{-1}	[84]
DLLME	Hippuric acid and methyl hippuric acid	Urine	120 μL	ChCl: 4-Chlorophenol	5 mL	≅ 2 min	HPLC-UV	0.024 and 0.033 μg mL^{-1}	[85]
SLM-EME	Drugs	Human plasma	4 μL	Coumarin:thymol	0.1 mL	20 min	UHPLC-MS/MS	–	[86]
EME and G-EME	Drugs	Human urine	4 μL	Coumarin:thymol	11 mL	15 min	HPLC-DAD	19–29 μg L^{-1}	[87]

PALME Parallel artificial liquid membrane extraction, *LPME* liquid-phase microextraction, *ALLME* Air-assisted liquid–liquid microextraction, *DLLME* dispersive liquid–liquid microextraction, *D-μ-SPE* Dispersive micro-solid phase extraction, *MSPE* magnetic solid-phase extraction, *UAE* ultrasound-assisted extraction, *UALLME* ultrasound-assisted liquid–liquid microextraction, *VALLME* Vortex-assisted liquid–liquid microextraction, *SLM-EME* supported liquid membrane for electromembrane extraction, *EME* electromembrane extraction, *G-EME* gel electromembrane extraction

10.7 Perspectives and Trends in Sample Preparation Using DESs

The evolution of DES-based analytical protocols has been remarkable and evident in the previous years. The use of sustainable solvents has increased, and the combination of green aspects and high extraction efficiency makes them a valuable alternative to replace the conventional solvents typically employed. In addition, some recent works have provided novel insights into the application of DESs in microextraction approaches. These studies investigated the possibilities of automating experimental workflows, as well as emphasized the proposal of microfluidic devices by exploiting the versatility of 3D printing. Moreover, studies on evaluating chemical interactions between DESs and certain analytes using computational simulations have also been reported.

Automation of analytical workflows is generally challenging in sample preparation; however, significant progress has been made recently using DESs as extraction phases. Lab-in-syringe and multiwell platforms are gaining attention as alternatives to promote more automated and independent configurations. Specifically, an efficient lab-in-syringe system has been proposed for LPME to determine Sudan dyes in chili-based sauces. In this approach, only 100 5 L of a NADES based on terpenes and fatty acids was required [68]. Additionally, the automation of an SDME approach using DESs has been proposed with a lab-in-syringe apparatus. This online system was successfully coupled with HPLC to determine fluoroquinolones in water [69]. Following the trends in automating sample preparation strategies using DESs, membrane-based approaches also offer promising possibilities for allowing multiple extractions. In this configuration, a series of polypropylene hollow fibers are filled with a DES that serves as the extraction solvent. Therefore, multiple samples can be analyzed simultaneously, substantially increasing the analysis throughput [34, 70]. Another recent tool being exploited in sample preparation is 3D printing [71]. This technique permits the production of specifically designed apparatus that can be employed in extraction/microextraction approaches [72, 73]. Due to the unique versatility and low cost of 3D printing, components with varied architectures and complexities can be rapidly produced. In this context, millifluidic or microfluidic flow cells have been printed to be employed to serve as support for liquid-phase microextraction using DESs as extraction phases [74]. Moreover, the 3D-printed devices can be attached or online coupled to analytical instruments, offering a more automated approach. Sample preparation systems produced through 3D printing, associated with sustainable solvents such as DESs, will certainly be reported in the coming years.

The use of computational systems to predict and understand the chemical interactions between DESs and analytes is also a subject that should be deeply investigated in the future. Computational tools, such as molecular dynamics (MD), can provide new insights into the system under investigation, not only by examining the physicochemical properties of DESs [75, 76] but also by defining the intermolecular mechanism and interactions involved in a microextraction procedure [65]. For example, these computational tools can be combined with experimental data to optimize the extraction conditions and enhance the detectability of an analytical methodology.

10.8 Conclusion

Green solvents such as DESs have gained enormous attention in extraction and microextraction approaches, not only due to their sustainable aspects but also because of their efficiency. As previously discussed, a broad applicability of DESs in varied matrices has been reported, which highlights their versatility as potential extraction phases in sample preparation strategies. The ability to extract analytes of varied chemical natures in complex samples makes them valuable alternatives to replace toxic organic solvents. This fact aligns with modern trends in analytical chemistry toward the proposal of more sustainable workflows with reduced environmental impact. Importantly, the development of magneto-active DESs has allowed for innovative applications in which phase separation is easily achieved through the application of magnetic fields. Therefore, the possibility of automation of the extraction or microextraction procedure using these solvents can be facilitated. Despite the growing number of studies reported in recent years, there are opportunities for innovation and improvements in DES-based analytical workflows, particularly related to automation and the proposal of alternative devices through 3D printing.

Declaration of Generative AI and AI-Assisted Technologies in the Writing Process During the preparation of this work, the authors employed AI tools to refine the clarity and readability of various statements in the manuscript. After using this tool/service, the authors reviewed and edited the content as needed and take full responsibility for the publication content.

Conflict of Interest Statement The authors declare no conflict of interest.

References

1. Wieczorek MN, Zhou W, Pawliszyn J (2024) Perspective on sample preparation fundamentals. Adv Sample Prep 10:100114. https://doi.org/10.1016/j.sampre.2024.100114
2. Merib J (2023) High-throughput platforms for microextraction techniques. Anal Bioanal Chem 415:3671–3681. https://doi.org/10.1007/s00216-022-04504-7
3. Wojnowski W, Tobiszewski M, Pena-Pereira F, Psillakis E (2022) AGREEprep—analytical greenness metric for sample preparation. TrAC Trends Anal Chem 149:116553. https://doi.org/10.1016/j.trac.2022.116553
4. Martínez-Pérez-Cejuela H, Gionfriddo E (2024) Evolution of green sample preparation: fostering a sustainable tomorrow in analytical sciences. Anal Chem 96(20):7840–7863. https://doi.org/10.1021/acs.analchem.4c01328
5. Carasek E, Bernardi G, do Carmo SN, CMS V (2019) Alternative green extraction phases applied to microextraction techniques for organic compound determination. Separations 6(3):35. https://doi.org/10.3390/separations6030035
6. Arthur CL, Pawliszyn J (1990) Solid phase microextraction with thermal desorption using fused silica optical fibers. Anal Chem 62(19):2145–2148. https://doi.org/10.1021/ac00218a019
7. Liu H, Dasgupta PK (1996) Analytical chemistry in a drop: solvent extraction in a microdrop. Anal Chem 68(6):1135–1140. https://doi.org/10.1021/ac960145h
8. Pedersen-Bjergaard S, Rasmussen KE (1999) Liquid−liquid−liquid microextraction for sample preparation of biological fluids prior to capillary electrophoresis. Anal Chem 71(14):2650–2656. https://doi.org/10.1021/ac990055n

9. Rezaee M, Assadi Y, Milani Hosseini MR, Aghaee E, Ahmadi F, Berijani S (2006) Determination of organic compounds in water using dispersive liquid–liquid microextraction. J Chromatogr 1116:1–9. https://doi.org/10.1016/j.chroma.2006.03.007
10. Carasek E, Merib J (2015) Membrane-based microextraction techniques in analytical chemistry: a review. Anal Chim Acta 880:8–25. https://doi.org/10.1016/j.aca.2015.02.049
11. Ballesteros-Gómez A, Rubio S (2023) Supramolecular solvents for multi-target and matrix-independent sample preparation. Anal Chim Acta 1150:100085. https://doi.org/10.1016/j.sampre.2023.100085
12. Monteiro SA, Scheid C, Machado CS, Machado C, Merib J (2024) A high-throughput sample preparation approach for the determination of nonsteroidal anti-inflammatory drugs in urine samples through membrane-based microextraction employing a green supramolecular solvent. Green Anal Chem 10:100142. https://doi.org/10.1016/j.greeac.2024.100142
13. Carasek E, Bernardi G, Morelli D, Merib J (2021) Sustainable green solvents for microextraction techniques: recent developments and applications. J Chromatogr A 1640:461944. https://doi.org/10.1016/j.chroma.2021.461944
14. Abbott AP, Capper G, Davies DL, Rasheed RK, Tambyrajah V (2003) Novel solvent properties of choline chloride/urea mixtures. Chem Commun 2003:70–71. https://doi.org/10.1039/B210714G
15. Salamat Q, Brandão JPC, Santos Junior AF, Soylak M (2025) Deep eutectic solvent-based ferrofluids: synthesis strategies, characteristics, and applications for trace analytes extraction. TrAC Trends Anal Chem 184:118126. https://doi.org/10.1016/j.trac.2024.118126
16. Andrade DC, Monteiro SA, Merib J (2022) A review on recent applications of deep eutectic solvents in microextraction techniques for the analysis of biological matrices. Adv Sample Prep 1:100007. https://doi.org/10.1016/j.sampre.2022.100007
17. Paiva A, Craveiro R, Aroso I, Martins M, Reis RL, Duarte ARC (2014) Natural deep eutectic solvents—solvents for the 21st century. ACS Sustain Chem Eng 2(5):1063–1071. https://doi.org/10.1021/sc500096j
18. Álvarez MS, Longo MA, Rodríguez A, Deive FJ (2024) The role of deep eutectic solvents in catalysis: a vision on their contribution to homogeneous, heterogeneous, and electrocatalytic processes. J Ind Eng Chem 132:36–49. https://doi.org/10.1016/j.jiec.2023.11.030
19. Di Carmine G, Abbott AP, D'Agostino C (2021) Deep eutectic solvents: alternative reaction media for organic oxidation reactions. React Chem Eng 6(4):582–598. https://doi.org/10.1039/D0RE00458H
20. Makoś-Chełstowska P (2023) VOCs absorption from gas streams using deep eutectic solvents—a review. J Hazard Mater 448:130957. https://doi.org/10.1016/j.jhazmat.2023.130957
21. Manafpour AA, Feyzi F, Rezaee M (2024) An environmentally friendly deep eutectic solvent for CO_2 capture. Sci Rep 14:19744. https://doi.org/10.1038/s41598-024-70761-4
22. Brett CMA (2024) Perspectives for the use of deep eutectic solvents in the preparation of electrochemical sensors and biosensors. Curr Opin Electrochem 45:101465. https://doi.org/10.1016/j.coelec.2024.101465
23. Cai Y (2019) Application of deep eutectic solvents in chromatography: a review. TrAC Trends Anal Chem 120:115623. https://doi.org/10.1016/j.trac.2019.115623
24. van Osch DJGP, Zubeir LF, van den Bruinhorst A, Rocha MAA, Kroon MC (2015) Hydrophobic deep eutectic solvents as water-immiscible extractants. Green Chem 17:4518–4521. https://doi.org/10.1039/C5GC01451D
25. Makoś-Chełstowska P, Słupek E, Gębicki J (2020) Hydrophobic deep eutectic solvents in microextraction techniques—a review. Microchem J 152:104384. https://doi.org/10.1016/j.microc.2019.104384
26. Bintanel-Cenis J, Fernández MA, Gómara B, Ramos L (2024) Critical overview on the use of hydrophobic (deep) eutectic solvents for the extraction of organic pollutants in complex matrices. Talanta 270:125599. https://doi.org/10.1016/j.talanta.2023.125599
27. Hu C, Feng J, Cao Y, Chen L, Li Y (2024) Deep eutectic solvents in sample preparation and determination methods of pesticides: recent advances and future prospects. Talanta 266:125092. https://doi.org/10.1016/j.talanta.2023.125092

28. Plastiras O-E, Samanidou V (2022) Applications of deep eutectic solvents in sample preparation and extraction of organic molecules. Molecules 27:7699. https://doi.org/10.3390/molecules27227699
29. Werner J, Zgoła-Grześkowiak A, Płatkiewicz J, Płotka-Wasylka J, Jatkowska N, Kalyniukova A, Zaruba S, Andruch V (2023) Deep eutectic solvents in analytical sample preconcentration part B: solid-phase (micro)extraction. Microchem J 191:108898. https://doi.org/10.1016/j.microc.2023.108898
30. Montagner CC, Vidal C, Acayaba RD (2017) Contaminantes emergentes em matrizes aquáticas do Brasil: cenário atual e aspectos analíticos, ecotoxicológicos e regulatórios. Quím Nova 40:1094–1110. https://doi.org/10.21577/0100-4042.20170091
31. Scheid C, Mello W, Buchner S, Benvenutti EV, Deon M, Merib J (2023) Efficient analysis of polycyclic aromatic hydrocarbons by dispersive-μ-solid-phase extraction using silica-based nanostructured sorbent phases coupled to gas chromatography-mass spectrometry. Adv Sample Prep 7:100070. https://doi.org/10.1016/j.sampre.2023.100070
32. Quintanilla I, Fontàs C, Anticó E (2024) Deep eutectic solvents incorporated in a polymeric film for organophosphorus pesticide microextraction from water samples. Anal Chim Acta 1318:342940. https://doi.org/10.1016/j.aca.2024.342940
33. Hristozova A, Vidal L, Aguirre MÁ, Simitchiev K, Canals A (2025) Natural deep eutectic solvent-based dispersive liquid-liquid microextraction of pesticides in drinking waters combined with GC-MS/MS detection. Talanta 282:126967. https://doi.org/10.1016/j.talanta.2024.126967
34. Morelli DC, Bernardi G, Morés L, Pierri ME, Carasek E (2020) A green-high throughput-extraction method based on hydrophobic natural deep eutectic solvent for the determination of emerging contaminants in water by high performance liquid chromatography–diode array detection. J Chromatogr A 1626:461377. https://doi.org/10.1016/j.chroma.2020.461377
35. Wang Y, Shen L, Yan Y, Gong B, Chen K, Zhu G, Li Z (2024) Ultrasound-assisted upper critical solution temperature type switchable deep eutectic solvent-based liquid-liquid microextraction for the determination of triazole in water. Anal Chim Acta 1328:343172. https://doi.org/10.1016/j.aca.2024.343172
36. Alves MS, Ferreira Neto LC, Scheid C, Merib J (2022) An overview of magnetic ionic liquids: from synthetic strategies to applications in microextraction techniques. J Sep Sci 45:258–281. https://doi.org/10.1002/jssc.202100599
37. Makoś Chełstowska P, Kaykhaii M, Płotka-Wasylka J, de la Guardia M (2022) Magnetic deep eutectic solvents—fundamentals and applications. J Mol Liq 365:120158. https://doi.org/10.1016/j.molliq.2022.120158
38. Shi R, Zhou F, Chen Y, Liu Z, Liu S, Mu T (2022) Magnetic deep eutectic solvents: formation and properties. Phys Chem Chem Phys 24:20073–20081. https://doi.org/10.1039/D2CP01592G
39. Mehrabi F, Ghaedi M, Alipanahpour Dil E, Tayebi L (2024) Deep eutectic solvent-based ferrofluid for highly efficient preconcentration and determination of metronidazole by vortex-assisted liquid-phase microextraction under experimental design optimization. Talanta 272:125705. https://doi.org/10.1016/j.talanta.2024.125705
40. Makoś-Chełstowska P, Kucharska K, Słupek E, Gębicki J, de la Guardia M (2023) Magnetic deep eutectic solvents as efficient media for extraction of furfural and 5-hydroxymethylfurfural from aqueous samples. J Mol Liq 390:122945. https://doi.org/10.1016/j.molliq.2023.122945
41. Ferreira KC, Rodrigues T, Isquibola G, Ayala-Durán SC, Merib J, de Lima Gomes PCF, Gonzalez MH, Anderson JL (2024) Determination of parabens and bisphenol A in sludge samples using hydrophobic deep eutectic solvents by gas chromatography coupled to mass spectrometry and sample introduction via pyrolizer. Microchem J 207:112057. https://doi.org/10.1016/j.microc.2024.112057
42. Zhao L, Zhang Z, Jiang H, Guo Y, Chen Z, Wang X, Jing X (2023) Hydrophilic and hydrophobic deep eutectic solvent-based extraction to determine parathion in cereals by digital image colorimetry integrated with smartphones. Talanta 265:124831. https://doi.org/10.1016/j.talanta.2023.124831

43. Anastassiades M, Lehotay SJ, Stajnbaher D, Schenck FJ (2003) Fast and easy multiresidue method employing acetonitrile extraction/partitioning and dispersive solid-phase extraction for the determination of pesticide residues in produce. J AOAC Int 86(2):412–431
44. González-Curbelo MA, Varela-Martínez DA, Riaño-Herrera DA (2022) Pesticide-residue analysis in soils by the QuEChERS method: a review. Molecules 27:4323. https://doi.org/10.3390/molecules27134323
45. Soltani S, Sereshti H (2022) A green alternative QuEChERS developed based on green deep eutectic solvents coupled with gas chromatography-mass spectrometry for the analysis of pesticides in tea samples. Food Chem 380:132181. https://doi.org/10.1016/j.foodchem.2022.132181
46. Della Posta S, Ascrizzi AM, Pietrangeli G, Terrigno V, De Arcangelis E, Ruggeri S, Fanali C (2024) Miniaturized matrix solid-phase dispersion assisted by deep eutectic solvent for acrylamide determination in bread samples. Anal Bioanal Chem 416:1117–1125. https://doi.org/10.1007/s00216-024-05315-8
47. Hao Y, Huo B, Wang F, Xie Q, Liang W, Jia L, Guo H, Wu Y, Wang Q (2024) Water-based green deep eutectic solvent: application in liquid–liquid microextraction of trace bisphenol a in edible oils. Talanta 286:127511. https://doi.org/10.1016/j.talanta.2024.127511
48. Mu G, Yan S, Pan F, Xu H, Jing X, Xue X (2024) Based on theoretical design simultaneous analysis of multiple neonicotinoid pesticides in beeswax by deep eutectic solvents extraction combined with UHPLC-MS/MS. Food Chem X 15:102073. https://doi.org/10.1016/j.fochx.2024.102073
49. Soylak M, Salamat Q, Sajjad S (2024) The usability of green deep eutectic solvents in hollow fiber liquid-phase microextraction for the simultaneous extraction of analytes of different natures: a comprehensive study. Spectrochim Acta A Mol Biomol Spectrosc 319:124552. https://doi.org/10.1016/j.saa.2024.124552
50. Wu Y-W, Zhang J, Ma X-R, Cao J (2025) Simultaneous extraction of flavonoids and saponins from functional food via basic deep eutectic solvent-assisted mechanochemical extraction. Food Chem 472:142925. https://doi.org/10.1016/j.foodchem.2025.142925
51. Wang M, Yue Y, Zhang J, Qin Y, Jia L, Jing X (2024) Dispersive liquid-liquid microextraction based on magnetic deep eutectic solvent for the determination of succinate dehydrogenase inhibitors in water, juice, wine, and vinegar. J Food Compos Anal 133:106411. https://doi.org/10.1016/j.jfca.2024.106411
52. Usman I, Hussain M, Imran A, Afzaal M, Saeed F, Javed M, Afzal A, Ashfaq I, Al Jbawi E, Saewan SA (2022) Traditional and innovative approaches for the extraction of bioactive compounds. Int J Food Prop 25:1215–1233. https://doi.org/10.1080/10942912.2022.2074030
53. Haq HU, Boczkaj G, Altunay N (2025) Magnetic deep eutectic solvents for vortex-assisted liquid phase microextraction method for selective and green extraction of patulin from fruits and fruit juices. Microchem J 208:112472. https://doi.org/10.1016/j.microc.2024.112472
54. Yang W, Wang R, Wang L, Tan Z (2025) Switchable deep eutectic solvents for the extraction of phlorizin and trilobatin from sweet tea (Lithocarpus litseifolius (Hance) Chun). Sep Purif Technol 285:129898. https://doi.org/10.1016/j.seppur.2024.129898
55. Zhang Y, Li H, Hai X, Guo X, Di X (2024) Designing green and recyclable switchable supramolecular deep eutectic solvents for efficient extraction of flavonoids from Scutellariae radix and mechanism exploration. J Chromatogr A 1730:465084. https://doi.org/10.1016/j.chroma.2024.465084
56. Jovanović MS, Milutinović M, Radan M, Krgović N, Ćujić Nikolić N, Menković N, Randjelović M, Kitić D, Šavikin K (2024) Ultrasound-assisted natural deep eutectic solvents (NaDES) extraction of gentiopicroside, isovitexin, and isogentisin from Gentiana asclepiadea L. biowaste. Sustain Chem Pharm 42:101808. https://doi.org/10.1016/j.scp.2024.101808
57. Plaza M, Marina M, Marina M (2025) Hydrophobic natural deep eutectic solvents for the sustainable extraction of carotenoids from persimmon peels. Microchem J 208:112421. https://doi.org/10.1016/j.microc.2024.112421
58. Mammana SB, Canizo BV, Elia-Dazat R, Gomez FJV, Silva MF, Silva MF (2024) Solid eutectic systems and natural deep eutectic solvents in matrix solid-phase dispersion for the extraction of phenolic compounds from lettuce samples. J Food Compos Anal 135:106661. https://doi.org/10.1016/j.jfca.2024.106661

59. Bouchouareb K, Combès A, Pichon V (2022) Determination of nerve agent biomarkers in human urine by a natural hydrophobic deep eutectic solvent-parallel artificial liquid membrane extraction technique. Talanta 249:123704. https://doi.org/10.1016/j.talanta.2022.123704
60. García-Atienza P, Armenta S (2024) The use of deep eutectic solvent-liquid phase microextraction as green analytical sample treatment for the analysis of drugs in urine. Green Anal Chem 9:100115. https://doi.org/10.1016/j.greeac.2024.100115
61. Akramipour R, Golpayegani MR, Gheini S, Fattahi N (2018) Speciation of organic/inorganic mercury and total mercury in blood samples using vortex assisted dispersive liquid-liquid microextraction based on the freezing of deep eutectic solvent followed by GFAAS. Talanta 186:17–23. https://doi.org/10.1016/j.talanta.2018.04.042
62. El-Deen AK, Elmansi H, Shimizu K (2022) Utilization of hydrophilic and hydrophobic deep eutectic solvents for dispersive liquid-liquid microextraction of bicalutamide from water and spiked human plasma. Sustain Chem Pharm 29:100825. https://doi.org/10.1016/j.scp.2022.100825
63. Khezeli T, Daneshfar A (2015) Dispersive micro-solid-phase extraction of dopamine, epinephrine and norepinephrine from biological samples based on green deep eutectic solvents and Fe_3O_4@MIL-100 (Fe) core−shell nanoparticles grafted with pyrocatechol. RSC Adv 5(80):65264–65273. https://doi.org/10.1039/C5RA08058D
64. Nishijo N, Hayama T, Tomita R, Fujioka T (2021) Deep eutectic solvent extraction of cortisol and cortisone from human saliva followed by LC-UV analysis. J Chromatogr B Analyt Technol Biomed Life Sci 1179:122828. https://doi.org/10.1016/j.jchromb.2021.122828
65. Andrade DC, Monteiro SA, Oliveira TE, Merib J (2025) Microextraction of steroidal hormones from urine samples using natural deep eutectic solvents: insights into chemical interactions using molecular dynamics simulations. Anal Bioanal Chem 417(7):1305–1314. https://doi.org/10.1007/s00216-024-05560-x
66. Norouzi F, Khoubnasabjafari M, Jouyban-Gharamaleki V, Soleymani J, Jouyban A, Farajzadeh MA, Afshar Mogaddam MR (2020) Determination of morphine and oxymorphone in exhaled breath condensate samples: application of microwave enhanced three-component deep eutectic solvent-based air-assisted liquid-liquid microextraction and derivatization prior to gas chromatography-mass spectrometry. J Chromatogr B Analyt Technol Biomed Life Sci 1152:122256. https://doi.org/10.1016/j.jchromb.2020.122256
67. Pasupuleti SB, Nanda Kumar A, Reddy PV, Rani PU, Reddy AR (2022) Eco-friendly magnetic solid-phase extraction and deep eutectic solvent for the separation of parabens in environmental water and urine samples. Microchem J 178:107330. https://doi.org/10.1016/j.microc.2022.107330
68. Shishov AY, Pochivalov AS, Dubrovsky IS, Bulatov AV (2023) Deep eutectic solvents with low viscosity for automation of liquid-phase microextraction based on lab-in-syringe system: separation of Sudan dyes. Talanta 255:124243. https://doi.org/10.1016/j.talanta.2022.124243
69. Yıldırım S, Cocovi-Solberg DJ, Uslu B, Solich P, Horstkotte B (2022) Lab-in-syringe automation of deep eutectic solvent-based direct immersion single drop microextraction coupled online to high-performance liquid chromatography for the determination of fluoroquinolones. Talanta 246:123476. https://doi.org/10.1016/j.talanta.2022.123476
70. Morelli DC, Morés L, Colombari E, Carasek E (2024) Quantification of phthalate esters in carbonated soft drinks by hollow fiber-microporous membrane liquid-liquid extraction and 96-well plate system using natural deep eutectic solvents based on fatty acids. J Sep Sci 47(24):1–9. https://doi.org/10.1002/jssc.70060
71. Mestre-Manrique F, Payà-Pou R, Beneito-Cambra M, Simó-Alfonso EF, Carrasco-Correa EJ (2023) Is 3D printing a good alternative to prepare novel devices for green analytical sample preparation? Adv Sample Prepar 6:100062. https://doi.org/10.1016/j.sampre.2023.100062
72. Li F, Rodas Ceballos M, Keshan Balavandy S, Fan J, Khataei MM, Yamini Y, Maya F (2020) 3D Printing in analytical sample preparation. J Sep Sci 43(9–10):1854–1866. https://doi.org/10.1002/jssc.202000035
73. Monteiro SA, Scheid C, Deon M, Merib J (2023) Fundamentals, recent applications, and perspectives of 3D printing in sample preparation approaches. Microchem J 195:109385. https://doi.org/10.1016/j.microc.2023.109385

74. Díaz-Álvarez M, Turiel E, Martín-Esteban A (2025) Natural deep eutectic solvent-based liquid phase microextraction in a 3D-printed millifluidic flow cell for the on-line determination of thiabendazole in juice samples. Anal Chim Acta 1339:343617. https://doi.org/10.1016/j.aca.2025.343617
75. Cheng K, Xu X, Song J, Chen Y, Kan Z, Li C (2024) Molecular dynamics simulations of choline chloride and ascorbic acid deep eutectic solvents: investigation of structural and dynamics properties. J Mol Graph Model 130:108784. https://doi.org/10.1016/j.jmgm.2024.108784
76. Panda DK, Bhargava BL (2022) Molecular dynamics investigation of non-ionic deep eutectic solvents. J Mol Graph Model 113:108152. https://doi.org/10.1016/j.jmgm.2022.108152
77. Lucci E, Falcinelli G, Antonelli L, Dal Bosco C, Felli N, De Cesaris MG, Gentili A (2024) Hydrophobic deep eutectic solvent-ferrofluid microextraction followed by liquid chromatography-mass spectrometry for the enantioselective determination of chiral agrochemicals in natural waters. Anal Bioanal Chem 417:1341. https://doi.org/10.1007/s00216-024-05619-9
78. Godunov PA, Gerasimova AA, Shishov AY, Bulatov AV (2024) Temperature-responsive deep eutectic solvent-based microextraction for the determination of bisphenols in beverages by HPLC-FLD. J Food Compos Anal 126:106569. https://doi.org/10.1016/j.jfca.2024.106569
79. Salamat Q, Soylak M (2024) Novel reusable and switchable deep eutectic solvent for extraction and determination of curcumin in water and food samples. Talanta 269:125401. https://doi.org/10.1016/j.talanta.2023.125401
80. Zhang H, Zhang Y, Wang Y, Wang X, Zhao F, Xu J, Han J (2024) Sustainable ultrasound-assisted extraction of osthole from Cnidium monnieri (L.) cuss based on natural deep eutectic solvents and switchable hydrophilicity solvents. Ind Crops Prod 222:120140. https://doi.org/10.1016/j.indcrop.2024.120140
81. Hai TC, Phan Van M, Hong Anh LT (2024) Optimization of ultrasound-assisted extraction of astaxanthin from black tiger shrimp (Penaeus monodon) shells using deep eutectic solvent and ethanol as a co-solvent. LWT Food Sci Technol 212:116965. https://doi.org/10.1016/j.lwt.2024.116965
82. Wu YW, Ma XR, Peng X, Cai X, Cao J, Ye LH (2024) Polyamide and deep eutectic solvent assisted matrix solid-phase dispersion microextraction for strong hydrophobic compounds. Sustain Chem Pharm 40:101679. https://doi.org/10.1016/j.scp.2024.101679
83. García-Atienza P, Martínez-Pérez-Cejuela H, Herrero-Martínez JM, Armenta S (2024) Liquid phase microextraction based on natural deep eutectic solvents of psychoactive substances from biological fluids and natural waters. Talanta 267:125277. https://doi.org/10.1016/j.talanta.2023.125277
84. Majidi SM, Hadjmohammadi MR (2020) Hydrophobic borneol-based natural deep eutectic solvents as a green extraction media for air-assisted liquid-liquid micro-extraction of warfarin in biological samples. J Chromatogr A 1621:461030. https://doi.org/10.1016/j.chroma.2020.461030
85. Damokhi A, Yousefinejad S, Jafari S, Soleimani E, Dehghani F (2023) Deep eutectic solvent-based dispersive liquid-liquid microextraction coupled with HPLC and central composite design for biological monitoring of hippuric acid and methyl hippuric acid in urine. J Mol Liq 386:122506. https://doi.org/10.1016/j.molliq.2023.122506
86. Hansen FA, Santigosa-Murillo E, Ramos-Payán M, Muñoz M, Øiestad EL, Pedersen-Bjergaard S (2020) Electromembrane extraction using deep eutectic solvents as the liquid membrane. Anal Chim Acta 1143:109–116. https://doi.org/10.1016/j.aca.2020.11.044
87. Abbasi H, Abbasi S, Haeri SA, Rezayati S, Kalantari F, Poor Heravi MR (2022) Electromembrane extraction using biodegradable deep eutectic solvents and agarose gel as green and organic solvent-free strategies for the determination of polar and non-polar basic drugs from biological samples: a comparative study. Anal Chim Acta 1221:339986. https://doi.org/10.1016/j.aca.2022.339986

Chapter 11
Current Status and Recent Advances in Hydrophobic Deep Eutectic Solvents for Drug Extraction and Separation

Nur Nadhirah Mohamad Zain, Nor Munira Hashim, Noorfatimah Yahaya, Hikmat, and Kartika A. Madurani

11.1 Introduction

Deep eutectic solvents (DESs) have recently attracted considerable attention as an environmentally conscious and innovative class of solvents, presenting a sustainable alternative to traditional organic solvents and selected ionic liquids (Fig. 11.1) [1, 2]. Initially reported by Abbott et al. in 2003, DESs result from mixing a hydrogen bond donor (HBD) with a hydrogen bond acceptor (HBA), forming a eutectic system whose melting point is significantly lower than that of its constituents [3, 4]. This decrease in melting point is attributed to expensive hydrogen bonding, imparting unique properties such as low volatility, adjustable polarity, high thermal resistance, and strong solubilization capabilities [2, 5].

The appeal of DESs lies not only in their physicochemical characteristics but also in their alignment with the green chemistry principles [2, 6]. Natural DESs (NADESs), in particular, are favoured for their ease of synthesis, low toxicity, and biodegradable nature, making them highly suitable for pharmaceutical, analytical,

N. N. M. Zain (✉) · N. M. Hashim · N. Yahaya
Department of Toxicology, Advanced Medical and Dental Institute (AMDI), Universiti Sains Malaysia, Kepala Batas, Penang, Malaysia
e-mail: nurnadhirah@usm.my; normunira@student.usm.my; noorfatimah@usm.my

Hikmat
Research Center for Chemistry, National Research and Innovation Agency (BRIN), Tangerang Selatan, Indonesia
e-mail: hikmat@brin.go.id

K. A. Madurani
Laboratory of Instrumentation and Analytical Sciences, Chemistry Department, Faculty of Science and Data Analytics, Institut Teknologi Sepuluh Nopember, Arief Rahman Hakim, Surabaya, Indonesia
e-mail: kartika.madurani@its.ac.id

K. Kuperkar et al. (eds.), *Deep Eutectic Solvents*, Environmental Footprints and Eco-design of Products and Processes,
https://doi.org/10.1007/978-3-032-03801-2_11

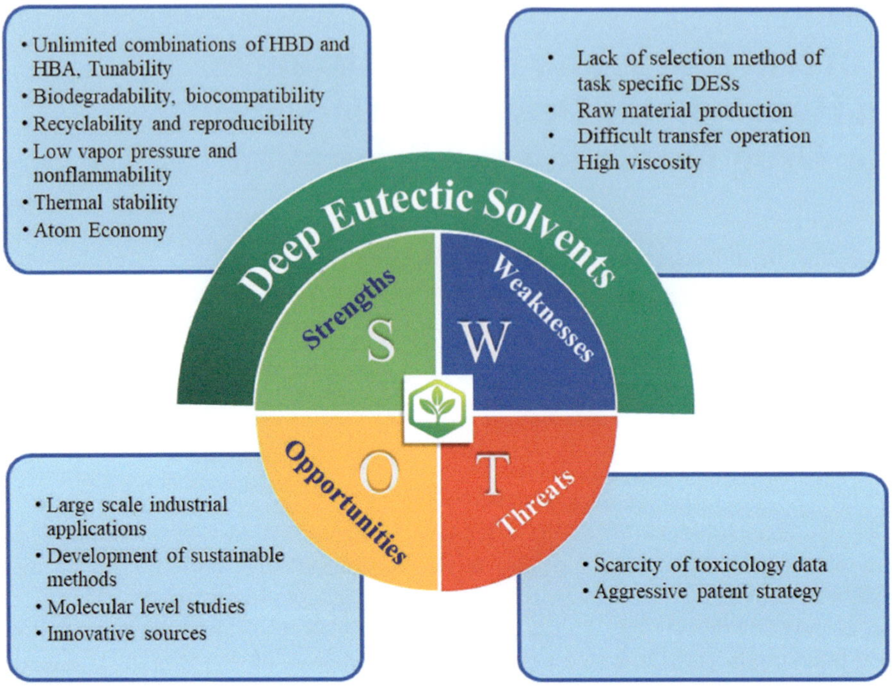

Fig. 11.1 Green features of deep eutectic solvents. Reproduced with permission from Reference [2]

biocatalytic, and environmental applications [7–14]. Their modular nature also allows customisation through deliberate selection of HBD and HBA combinations, as well as molar ratios, to fine-tune performance [5, 15].

DESs can be broadly classified into hydrophilic and hydrophobic categories based on their water compatibility, which directly influences their suitability for different extraction environments. Hydrophilic DESs have demonstrated efficacy in extracting polar analytes from aqueous and plant-based matrices, while hydrophobic DESs (HDESs) have emerged more recently as effective media for extracting nonpolar or lipophilic compounds from complex matrices [16, 17]. This chapter elaborates on the synthesis, classification, and diverse uses of both types, emphasising pharmaceutical extractions and separations. Additionally, it addresses existing limitations, technological developments, and prospective research directions for establishing DESs as mainstream green solvents.

11.2 Classification and Types of Deep Eutectic Solvents (DESs)

Deep eutectic solvents (DESs) are a category of green solvents created by blending a hydrogen bond donor (HBD) and a hydrogen bond acceptor (HBA), leading to a notable substantial drop in melting point relative to the separate constituents.

Originally reported by Abbott et al. using a 1:2 molar ratio of choline chloride and urea, DESs exhibit remarkable versatility based on their composition and physicochemical traits [3]. DESs are categorized into distinct classifications according to their composition, physicochemical characteristics, and applications. These solvents are transforming sectors such as medicines, green chemistry, and biotechnology through their customisation and sustainability. DESs are widely used in biocatalysis, material synthesis, and separation processes [18].

According to their composition and the characteristics of their components, DESs are mainly classified as:

11.2.1 Natural DESs (NADESs)

Natural deep eutectic solvents (NADESs) represent a subclass of DESs developed from naturally occurring substances like sugars, amino acids, polyols, and organic acids, making them suitable for green applications [19, 20]. Because of their low toxicity, high degradability, and renewable origin, NADESs offer a strong alternative to conventional organic solvents. Typical HBAs used in NADES preparation include quaternary ammonium compounds (e.g., choline chloride) and alcohols, while HBDs consist of biomolecules like glucose, fructose, citric acid, lactic acid, proline, and arginine. Strong hydrogen bonding among these constituents facilitates lower melting points and high solvating power [21]. Their physical features include low vapour pressure, high viscosity, modifiable polarity, excellent solubility for bioactive substances, and good biocompatibility [22]. Applications span pharmaceutical and cosmetic extractions, food and nutraceutical formulation, antimicrobial treatments, and biomass transformation [23–27]. However, the high viscosity of NADESs can be a limitation, potentially reducing their extraction efficiency [28].

11.2.2 Hydrophilic and Hydrophobic DESs

Hydrophilic DESs exhibit excellent solubility in water, whereas hydrophobic DESs are employed in scenarios where the exclusion of water is essential [29]. The DESs are categorized according to their water affinity into hydrophilic DESs (HDESs) and hydrophobic DESs (HDESs). The difference between these two categories greatly impacts their use across multiple sectors, such as pharmaceuticals, food chemistry, environmental sciences, and material processing.

11.2.2.1 Hydrophilic DESs

Hydrophilic DESs are generally composed of HBAs like choline chloride, urea, or amino acids, paired with HBDs such as sugars, polyols, or weak organic acids (e.g., citric or lactic acid). They dissolve easily in water, making them efficient for extracting polar molecules [16]. These solvents are environmentally benign, thermally stable, and exhibit low volatility, supporting their use in pharmaceutical formulations, catalysis, environmental remediation, and sustainable chemistry [16, 30–32].

11.2.2.2 Hydrophobic DESs

Hydrophobic DESs are formed using nonionic or water-repelling components such as menthol, thymol, and long-chain quaternary ammonium salts as HBAs, and HBDs include fatty acids like decanoic acid or natural terpenes [33]. These systems do not mix with water and show high affinity for nonpolar compounds, offering enhanced extraction efficiency for lipophilic targets. They are biodegradable, exhibit low toxicity, and are widely used in fields such as environmental science, food safety, and renewable energy [30, 34–36].

11.2.2.3 Comparison of Hydrophilic Vs Hydrophobic DESs

Hydrophilic and hydrophobic DESs vary notably in their solubility, selectivity, and application scope, as summarized in Table 11.1.

The stability and scalability challenges of both types arise from their sensitivity to water absorption, necessitating the optimization of formulations for broader industrial applications [17, 37].

11.2.3 Hydrogen Bond Donor (HBD) and Acceptor (HBA) Systems

DESs are categorized based on the nature of their hydrogen bond donors and acceptors into five primary types, labelled Type I to Type V. Each type features distinct structural and functional traits suited for specific applications.

Table 11.1 Comparison of hydrophilic vs hydrophobic DESs

Feature	Hydrophilic DESs	Hydrophobic DESs
Water solubility	High	Low
Best for	Polar compounds	Nonpolar compounds
Applications	Drug delivery, food, and catalysis	Bioactive extraction, VOC removal, green energy
Examples	Choline chloride + urea	Menthol + decanoic acid

11.2.3.1 Type I

This type involves a mixture of quaternary ammonium salts with anhydrous metal salts. Such combinations typically lead to the formation of ionic complexes that exhibit significant melting point depression due to charge delocalization. Examples include choline chloride (HBA) mixed with $ZnCl_2$, $AlCl_3$, or $SnCl_2$ (HBD) [38, 39]. Type I DESs are characterized by high ionic conductivity and notable thermal stability, and their properties can be fine-tuned by adjusting the metal salt types or molar ratio. Applications of this class of DESs include electroplating, catalysis, and energy storage [38, 41–43].

11.2.3.2 Type II

This variant combines quaternary ammonium salts with hydrated metal salts. The water molecules in these salts assist in forming strong hydrogen bonds and stabilize ionic interactions, enhancing solubility and thermal performance. Typical components include choline chloride (HBA) and metal salts like $FeCl_3{\cdot}6H_2O$ and $CuSO_4{\cdot}5H_2O$ (HBD) [38]. They are primarily used in catalysis, electrochemical reactions, and green extraction processes [40, 44–46].

11.2.3.3 Type III

This category pairs quaternary ammonium salts as HBAs with organic HBDs such as alcohols, amides, or carboxylic acids. Choline chloride or tetrabutylammonium chloride are common HBAs, while HBDs may include lactic acid, glycerol, or urea [47]. These DESs offer significant flexibility due to their low melting points and strong hydrogen bonding, making them ideal for extraction, biodegradation, and green chemical processes [42, 48–53].

11.2.3.4 Type IV

Type IV DESs consist of metal salts, often transition metal salts, which act as hydrogen bond acceptors, while organic compounds like amides or alcohols serve as hydrogen bond donors. The systems demonstrate unique coordination interactions among their components, making them suitable for various advanced applications. The metal salts commonly employed in the synthesis of type IV DESs comprise $ZnCl_2$, $FeCl_3$, and $AlCl_3$. Currently, the applied HBD comprises organic compounds such as urea, amides, or alcohols, including ethylene glycol. The establishment of type IV DESs is significantly dependent on the interaction between metal ions and HBD molecules [48]. This process involves hydrogen bonding and coordination between metal salts and organic hydrogen bond donors, leading to a reduced melting point and the creation of a liquid [54]. The main characteristics displayed by this class are:

- Thermal Stability: high stability, suitable for processes requiring elevated temperatures.
- Customizability: properties like viscosity and polarity can be tuned by varying the HBA/HBD ratio.
- Coordination Strength: the nature of metal-ligand bonds defines the solvent's function [55].

Type IV DESs have been utilized in extractive desulfurization, catalysis, metal recovery, and electrochemical applications [46, 48, 54, 56].

11.2.3.5 Type V

Distinct from other types, Type V DESs are formed solely from nonionic, hydrophobic substances. Common HBAs include camphor and lidocaine, while HBDs can be menthol, thymol, or various carboxylic acids. These solvents lack ionic species, making them water-immiscible and well-suited for extracting nonpolar analytes. Their low melting points and hydrophobic nature support applications in pesticide removal, green media, and solvent extraction [57–59]. Type V DESs have been utilized in a variety of applications, such as solvent extraction for metal recovery and pesticide removal, electromembrane extraction (EME), catalysis and reaction media, green chemistry, and microemulsion [60–64].

11.2.4 Organic Acid DESs (OA-DESs)

Organic acid-based deep eutectic solvents (OA-DESs) incorporate organic acids as HBDs, providing strong antibacterial and therapeutic traits [65]. These systems often use choline chloride, sugars, or urea as HBAs, paired with glycolic, malic, or citric acids [20]. Advantages of this DESs class include high biocompatibility, tunable polarity, and biodegradability [66, 67]. Organic acid DESs have found applications in various fields, including biomedical and pharmaceutical sectors, as well as in antimicrobial and antibacterial uses, lignocellulose processing, biomass conversion, catalysis, and green chemistry [20, 68–70]. The challenges remain with toxicity control and upscaling for industrial use [20, 68, 71].

11.2.5 Therapeutic DESs (THEDESs)

Therapeutic deep eutectic solvents (THEDESs) are specialised DESs that incorporate active pharmaceutical ingredients (APIs) as one or more components. These solvents improve drug solubility, bioavailability, and delivery efficiency [72, 73].

Formulations often include choline chloride, citric acid, and APIs acting as either HBA and HBD [74–76]. Therapeutic DESs have been employed in enhancing drug solubility, facilitating transdermal and liposomal drug delivery, and in applications related to antimicrobial and anticancer treatments, as well as in managing diabetes and metabolic diseases [72, 77–80]. The challenges and future directions for therapeutic DESs include issues of toxicity and biocompatibility, as well as stability and scalability, along with the need for formulation optimization [75, 76].

11.2.6 Biocompatible DESs

Biocompatible DESs are tailored for safe use in biological environments and often include naturally derived or physiologically benign constituents. HBAs commonly include choline chloride, betaine, and amino acids, while HBDs range from sugars and polyols to weak acids like lactic acid [81]. Their properties, such as low toxicity, biodegradability, and high solubilising capability, make them suitable for drug delivery, enzyme stabilisation, biosensors, and biomedical coatings. These solvents contribute significantly to the advancement of biosafe and eco-friendly chemical technologies [78, 81–84].

11.3 Hydrophilic Deep Eutectic Solvents (DESs)

Innovative solvent systems known as deep eutectic solvents (DESs) are created from two or more components, usually a hydrogen bond donor (HBD) and a hydrogen bond acceptor (HBA) [85]. They interact to form a liquid mixture with a significantly lower melting point than the individual constituents. Stupar et al. (2021) introduced a switchable-hydrophilicity eutectic solvent system as a successful way to separate carotenoids previously extracted from pumpkin (*Cucurbita maxima*). They used a natural deep eutectic solvent (NADES, composed of 4 mol NH_4OH and 1 mol of C8:C10 (Caprylic acid: Capric acid (3:1)). The switchable-hydrophilicity eutectic solvent system successfully recovered 90% of the total carotenoids in the extract [86]. In other research, Babu Balaraman et al. (2022), purified thrombolytic cysteine protease from Carica papaya peels using ultrasound-assisted dispersive liquid-phase microextraction using eutectic switchable solvents (ChCl/Octanoate) and CO_2 for phase dispersion [87]. Chen et al. (2023) and Yue et al. (2023) have used the introduction of CO_2 into DESs or NADESs as a transition mechanism to automatically assign bioactive compounds to the hydrophilic or hydrophobic layer of the mixture based on their polarity [88, 89]. This makes the recovery of target compounds much more effective and gives higher results. This is due to the reaction between the H_2CO_3 generated by the addition of CO_2 and the weak acid ions of HBD or HBA, which generates an imbalance in the eutectic system by breaking the original hydrogen bond between HBA and HBD [89]. Hydrophobic DESs or NADESs have nonpolar properties, while hydrophilic DESs or NADESs are polar subjects.

The main components of hydrophilic DESs are carboxylic acids, amines, and natural sugars [90]. Their composition supports their biodegradable nature and is consistent with green chemistry principles, which promote the use of eco-friendly materials in chemical reactions [91]. These solvents' nontoxic nature reduces possible health hazards to people and the environment, making them good substitutes for conventional organic solvents, which are usually poisonous or dangerous [85, 92]. Hydrophilic DESs have distinct qualities, such as low vapor pressure and customizable physicochemical properties, that allow for a variety of extraction techniques while also supporting sustainability. One of the most important properties of these solvents is their capacity to increase the solubility of polar molecules and aqueous solutes at ambient temperatures. This property eliminates the significant energy expenditures associated with solvent evaporation and heating. The lower temperature not only saves energy but also stabilizes delicate bioactive chemicals during extraction [93]. As a result, hydrophilic DESs have received attention for their novel uses in green extraction procedures.

Research has demonstrated that the solvation behavior of hydrophilic DESs is largely determined by the nature and content of the HBDs and HBAs utilized, allowing for greater flexibility in developing solvents that target certain bioactive molecules [94–96]. For example, employing choline chloride as an HBA in conjunction with other HBDs (such as urea or citric acid) can aid in the efficient extraction of a wide range of bioactive chemicals while retaining safety and efficacy [94, 97]. This versatility is especially advantageous for sectors seeking greener manufacturing and more efficient resource utilization.

11.3.1 Interaction Mechanisms and Extraction Efficiency

In practice, hydrophilic DESs have been implemented in a variety of extraction applications, which have served to illustrate their effectiveness in the isolation of valuable compounds from a wide range of matrices, such as plant materials and agricultural products [94]. For example, researchers have used DESs to improve phenolic chemicals, antioxidants, and essential oils extraction from plants, demonstrating their utility in food science and biomedical chemistry [98]. The tuning of extraction parameters, such as the molar ratio of DES components and water content, has increased extraction [99]. The development and application of hydrophilic deep eutectic solvents not only pave the way for more sustainable extraction operations but also uphold the principles of efficiency and environmental accountability in chemical research and industry [99, 100].

The extraction effectiveness of hydrophilic DESs is intrinsically linked to the specific interactions of solvent components and solutes. When generating a hydrophilic DES, strong hydrogen bonding networks occur between the HBD and HBA, lowering the melting point and allowing the combination to remain liquid at room temperature. These hydrogen bonds greatly boost the solvent's capacity to dissolve polar solutes such as organic acids, sugars, and phenolic compounds, resulting in higher extraction yields compared to traditional solvents [85].

The interaction specificity between solutes and hydrophilic DES components can also be modified by adjusting the composition of the DES. For example, adjusting the molar ratios of HBDs and HBAs can improve the solvent's polarity and viscosity, hence changing the extraction performance. El Achkar et al. (2021) highlight the need to achieve a balance in these ratios to enable maximum solvation of specific biomolecules while minimizing viscosity impediment. Furthermore, researchers discovered that adding water to hydrophilic DESs drastically alters solvation dynamics and extraction efficiency, resulting in improved dissolution capacities for particular bioactive chemicals [98].

Moreover, the impact of temperature and pH on the extraction process cannot be overstated. These parameters play critical roles in the ionization state of solutes, subsequently affecting their solubility in hydrophilic DESs [99]. For example, increasing the pH may lead to the deprotonation of acidic compounds, enhancing their solubility and extraction potential. Recent advances in computational modeling and laboratory experiments have provided novel insights into how solutes interact with DESs at the molecular level, leading to the development of more effective extraction strategies [101, 102]. Understanding the fundamental principles of extraction with hydrophilic DESs is critical for establishing optimum methods for specific applications. Identifying the appropriate physicochemical conditions for extraction could result in considerable increases in the yield and purity of extracted chemicals, leading to superior quality products in medicines and nutraceuticals, among other uses.

11.3.2 Customization and Optimization of Hydrophilic DESs

The customization and modification of hydrophilic DESs are critical to assuring their efficacy in extraction applications. Researchers can formulate DESs with specialized characteristics designed to extract specific bioactive chemicals. This unique technique frequently entails selecting HBDs that are compatible with the desired solute, allowing for maximum solvation and extraction efficiency [103]. For example, employing amino acids like glycine or citric acid as HBDs in conjunction with other HBAs, such as choline chloride, can produce DESs that are more effective at dissolving a variety of phytochemicals due to their specific hydrogen bonding capabilities [97].

Optimization strategies for hydrophilic DESs frequently involve altering parameters such as component molar ratios, the presence of additives (water or salts), and operational variables such as temperature and time [104, 105]. Extensive research has demonstrated that even small changes in these variables can significantly affect extraction yields and solvent performance. For example, adding water to a hydrophilic DES can greatly increase the dissolving capacity of certain chemicals while also influencing viscosity and overall extraction kinetics, which is critical for scaling up extraction procedures in industrial settings. Furthermore, systematic optimization techniques such as response surface methodology (RSM) or factorial design allow researchers to quickly discover the ideal operational parameters for each application [106, 107]. These approaches can reveal interactions between many parameters, providing insights into how to enhance extraction efficiencies while producing desired product quality. Research indicates that applying these advanced methodologies can lead to a more data-driven knowledge of how personalization affects the performance of hydrophilic DESs in real-world applications.

As the field moves forward, continued innovations in the customization and optimization of hydrophilic DESs are expected. Ongoing research aims to discover additional HBDs and HBAs, increasing the range of hydrophilic DES formulations and their applicability across many sectors while keeping sustainability as a top priority.

11.3.3 Application of Hydrophilic DESs

Numerous case studies have demonstrated the versatility and usefulness of hydrophilic DESs in a variety of applications across sectors. One significant area of application is the pharmaceutical industry, where DESs have been successfully used to extract bioactive chemicals from plants. Meanwhile, research has shown that employing a DES from choline chloride and urea resulted in better extraction rates of flavonoids and alkaloids than traditional organic solvents [108, 109]. This not only improves the active pharmaceutical ingredients (APIs) purity but also helps to ensure the overall extraction process sustainability by reducing the usage of harmful solvents.

In the food industry, hydrophilic DESs are being widely implemented to improve the extraction of functional ingredients, such as phenolic compounds from various fruits and vegetables. Research indicates that these solvents can effectively preserve the antioxidant properties of extracted compounds, provide health benefits while also enhancing food quality [110, 111]. For example, using DESs composed of natural components has shown promising results in extracting antioxidants from olive oil, yielding extracts with significant health-promoting properties [112]. This approach not only supports sustainable food processing methods but also aligns with consumer preferences for health-conscious products.

The application of hydrophilic DESs extends into the cosmetics industry as well. They serve to extract and stabilize bioactive compounds that possess beneficial properties for skin health. Studies have illustrated how DES formulations can effectively solubilize vitamins, essential oils, and antioxidants while maintaining their bioactivity and ensuring compatibility with cosmetic formulations [113]. This has led to increased interest among cosmetic manufacturers in utilizing hydrophilic DESs as safe and effective alternatives to traditional extraction methods involving harsh chemicals.

Researchers highlight the potential of hydrophilic DESs as adaptable and eco-friendly solvents across a wide range of industries by investigating their innovative uses. These case studies demonstrate the efficacy of DESs in extracting valuable biocompounds while meeting consumer demand for sustainable and nontoxic formulations, thereby encouraging the wider use of hydrophilic DESs in a variety of fields, including pharmaceuticals, food, and cosmetics.

11.4 Hydrophobic DES Synthesis (Common HBA and HBD)

Hydrophobic deep eutectic solvents (Hydrophobic DESs) emerged as a response to the limitations of hydrophilic deep eutectic solvents, which, despite their widespread application in extraction, catalysis, and green chemistry, faced obstacles related to water solubility and selectivity [17, 33]. Traditional hydrophilic DESs tend to dissolve in aqueous systems, leading to difficulties in phase separation and limiting their efficiency in extracting hydrophobic compounds [16, 114]. These drawbacks motivated researchers to develop HDES, which maintains the low toxicity, tunability, and biodegradability of DES while offering water immiscibility and enhanced selectivity for nonpolar and hydrophobic analytes [67]. The first reports on HDES by van Osch et al. (2015) primarily focused on their use in liquid–liquid extraction and environmental applications [115]. Since then, HDES have been increasingly explored for their potential in sample preparation, pollutant removal, and pharmaceutical analysis, providing a more versatile and efficient alternative to their hydrophilic counterparts [7, 8, 116, 117].

The synthesis of hydrophobic DES plays a critical role in defining their physicochemical properties, extraction efficiency, and overall applicability in drug extraction and separation. The preparation of hydrophobic DES typically involves the straightforward combination of a hydrogen bond acceptor (HBA) and a hydrogen bond donor (HBD), which self-assemble via hydrogen bonding interactions to form a eutectic mixture with a significantly lower melting point than the individual components [17, 118]. The concept of DES was first introduced by Abbott et al. in 2003 using hydrophilic DES, but their limited ability to extract nonpolar and hydrophobic compounds from aqueous matrices prompted researchers to shift towards the development of hydrophobic DES [3]. This innovation addresses the challenge of extracting lipophilic drugs and environmental pollutants that are poorly soluble in water.

The synthesis of hydrophobic DES typically involves the careful selection of hydrophobic HBAs and/or hydrophobic HBDs to achieve both low water solubility and strong extractive performance for nonpolar compounds [114, 119]. The HBAs commonly used in HDES include tetrabutylammonium bromide (TBAB) and tetraethylammonium chloride (TEAC), which exhibit low polarity and high thermal stability [120, 121]. Meanwhile, the HBDs are often composed of long-chain fatty acids (e.g., decanoic acid, dodecanoic acid), natural monoterpenes (e.g., menthol, thymol), and hydrophobic alcohols (e.g., 1-undecanol, octanol), which contribute to the overall hydrophobic nature of the solvent. The eutectic mixture is typically prepared by mixing the HBA and HBD at specific molar ratios (1:1 to 1:4) under mild heating (40–80 °C) with continuous stirring until a homogeneous liquid is formed [122–125].

Overall, the synthesis of hydrophobic DES provides a versatile and tunable platform for developing green extraction solvents with enhanced selectivity, efficiency, and sustainability. The choice of HBA, HBD, and their molar ratio significantly influences the viscosity, polarity, and hydrophobicity of the solvent, offering researchers a powerful tool to design customized extraction systems for various pharmaceutical and environmental applications. The growing interest in HDES highlights their potential as a green alternative to conventional organic solvents, aligning with green chemistry principles and green analytical techniques.

11.5 Application of HDES in Drug Extraction and Separation

11.5.1 Liquid–Liquid Extraction

Diverse types of hydrophobic DES have been utilized for the extraction of pharmaceuticals from diverse sample matrices, including food, water, and biological fluids, using liquid–liquid extraction (LLE) techniques (Table 11.2). In the context of pharmaceutical extraction from food samples, Sereshti et al. (2020) utilized a hydrophobic DES composed of octanoic acid as an HBA and thymol as an HBD in dispersive liquid–liquid microextraction (DLLME) paired with high-performance liquid chromatography-ultraviolet detection (HPLC-UV) for the extraction and quantification of tetracycline in milk. Their study achieved low limits of detection (LOD) and quantification (LOQ), varying from 1.5 to 8.5 μg/kg and 5.1 to 28.4 μg/kg, respectively [7]. Similarly, thymol was utilized by Qiao et al. (2021) in combination with guanidinium chloride as an HBA for the extraction of non-steroidal anti-inflammatory drugs (NSAIDs) from water and milk samples, achieving LODs between 0.5 and 1.0 μg/kg (Fig. 11.2) [8].

Additionally, thymol has been employed as an HBD for the extraction of fluoroquinolones from milk and shrimp (Barbayanov, Timofeeva & Bulatov, 2022) and sulfonamides from milk (Pochivalov et al., 2021). In contrast, Sereshti et al. (2022) applied thymol as an HBA for the extraction of tetracyclines from infant formula, highlighting its versatile role in HDES formulations. The widespread use of thymol as both an HBA and an HBD can be attributed to its ability to form intermolecular interactions with pharmaceutical compounds through its aromatic ring structure [126–128].

Beyond thymol-based HDES, Li et al. (2023) developed an extraction method using methyltrioctylammonium chloride as the HBA and capric acid as the HBD to preconcentrate sulfonamides from fish samples [129]. Similarly, Taşpınar and Altunay (2023) employed tetrabutylammonium bromide, a long-chain natural fatty

Table 11.2 Application of HDES-based liquid–liquid extraction in extracting pharmaceuticals from different sample matrices

HDES							
HBA	HBD	Analytical method	Pharmaceuticals	Sample matrix	LOD/LOQ (μg/kg)	Recovery (%)	Ref.
Octanoic acid	Thymol	DLLME/HPLC-UV	Tetracyclines	Milk	1.5–8.5/5.1–28.4	70.0–102.0	[7]
Guanidinium chloride	Thymol	UA-DLLME/HPLC-UV	NSAIDs	Water and milk	0.5–1.0/1.0–5.0	79.4–107.5	[8]
Tetramethylammonium chloride	Alpha terpineol	LLME/HPLC-MS	β-Blockers	Human plasma	0.130–0.205/0.435–0.692	72.0–86.0	[116]
Thymol	Acetic acid	VA-DLLME/UHPLC-DAD	Sulfonamides	Water	0.78–3.42/2.38–10.37	78.2–100.7	[131]
Methyltrioctylammonium chloride	Capric acid	VA-DLLME/HPLC-PDA	Sulfonamides	Fish	5.3–9.0	96.6–113.5	[129]
Nonanoic acid	Dodecanoic acid	USAEME/Spectrofluorimeter	Hepatitis C antiviral drug	Human plasma	1.42/4.30	92.38–93.82	[146]
Tetrabutylammonium bromide	1-dodecanol	UA-DLLME/UV–VIS spectrophotometer	Sulfadiazine (sulfonamide)	Dairy products and honey	0.60/2.00	89.0–98.0	[122]
Benzyltriethylammonium chloride	Thymol	DLLME/HPLC-PDA	Sulfonamides	Environmental water	0.30–1.10/1.10–3.80	84.0–112.0	[123]
Ethyl n-phenylcarbamate	Thymol	VA-DLLME/HPLC-UV	Famciclovir	Environmental water	0.20/0.70	86.0–118.0	[132]
Dodecanoic acid	DL-menthol	LLME/HPLC-UV	Sex hormones	Cosmetics	2.33–3.30/6.99–9.90	85.7–101.3	[147]
Heptanoic	Thymol	HLLME/LC-MS/MS	Quinolones and sulfonamides	Wastewater	0.0021–0.0334/0.0073–0.1114	70.1–115.3	[133]
4-methoxybenzaldehyde	Thymol	LLME/HPLC-UV	Isoniazid	Human plasma	20.0/60.0	95.0	[136]
Methyltrioctylammonium bromide	Capryl alcohol	DLLME-MPAR/LC-UV	Antidepressants	Human plasma and urine	0.020–0.028/0.068–0.092	93.8–111.6	[138]
Heptanoic acid	Thymol	EA-DLLME/HPLC-FLD	Fluoroquinolones	Milk and shrimp	0.03–0.6	86.0–99.0	[126]

(continued)

Table 11.2 (continued)

HDES							
HBA	HBD	Analytical method	Pharmaceuticals	Sample matrix	LOD/LOQ (μg/kg)	Recovery (%)	Ref.
Nonanoic acid	Thymol	LLME/LC-FLD	Fluoroquinolones	Environmental water	0.01–0.04/0.03–0.12	94.0–97.0	[134]
Menthol	Hexanoic acid	DLLME/ HPLC-FLD	Enrofloxacin	Milk	7.00/20.00	85.0	[130]
Octanoic acid	Thymol	LLME/LC-UV	Sulfonamides	Milk	1.00–5.00	77.0–99.0	[127]
Menthol	n-octanol	AA-LLME/UV–vis spectrophotometer	Anti-conjunctivitis olopatadine	Eye drops	46.00/100.00	95.52–104.95	[140]
Thymol	Ethylene glycol and benzyl alcohol	VA-ELLME/ HPLC- UV	Tetracyclines	Infant formulas	0.88–2.74/3.00–9.32	68.9–102.0	[128]
L-menthol	Camphor sulphonic acid	LLME/HPLC-UV	Morphine and codeine	Human whole blood	0.5/1.5	83.0–166.0	[139]
Menthol	Thymol	DLLME/LC-UV	Pacritinib	Human blood serum	25.0/100.0	100.88–111.01	[137]
Formic acid	L-menthol	HF-LPME/ HPLC-UV	Sulfonamides	Environmental water	0.61–3.23/2.05–10.78	53.0–121.0	[135]

DLLME Dispersive liquid–liquid microextraction, *HPLC-UV* High-performance liquid chromatography-ultraviolet detector, *UA-DLLME* Ultrasonication assisted-dispersive liquid–liquid microextraction, *LLME* Liquid–liquid microextraction, *HPLC-MS* High-performance liquid chromatography-mass spectrometry, *VA-DLLME* Vortex-assisted dispersive liquid–liquid microextraction, *UHPLC-DAD* Ultra-high-performance liquid chromatography-diode array detector, *HPLC-PDA* High-performance liquid chromatography-photodiode array, *USAEME* Ultrasound-assisted emulsification microextraction, *SF-μSPE* Syringe filter-based micro-solid phase extraction, *HLLME* Homogenous liquid–liquid microextraction, *LC-MS/MS* Liquid chromatography-triple quadrupole mass spectrometry, *DLLME-MPAR* dispersed liquid–liquid microextraction with magnetic particles-assisted retrieval, *EA-DLLME* effervescence-assisted dispersive liquid–liquid microextraction, HPLC-FLD High-performance liquid chromatography-fluorometric detection, *LLE* liquid–liquid extraction, *HF-LPME* Hollow fiber liquid-phase microextraction

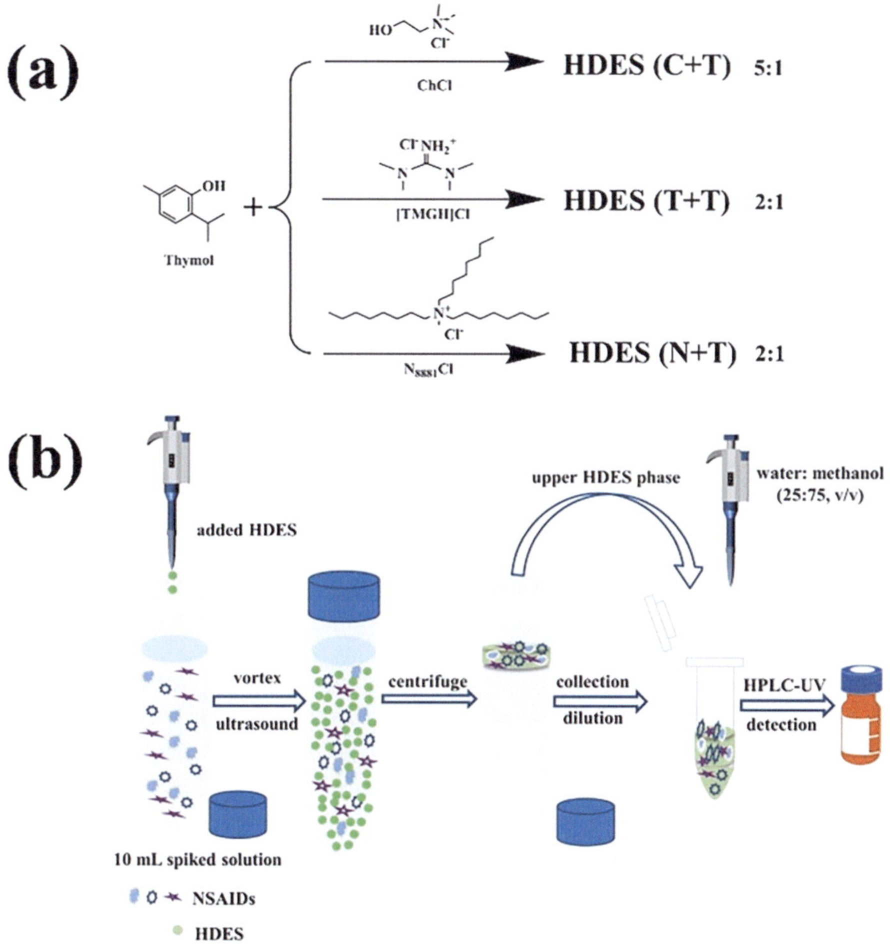

Fig. 11.2 Schematic illustrations of the preparation and chemical structural formula of HDESs (**a**) and the experimental procedure of UA-HDES-DLLME (**b**). Reproduced with permission from Reference [8])

acid, as the HBA in combination with 1-dodecanol as the HBD for the extraction of sulfadiazine, a sulfonamide antibiotic, from dairy products and honey [122]. In addition, Kurashov et al. (2024) implemented a hydrophobic DES composed of menthol and hexanoic acid in a supramolecular solvent (SUPRAS)-based extraction for the determination of enrofloxacin in milk [130]. These studies collectively demonstrate the effectiveness of hydrophobic DES in pharmaceutical extraction from food matrices, with variations in HBA and HBD combinations tailored to enhance extraction efficiency for specific drug classes.

In water matrices, thymol is the main component in hydrophobic DES formulations, functioning as either an HBA or an HBD. Mostafa et al. (2022) employed thymol as the HBA in combination with acetic acid for the extraction of sulfonamides from water samples [131]. Conversely, thymol was utilized as an HBD in several studies, including those conducted by Wu et al. (2023), Li et al. (2024), Zhang et al. (2024), and Pochivalov et al. (2023), for the extraction of sulfonamides and famciclovir from environmental water [123, 132–134]. In addition to thymol-based hydrophobic DES, Díaz-Álvarez and Martín-Esteban applied a different hydrophobic DES system composed of formic acid and menthol in hollow fiber liquid-phase microextraction (HF-LPME) for the preconcentration of sulfonamides from environmental water, demonstrating the versatility of hydrophobic DES in pharmaceutical extraction from aqueous matrices [135].

Many researchers have applied hydrophobic DES in the LLE method for extracting pharmaceuticals from biological samples such as human plasma, serum, and urine. Jouyban et al. (2021) utilized a combination of tetramethylammonium chloride and alpha-terpineol as a hydrophobic DES to extract β-blockers from human plasma [116]. Additionally, thymol has been a preferred component in hydrophobic DES formulations for pharmaceutical extraction from biological samples. Meshcheva et al. (2024) employed a mixture of 4-methoxybenzaldehyde and thymol as a hydrophobic DES to extract isoniazid, an antibiotic used for tuberculosis treatment, from human plasma matrices [136]. Treder et al. (2024) used a combination of menthol and thymol in dispersive liquid–liquid microextraction to preconcentrate pacritinib, an anticancer drug, from human blood serum [137]. Similarly, Han et al. (2024) utilized methyltrioctylammonium bromide and capryl alcohol as a hydrophobic DES to preconcentrate antidepressants from human plasma and urine [138]. Binjawhar and Mohammedsaeed (2024) applied a combination of L-menthol and camphor sulfonic acid to extract morphine and codeine from human whole blood [139]. Notably, only one study has explored the application of hydrophobic DES in a matrix other than food, environmental water, or biological samples by Magdy Saad et al. (2024), where they used a hydrophobic DES composed of menthol and n-octanol to extract olopatadine, an anti-conjunctivitis medication, from eye drop samples [140].

11.5.2 Solid-Phase Extraction

Several studies have employed hydrophobic deep eutectic solvents (HDES) in solid-phase extraction (SPE) methods for the analysis of biological and environmental water samples (Table 11.3). Pourvali Abarbakouh et al. (2024) synthesized magnetic ferrofluid using hydrophobic DES composed of benzyltriphenylphosphonium bromide and 1-undecanol, which was thereafter applied in dispersive solid-phase microextraction (DSPME) paired with liquid chromatography-ultraviolet detector (LC-UV) for the determination of tamoxifen and its metabolites in human plasma [141]. Guo et al. (2024) utilized hydrophobic DES based on nonionic surfactant Brij 35 and hexafluoroisopropanol in the preparation of molecularly imprinted polymers (MIPs). The MIP was applied in SPE combined with dispersive liquid–liquid microextraction (DLLME) for the preconcentration of estrogens from human serum samples, achieving limits of detection between 0.08 and 0.40 μg/kg [142].

Table 11.3 Application of HDES-based solid-phase extraction in extracting pharmaceuticals from different sample matrices

HDES		Sorbents/ analytical method	Pharmaceuticals	Sample matrix	LOD/LOQ (μg/kg)	Recovery (%)	Ref.
HBA	HBD						
Benzyltriphenylphosphonium bromide	1-undecanol	Magnetic ferrofluid/ DSPME/LC-UV	Tamoxifen and its metabolites	Human plasma	0.20–200	94.3–105.1	[141]
Nonionic surfactant Brij 35	Hexafluoroisopropanol	MIP/SPE combined with DLLME/ HPLC-UV	Estrogens	Human serum	0.08–0.4/0.3–1.5	94.6–104.4	[142]
Tetraethylammonium chloride	Thymol	Ferrofluid/ DSPME/ HPLC-UV	Clozapine	Human urine	2.8–3.1/9.2–10.2	91.4–98.4	[143]
Menthol	Dodecanoic acid	MNP/SBME/ HPLC-DAD	Steroid hormones	Human urine and environmental water	0.269/0.899	91.70–98.58	[148]
Thymol	Menthol	Electrospun nanofibers' mat/ SF-μSPE/ HPLC-UV	Anti-diabetic drugs	Plasma, serum, and wastewater	0.03–0.09/0.10–0.25	97.44–99.33	[145]

DSPME Dispersive solid phase microextraction, *LC-UV* Liquid chromatography- ultraviolet detector, *MIP* Molecular imprinted polymer, *HPLC-UV* High performance liquid chromatography-ultraviolet detector, *MNP* Magnetic nanoparticles, *SBME* Stir bar microextraction, *HPLC-DAD* High performance liquid chromatography-diode array detector

Nia and Hadjmohammadi (2023) employed a hydrophobic DES-based ferrofluid composed of tetraethylammonium chloride and thymol for the extraction of clozapine from human urine samples [143]. Al-Hashimi et al. (2024) applied magnetic nanoparticles (MNPs) functionalized with hydrophobic DES prepared from menthol and dodecanoic acid to extract steroid hormones from human urine and environmental water matrices [144]. Khodayari and Ebrahimzadeh (2023) incorporated HDES composed of thymol and menthol in the fabrication of electrospun nanofiber mats, which served as sorbents for the extraction of antidiabetic drugs from human plasma, serum, and wastewater samples (Fig. 11.3) [145]. These studies demonstrate the remarkable versatility of hydrophobic DES-based materials in solid-phase extraction techniques, highlighting their potential as green, efficient, and modifiable sorbents for the extraction of various pharmaceuticals from complex biological and environmental matrices.

11.6 Challenges, Opportunities and Perspectives

Hydrophobic deep eutectic solvents (HDESs) have recently obtained considerable recognition as an advanced subclass of DESs, addressing many of the critical limitations associated with traditional hydrophilic DESs. While hydrophilic DESs were initially favored for their biodegradability, low toxicity, and tunability, their high water miscibility and limited selectivity toward nonpolar analytes have restricted their efficiency in applications requiring the extraction of hydrophobic compounds from aqueous or complex biological matrices.

In contrast, HDESs—typically composed of hydrophobic hydrogen bond donors and acceptors such as menthol, thymol, fatty acids, or long-chain quaternary ammonium salts—exhibit low water solubility, excellent phase separation, and strong affinity for nonpolar pharmaceutical residues. This immiscibility with water has proven to be a decisive advantage, particularly in dispersive liquid–liquid microextraction (DLLME) and other microextraction techniques, where HDESs enable efficient preconcentration of analytes without the need for additional dispersing solvents.

Nevertheless, several challenges must still be addressed to fully exploit the potential of HDESs. Although their viscosity is generally less than that of hydrophilic DESs, some formulations remain too viscous for high-throughput or automated systems, necessitating careful selection or modification of components. Furthermore, their physicochemical behaviour, particularly in multicomponent or bioactive systems, remains underexplored at the molecular level. Unlike traditional DESs, HDESs often lack comprehensive toxicological profiles, especially when used in food, clinical, or pharmaceutical applications. Therefore, further assessment of their long-term environmental impact and bioaccumulation is essential.

Despite these concerns, HDESs offer a wide range of opportunities that position them at the forefront of green and sustainable extraction technologies. They have demonstrated superior selectivity and enrichment factors in the extraction of antibiotics, non-steroidal anti-inflammatory drugs (NSAIDs), sulfonamides, and hormone residues from milk, blood, urine, water, and cosmetics. This performance has surpassed that of hydrophilic DESs in several analytical scenarios, reinforcing the notion that HDESs are not merely complementary, but in many cases, a functional replacement.

Moreover, recent innovations have expanded the utility of HDESs beyond classical DLLME to include solid-phase extraction (SPE), supramolecular solvent systems (SUPRAS), and nanomaterial-functionalized platforms. The versatility of HDESs is further enhanced by their compatibility with green synthesis methods, their ability to incorporate bio-based components, and their suitability for miniaturized and portable analytical systems.

Looking ahead, key opportunities lie in designing task-specific HDESs tailored to target compound classes, especially those with low aqueous solubility. The integration of computational design tools, machine learning, and molecular modeling can accelerate the rational design of HDESs with optimized extraction efficiency, viscosity, and recyclability. Additionally, regulatory and industrial acceptance of HDESs will benefit from deeper toxicological evaluations and the development of standardized protocols for synthesis and reuse.

In summary, HDESs have proven to be not only a solution to the inherent drawbacks of hydrophilic DESs but also a promising platform for next-generation green analytical solvents. Their hydrophobic nature, combined with tunability and eco-compatibility, supports their continued advancement in pharmaceutical, environmental, and food safety applications.

11.7 Conclusion

Deep eutectic solvents (DESs) represent a major change in solvent chemistry, offering a highly flexible, environmentally friendly, and cheap alternative to traditional organic solvents. Their unique ability to form eutectic mixtures with reduced melting points through hydrogen bonding has enabled their application across a wide variety of scientific disciplines. The versatility of DESs—spanning hydrophilic and hydrophobic systems, natural and synthetic formulations, and therapeutic uses—shows their strong potential in analytical, pharmaceutical, and industrial chemistry. Hydrophilic DESs have been deeply studied for the extraction of polar compounds, while the development of hydrophobic DESs has addressed the need for efficient extraction of nonpolar analytes from aqueous and lipid-rich mixtures. The use of DESs in both liquid–liquid and solid-phase extraction platforms has demonstrated their better performance in drug recovery, environmental analysis, and biomolecule isolation. However, the continued growth of DESs requires focused research to solve current challenges, particularly regarding viscosity control, phase stability, toxicity, and following safety rules. The future path of DESs lies in the combination of green chemistry ideas with better material design, computer modeling, and sustainable engineering methods. In conclusion, DESs are steadily playing an

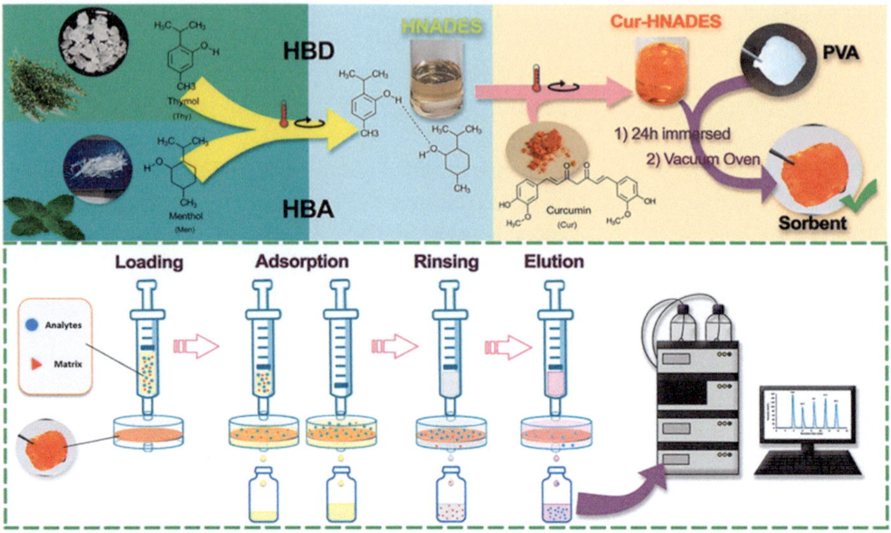

Fig. 11.3 Summarization of QuEChERS syringe filter-based micro-solid phase extraction for the extraction of antidiabetic drugs from human plasma, serum, and wastewater samples. Reproduced with permission from Reference [145]

important role in the future of green analytical chemistry, drug formulation, and sustainable separation methods. Their natural flexibility, environmental safety, and growing uses make them valuable tools in the search for greener and more effective chemical processes.

Declaration of Generative AI and AI-Assisted Technologies in the Writing Process While preparing this work, the authors utilized AI-assisted tools to enhance the language and readability of several statements in the manuscript. After using this tool/service, the authors reviewed and edited the content as needed and take full responsibility for the publication content.

Conflict of Interest Statement *The authors declare no conflict of interest.*

References

1. Lu W, Chen H (2022) Application of deep eutectic solvents (DESs) as trace level drug extractants and drug solubility enhancers: state-of-the-art, prospects and challenges. J Mol Liq 349:118105
2. Prabhune A, Dey R (2023) Green and sustainable solvents of the future: deep eutectic solvents. J Mol Liq 379:121676
3. Abbott AP, Capper G, Davies DL, Rasheed RK, Tambyrajah V (2003) Novel solvent properties of choline chloride/urea mixtures. Chem Commun 1:70–71
4. El Achkar T, Fourmentin S, Greige-Gerges H (2019) Deep eutectic solvents: An overview on their interactions with water and biochemical compounds. J Mol Liq 288:111028

5. Shafique S, Belousov AS, Rashid R, Shafiq I, Aziz KHH, Riaz N, Khan MS, Shaheen A, Ishaq M, Akhter P, Hussain M (2025) Deep eutectic solvents (DES): structure, properties, and cutting-edge applications in green catalysis. J Mol Liq 419:126769
6. Santana-Mayor Á, Rodríguez-Ramos R, Herrera-Herrera AV, Socas-Rodríguez B, Rodríguez-Delgado MÁ (2021) Deep eutectic solvents. The new generation of green solvents in analytical chemistry. TrAC Trends Anal Chem 134:116108
7. Sereshti H, Semnani Jazani S, Nouri N, Shams G (2020) Dispersive liquid–liquid microextraction based on hydrophobic deep eutectic solvents: application for tetracyclines monitoring in milk. Microchem J 158:105269
8. Qiao L, Sun R, Yu C, Tao Y, Yan Y (2021) Novel hydrophobic deep eutectic solvents for ultrasound-assisted dispersive liquid-liquid microextraction of trace non-steroidal anti-inflammatory drugs in water and milk samples. Microchem J 170:106686
9. Panić J, Papović S, Šarac B, Cerc Korošec R, Rapaić M, Gadžurić S, Bešter-Rogač M, Vraneš M (2023) Active pharmaceutical ingredient ionic liquid in natural deep eutectic solvents—influence of solvent composition. J Mol Liq 384:122282
10. Ganorkar SB, Hadole PM, Patil MR, Pardeshi CV, Bobade PS, Shirkhedkar AA, Vander Heyden Y (2025) Deep eutectic solvents in analysis, delivery and chemistry of pharmaceuticals. Int J Pharm 672:125278
11. Putra SSS, Basirun WJ, Elgharbawy AAM, Hayyan A, Hayyan M, Mohammed MA (2022) Nanocellulose and natural deep eutectic solvent as potential biocatalyst system toward enzyme immobilization. Mol Catalysis 528:112422
12. Zhang Y, Wang Q, Lin Z, Xia J, Suo H, Huang R, Zou B (2025) Efficient synthesis of isoamyl acetate in deep eutectic solvent utilizing in-situ aqueous phase immobilized lipase biocatalyst. Food Biosci 64:105963
13. Rozas S, Martín-Martel S, Aguilar N, Gutiérrez A, Marcos PA, Bol A, Atilhan M, Mele A, Aparicio S (2025) An in silico study on the carbon capture performance of menthol-based natural deep eutectic solvent. Fuel 392:134784
14. Qian J, Li Y, Su F, Liu S, Wang J, Li W (2025) High-performance adsorbents from spent coffee grounds modified with natural deep eutectic solvents. J Environ Chem Eng 13(3):116206
15. Omar KA, Sadeghi R (2022) Physicochemical properties of deep eutectic solvents: a review. J Mol Liq 360:119524
16. Tang W, An Y, Row KH (2021) Emerging applications of (micro) extraction phase from hydrophilic to hydrophobic deep eutectic solvents: opportunities and trends. TrAC Trends Anal Chem 136:116187
17. Zainal-Abidin MH, Hayyan M, Wong WF (2021) Hydrophobic deep eutectic solvents: current progress and future directions. J Ind Eng Chem 97:142–162
18. Zhang Q, De Oliveira Vigier K, Royer S, Jérôme F (2012) Deep eutectic solvents: syntheses, properties and applications. Chem Soc Rev 41(21):7108–7146
19. Ling JK, Hadinoto K (2022) Deep eutectic solvent as green solvent in extraction of biological macromolecules: a review. Int J Mol Sci 23:3381
20. Swebocki T, Barras A, Abderrahmani A, Haddadi K, Boukherroub R (2023) Deep eutectic solvents comprising organic acids and their application in (bio)medicine. Int J Mol Sci 24:8492
21. González-Laredo RF, Sayago-Monreal VI, Moreno-Jiménez MR, Rocha-Guzmán NE, Gallegos-Infante JA, Landeros-Macías LF, Rosales-Castro M (2023) Natural deep eutectic solvents (NaDES) as an emerging technology for the valorisation of natural products and agro-food residues: a review. Int J Food Sci Technol 58(12):6660–6673
22. Satija P, Chambyal A, Singh G, Ritvik G, Singh S, Devi JS (2024) Natural deep eutectic solvents (NADES): manufacture, characteristics, and their significance as designer solvents. ChemistrySelect 9(22):e202401212
23. Al-Akayleh F, Khalid RM, Hawash D, Al-Kaissi E, Al-Adham ISI, Al-Muhtaseb N, Jaber N, Al-Remawi M, Collier PJ (2022) Antimicrobial potential of natural deep eutectic solvents. Lett Appl Microbiol 75(3):607–615

24. Jauregi P, Esnal-Yeregi L, Labidi J (2024) Natural deep eutectic solvents (NADES) for the extraction of bioactives: emerging opportunities in biorefinery applications. PeerJ Anal Chem 6:e32
25. Villa C, Caviglia D, Cuna FSR d, Zuccari G, Russo E (2024) NaDES application in cosmetic and pharmaceutical fields: an overview. Gels 10:107
26. Wils L, Hilali S, Boudesocque-Delaye L (2021) Biomass valorization using natural deep eutectic solvents: what's new in France? Molecules 26:6556
27. Wu K, Ren J, Wang Q, Nuerjiang M, Xia X, Bian C (2022) Research Progress on the preparation and action mechanism of natural deep eutectic solvents and their application in food. Foods 11:3528
28. Jurić T, Uka D, Holló BB, Jović B, Kordić B, Popović BM (2021) Comprehensive physicochemical evaluation of choline chloride-based natural deep eutectic solvents. J Mol Liq 343:116968
29. Mu L, Gao J, Zhang Q, Kong F, Zhang Y, Ma Z, Sun C, Lv S (2023) Research progress on deep eutectic solvents and recent applications. Processes 11:1986
30. Mastellone G, Abbasi NM, Cagliero C, Anderson JL (2023) New class of tunable choline bromide-based hydrophobic deep eutectic solvents for the extraction of bioactive compounds of varying polarity from a plant matrix. ACS Sustain Chem Eng 11(17):6665–6675
31. Sereshti H, Karami F, Nouri N (2021) A green dispersive liquid-liquid microextraction based on deep eutectic solvents doped with β-cyclodextrin: application for determination of tetracyclines in water samples. Microchem J 163:105914
32. Shishov A (2022) Deep eutectic solvents in chemical analysis. Opportunities and limitations, transactions of the Kola science Centre of RAS. Ser Eng Sci 13:278–282
33. Cao J, Su E (2021) Hydrophobic deep eutectic solvents: the new generation of green solvents for diversified and colorful applications in green chemistry. J Clean Prod 314:127965
34. Boateng ID (2022) A critical review of emerging hydrophobic deep eutectic solvents' applications in food chemistry: trends and opportunities. J Agric Food Chem 70(38):11860–11879
35. Chen C-C, Huang Y-H, Fang J-Y (2022) Hydrophobic deep eutectic solvents as green absorbents for hydrophilic VOC elimination. J Hazard Mater 424:127366
36. Salehi HS, Celebi AT, Vlugt TJH, Moultos OA (2021) Thermodynamic, transport, and structural properties of hydrophobic deep eutectic solvents composed of tetraalkylammonium chloride and decanoic acid. J Chem Phys 154(14):144502
37. Kivelä H, Salomäki M, Vainikka P, Mäkilä E, Poletti F, Ruggeri S, Terzi F, Lukkari J (2022) Effect of water on a hydrophobic deep eutectic solvent. J Phys Chem B 126(2):513–527
38. Mishra DK, Gopakumar G, Pugazhenthi G, Rao CVSB, Nagarajan S, Banerjee T (2021) Molecular and spectroscopic insights into a metal salt-based deep eutectic solvent: a combined quantum theory of atoms in molecules, noncovalent interaction, and density functional theory study. Chem A Eur J 125(44):9680–9690
39. Shi R, Yu D, Zhou F, Yu J, Mu T (2022) An emerging deep eutectic solvent based on halogen-bonds. Chem Commun 58(29):4607–4610
40. Karim WO, Juma JA, Omer KM, Aziz KHH, Salah YM, Aziz SB (2021) Effect of copper ion and water on anodic dissolution of metallic copper in a deep eutectic solvent (DES). Electrochemistry 89(1):71–74
41. Elgammal RA, Fraenza CC, Greenbaum S, Zawodzinski TA (2022) (Digital presentation) Fundamentals of deep eutectic solvents As electrolytes for lithium-ion batteries, Electrochemical Society Meeting Abstracts 242. The Electrochemical Society, Inc., pp 1719–1719
42. Shi M, Xiong W, Tu Z, Zhang X, Hu X, Wu Y (2021) Task-specific deep eutectic solvents for the highly efficient and selective separation of H2S. Sep Purif Technol 276:119357
43. Zhang M, Hao S, Chenghang Z, Shaojun L, Zhenglong L, Yi L, Weihong WX, Gao (2021) Highly efficient selective extraction of Mo with novel hydrophobic deep eutectic solvents. J Air Waste Manage Assoc 71(12):1492–1501

44. D. Makhota, O. Sukhatskyi, T. Butyrina, & , V. Protsenko, Application of deep eutectic solvents to prepare Electrocatalysts for green hydrogen production Materials science and surface engineering (2023) 18–22
45. Rashid SN, Hizaddin HF, Hayyan A, Chan SE, Hasikin K, Razak SA, Mokhtar MI, Azizan MM (2024) A kinetic study of ex-situ soil remediation by nickel extraction using natural deep eutectic solvent. Environ Technol 45(23):4820–4833
46. Tang N, Liu L, Yin C, Zhu G, Huang Q, Dong J, Yang X, Wang S (2021) Environmentally benign hydrophobic deep eutectic solvents for palladium (II) extraction from hydrochloric acid solution. J Taiwan Inst Chem Eng 121:92–100
47. Paveglio GC, Milani FA, Sauer AC, Roman D, Meyer AR, Pizzuti L (2021) Structure-physical properties relationship of eutectic solvents prepared from benzyltriethylammonium chloride and carboxylic acids. J Braz Chem Soc 32(3):542–551
48. Abbasi NM, Farooq MQ, Anderson JL (2021) Modulating solvation interactions of deep eutectic solvents formed by ammonium salts and carboxylic acids through varying the molar ratio of hydrogen bond donor and acceptor. J Chromatogr A 1643:462011
49. Qu W, Qader IB, Abbott AP (2022) Controlled release of pharmaceutical agents using eutectic modified gelatin. Drug Deliv Transl Res 12(5):1187–1194
50. Aghapoor K, Mohsenzadeh F, Darabi HR, Sayahi H (2022) Choline chloride-based eutectic mixtures for greener synthesis of Quinoxaline-2, 3-diol derivatives and Terephthalaldehyde bis-(2-Aminophenylimine). Org Prep Proced Int 54(2):157–166
51. Bowen H, Durrani R, Delavault A, Durand E, Chenyu J, Yiyang L, Lili S, Jian S, Weiwei H, Fei G (2022) Application of deep eutectic solvents in protein extraction and purification. Front Chem 10:912411
52. Elgammal RA, Fraenza CC, Zawodzinski TA Jr, Greenbaum S (2023) Lithium ion battery electrolytes based on deep eutectic solvents, electrochemical society meeting abstracts 243, vol MA2023-01. The Electrochemical Society, Inc., pp 547–547
53. Majová V, Jablonský M, Lelovský M (2021) Delignification of unbleached pulp by ternary deep eutectic solvents. Green Process Synthesis 10(1):666–676
54. Xu L, Luo Y, Liu H, Yin J, Li H, Jiang W, Zhu W, Li H, Ji H (2021) Extractive desulfurization of diesel fuel by amide-based type IV deep eutectic solvents. J Mol Liq 338:116620
55. Malinowski A, Śmiechowski M (2024) Computational study of molecular interactions in ZnCl2(urea)2 crystals as precursors for deep eutectic solvents. Crystals 14:641
56. Geng T, Zeller SJ, Kibler LA, Ceblin MU, Jacob T (2022) Electrodeposition of cu onto au(111) from deep eutectic solvents: molar ratio of salt and hydrogen bond donor. ChemElectroChem 9(7):e202101283
57. Boventi M, Mauri M, Castiglione F, Simonutti R (2024) Exploring the structure of type V deep eutectic solvents by xenon NMR spectroscopy. Faraday Discuss 253(0):118–128
58. Srivastava S, Sinha S, Bhattacharjee S, Seth D (2024) Solute dynamics of a hydrophobic molecule in a menthol–thymol based type-V deep eutectic solvent: effect of composition of the components. Phys Chem Chem Phys 26(16):12638–12651
59. Teixeira G, Abranches DO, Ferreira O, Coutinho JAP (2023) Estimating the melting temperatures of type V deep eutectic solvents. Ind Eng Chem Res 62(36):14638–14647
60. Dhingra D, Behera K, Bhawna SP (2021) Formation of water-in-oil microemulsions within a hydrophobic deep eutectic solvent. Phys Chem Chem Phys 23(17):10629–10635
61. Hansen FA, Santigosa-Murillo E, Ramos-Payán M, Muñoz M, Leere Øiestad E, Pedersen-Bjergaard S (2021) Electromembrane extraction using deep eutectic solvents as the liquid membrane. Anal Chim Acta 1143:109–116
62. Kulshrestha A, Pancha I, Mishra S, Kumar A (2021) Deep eutectic solvents and ionic liquid assisted hydrolysis of microalgal biomass: a promising approach towards sustainable biofuel production. J Mol Liq 335:116264
63. Soltani S, Sereshti H, Nouri N (2021) Deep eutectic solvent-based clean-up/vortex-assisted emulsification liquid-liquid microextraction: application for multi-residue analysis of 16 pesticides in olive oils. Talanta 225:121983

64. Vargas SJR, Pérez-Sánchez G, Schaeffer N, Coutinho JAP (2021) Solvent extraction in extended hydrogen bonded fluids—separation of Pt(iv) from Pd(ii) using TOPO-based type V DES. Green Chem 23(12):4540–4550
65. Swebocki T, Kocot AM, Barras A, Arellano H, Bonnaud L, Haddadi K, Fameau A-L, Szunerits S, Plotka M, Boukherroub R (2024) Comparison of the antibacterial activity of selected deep eutectic solvents (DESs) and deep eutectic solvents comprising organic acids (OA-DESs) toward gram-positive and gram-negative species. Adv Healthc Mater 13(14):2303475
66. Affat S (2024) A review of deep eutectic solvents (DESs), preparation, classification, physicochemical properties, advantages and disadvantages, University of Thi-Qar. J Science 11(1):167–175
67. Lomba L, Ribate MP, Sangüesa E, Concha J, Garralaga M, Errazquin D, García CB, Giner B (2021) Deep eutectic solvents: are they safe? Appl Sci 11:10061
68. Fan K, Yang B, Yu S, Yang R, Zhang L, Chi W, Yin M, Wu H, Guo J (2023) Ternary choline chloride/benzene sulfonic acid/ethylene glycol deep eutectic solvents for oxidative desulfurization at room temperature. RSC Adv 13(37):25888–25894
69. Lim W-L, Gunny AAN, Kasim FH, Gopinath SCB, Kamaludin NHI, Arbain D (2021) Cellulose nanocrystals from bleached rice straw pulp: acidic deep eutectic solvent versus sulphuric acid hydrolyses. Cellulose 28(10):6183–6199
70. Lomba L, García CB, Ribate MP, Giner B, Zuriaga E (2021) Applications of deep eutectic solvents related to health, synthesis, and extraction of natural based chemicals. Appl Sci 11:10156
71. Lima F, Branco LC, Lapa N, Marrucho IM (2021) Beneficial and detrimental effects of choline chloride–oxalic acid deep eutectic solvent on biogas production. Waste Manag 131:368–375
72. Bakr El-Nassan H (2024) Applications of therapeutic deep eutectic solvents (THEDESs) as antimicrobial and anticancer agents. Pharm Dev Technol 29(10):1084–1092
73. Kalantri SA, Vora (2024) Eutectic solutions for healing: a comprehensive review on therapeutic deep eutectic solvents (TheDES). Drug Dev Ind Pharm 50(5):387–400
74. Alkhawaja B, Al-Akayleh F, Nasereddin J, Kamran M, Woodman T, Al-Rubaye Z, Qinna N, Al-Remawi M, Olaimat AR (2024) Structural insights into novel therapeutic deep eutectic systems with capric acid using 1D, 2D NMR and DSC techniques with superior gut permeability. RSC Adv 14(21):14793–14806
75. Cysewski P, Jeliński T, Przybyłek M, Mai A, Kułak J (2024) Experimental and machine-learning-assisted Design of Pharmaceutically Acceptable Deep Eutectic Solvents for the solubility improvement of non-selective COX inhibitors ibuprofen and Ketoprofen. Molecules 29:2296
76. García CB, Concha J, Culleré L, Lomba L, Sangüesa E, Ribate MP (2023) Has the toxicity of therapeutic deep eutectic systems been assessed? Appl Sci 13:5980
77. Huang C, Chen X, Wei C, Wang H, Gao H (2021) Deep eutectic solvents as active pharmaceutical ingredient delivery Systems in the Treatment of metabolic related diseases. Front Pharmacol 12:794939
78. Oyoun F, Toncheva A, Henríquez LC, Grougnet R, Laoutid F, Mignet N, Alhareth K, Corvis Y (2023) Deep eutectic solvents: An eco-friendly Design for Drug Engineering. ChemSusChem 16(20):e202300669
79. Song J-E, Jun S-H, Ryoo J-Y, Kang N-G (2024) Formulation of ascorbic acid and betaine-based therapeutic deep eutectic system for enhanced transdermal delivery of ascorbic acid. Pharmaceutics 16:687
80. Wu J, Yang S, Yin T, Wang X (2021) Eutectic-based liposome as a potential delivery system of paeonol. RSC Adv 11(62):39343–39348
81. Pedro SN, Freire CSR, Silvestre AJD, Freire MG (2021) Deep eutectic solvents and pharmaceuticals. Encyclopedia 1:942–963
82. Durante-Salmerón DA, Fraile-Gutiérrez I, Gil-Gonzalo R, Acosta N, Aranaz I, Alcántara AR (2024) Strategies to prepare chitin and chitosan-based bioactive structures aided by deep eutectic solvents: a review. Catalysts 14:371

83. Svigelj R, Dossi N, Grazioli C, Toniolo R (2021) Deep eutectic solvents (DESs) and their application in biosensor development. Sensors 21:11060
84. Yadav N, Venkatesu P (2022) Current understanding and insights towards protein stabilization and activation in deep eutectic solvents as sustainable solvent media. Phys Chem Chem Phys 24(22):13474–13509
85. El Achkar T, Greige-Gerges H, Fourmentin S (2021) Understanding the basics and properties of deep eutectic solvents. In: Deep eutectic solvents for medicine, gas solubilization and extraction of natural substances. Springer International Publishing
86. Stupar A, Šeregelj V, Ribeiro BD, Pezo L, Cvetanović A, Mišan A, Marrucho I (2021) Recovery of β-carotene from pumpkin using switchable natural deep eutectic solvents. Ultrason Sonochem 76:105638
87. Babu Balaraman H, Viswanathan G, Muniasamy R, Gayatri T, Kumar Rathnasamy S (2022) Sustainable valorization of papaya peels for thrombolytic cysteine protease isolation by ultrasound assisted disruptive liquid phase microextraction with task specific switchable natural deep eutectic solvents. Microchem J 175:107118
88. Chen J, Xu Y, Tan Z (2023) CO2-triggered switchable hydrophilicity solvent as a recyclable extractant for ultrasonic-assisted extraction of Polygonatum sibiricum polysaccharides. Food Chem 402:134301
89. Yue W, Li X, Jing J, Qi J, Dai S, Lu H, Huang Z (2023) The green CO2-controllable fatty acid-based deep eutectic solvents for sustainable oil separation. J Mol Liq 378:121590
90. Płotka-Wasylka J, de la Guardia M, Andruch V, Vilková M (2020) Deep eutectic solvents vs ionic liquids: similarities and differences. Microchem J 159:105539
91. Anastas PT (2003) Meeting the challenges to sustainability through green chemistry Paul T, Anastas of the Chemistry Department, University of Nottingham in the UK discusses green chemistry as an approach toward meeting the goals of sustainability. Green Chem 5(2):G29–G34
92. Kudłak B, Owczarek K, Namieśnik J (2015) Selected issues related to the toxicity of ionic liquids and deep eutectic solvents—a review. Environ Sci Pollut Res 22(16):11975–11992
93. Zain NNM, Yahaya N, Madurani KA, Shahabuddin S, Sambasevam KP, Baharin SNA, Jeerapan I (2024) Green techniques: revolutionizing deep eutectic solvents-based modified electrodes for electrochemical sensing of natural antioxidant. Microchem J 206:111491
94. El-Deen AK, Elmansi H, Shimizu K (2022) Utilization of hydrophilic and hydrophobic deep eutectic solvents for dispersive liquid-liquid microextraction of bicalutamide from water and spiked human plasma. Sustain Chem Pharm 29:100825
95. Qi J, Liu J, Shui L, Hao Z, Yan G, Wang N, Lu H, Xiong Y (2024) Switchable hydrophilicity solvents incorporating hydrophilic deep eutectic solvents: a sustainable alternative for oil-based drilling cuttings washing. J Clean Prod 442:140976
96. Yan Y, Shen L, Wang Y, Gong B, Li Z, Qiu H (2025) Development of switchable deep eutectic solvents: applications in extraction of natural products. Chin Chem Lett 110845:110845
97. Gonzalez–Diaz A, García–Núñez JA (2023) Hydrophilic deep eutectic solvents: a new generation of green and safe extraction systems for bioactive compounds obtaining from natural oil & fats—a review. Sustain Chem Pharm 36:101278
98. El Achkar T, Greige-Gerges H, Fourmentin S (2021) Basics and properties of deep eutectic solvents: a review. Environ Chem Lett 19(4):3397–3408
99. Liu J, Yan Y, Di X, Zhao L (2024) Investigation of the extraction efficiency and antioxidant capacity of deep eutectic solvent-enhanced Paeonia lactiflora. J Mol Liq 413:125871
100. Dong X, Cai Z, Wang Y, Zhou Q, Wang L, Fu Y (2025) Simultaneous extraction and separation of different polarity phytochemicals from in pigeon pea leaves with biphasic temperature-switchable deep eutectic solvent system. Microchem J 208:112418
101. Khajavian M, Ismail S (2025) Deep eutectic solvent-modified polyvinyl alcohol/chitosan thin film membrane for dye adsorption: machine learning modeling, experimental, and density functional theory calculations. Int J Biol Macromol 294:139479

102. Vorobyova V, Skiba M, Vinnichuk K, Linyucheva O, Vasyliev G (2025) Influence of hydrogen bond donor compound type on the properties of betaine based deep eutectic solvents: a computational and experimental approach. J Mol Struct 1323:140743
103. Bide Y, Shahvelayati AS, Kherad EM, Mohammadpour Z, Mahyari M (2025) Emerging frontiers in poly (ionic liquids) and poly (deep eutectic solvents). J Mol Liq 425:127171
104. Cao Y, Wang H, Jian Y, Hao G, Di D, Liu J (2024) Temperature-switchable deep eutectic solvent extraction of polysaccharides from wolfberry fruits: process optimization, structure characterization, and antioxidant activity. J Mol Liq 398:124352
105. Peng W, Wang X, Wang W, Wang Y, Huang J, Zhou R, Bo R, Liu M, Yin S, Li J (2024) Comparison, optimization and antioxidant activity of ultrasound-assisted natural deep eutectic solvents extraction and traditional method: a greener route for extraction of flavonoid from Moringa oleifera lam. Leaves. Ultrason Sonochem 109:107003
106. Lasu CS, Muhamad M, Kamal NNSNM, Yahaya N, Raoov M, Zain NNM (2024) Assessing the performance of non-toxic nanomagnetic modified deep eutectic solvents for tetracycline adsorption: Taguchi approach for optimization. J Mol Liq 399:124374
107. Villegas M, Toledo-Neira C, Merlet G, Cabezas R, Romero J, Villarroel-Utreras E, Quijada-Maldonado E (2025) Sustainable framework to develop deep eutectic solvent-based DLLME methodologies using COSMO-RS and RSM: application to PCB-77 in greywater. J Mol Liq 419:126790
108. Majid Q, Richu AK (2021) Effect of choline chloride and urea based deep eutectic solvent on the physicochemical properties of salicylic acid and salicylamide at T = (288.15 to 313.15) K. J Mol Liq 334:116500
109. Monem A, Habibi D, Goudarzi H, Mahmoudiani-Glian M, Benrashid A, Alshablawi Z (2024) The choline chloride-based DES as a capable and new catalyst for the synthesis of benzopyranophenazinecarbonitriles. Catal Commun 187:106913
110. Suthar P, Kaushal M, Vaidya D, Thakur M, Chauhan P, Angmo D, Kashyap S, Negi N (2023) Deep eutectic solvents (DES): An update on the applications in food sectors. J Agric Food Res 14:100678
111. Ortizo RGG, Sharma V, Tsai M-L, Nargotra P, Wang J-X, Sun P-P, Chen C-W, Dong C-D (2024) Exploring the potential of magnetic deep eutectic solvents and DES-functionalized nanomaterials for food analysis: advancements and current trends. Food Biosci 61:104764
112. Abdullahi AB, Ismail S, Alshana U (2023) Edible oil-based switchable-hydrophilicity solvent liquid–liquid microextraction for the determination of lead in food samples using flame-atomic absorption spectrometry. J Food Compos Anal 118:105189
113. Altunay AÖ, Elik A (2022) A novel sonication assisted dispersive liquid-liquid microextraction method for methylparaben in cosmetic samples using deep eutectic solvent. Sustain Chem Pharm 29:100781
114. Makoś P, Słupek E, Gębicki J (2020) Hydrophobic deep eutectic solvents in microextraction techniques–a review. Microchem J 152:104384
115. van Osch DJGP, Zubeir LF, van den Bruinhorst A, Rocha MAA, Kroon MC (2015) Hydrophobic deep eutectic solvents as water-immiscible extractants. Green Chem 17(9):4518–4521
116. Jouyban A, Ali Farajzadeh M, Afshar Mogaddam MR, Khodadadeian F, Nemati M, Khoubnasabjafari M (2021) In-situ formation of a hydrophobic deep eutectic solvent based on alpha terpineol and its application in liquid-liquid microextraction of three β-blockers from plasma samples. Microchem J 170:106687
117. Zhang Y, Hunter JR, Ullah A, Shao Q, Shi J (2024) Lignin derived hydrophobic deep eutectic solvents for the extraction of nanoplastics from water. J Hazard Mater 467:133695
118. Gao Y, Fan M, Cheng X, Liu X, Yang H, Ma W, Guo M, Li L (2024) Deep eutectic solvent: synthesis, classification, properties and application in macromolecular substances. Int J Biol Macromol 278:134593
119. Lee J, Jung D, Park K (2019) Hydrophobic deep eutectic solvents for the extraction of organic and inorganic analytes from aqueous environments. TrAC Trends Anal Chem 118:853–868

120. Yusof R, Abdulmalek E, Sirat K, Rahman MB (2014) Tetrabutylammonium bromide (TBABr)-based deep eutectic solvents (DESs) and their physical properties. Molecules 19:8011–8026
121. Li X, Wang X, Chen Q, Wu H, Huang W, Zhang W, Ren S, Ouyang R, Jiang Y, Yao C (2025) Oil-phenol separation by tetraethylammonium chloride-based deep eutectic solvents: microscopic contribution of basic sites. Sep Purif Technol 358:130296
122. Taşpınar H, Altunay N (2023) A chemometric-based sonication-assisted liquid-liquid microextraction for analysis of sulfadiazine in dairy products and honey using low viscous hydrophobic deep eutectic solvents. Microchem J 193:109024
123. Wu Y, Zhou M, Yu Q, Geng X, Chen C, Wang X (2023) Dispersive liquid-liquid microextraction based on deep eutectic solvent doped with hydroxyethyl-β-cyclodextrin for the determination of sulfonamides in environmental waters. Microchem J 193:109029
124. Yahaya N, Mohamed AH, Sajid M, Zain NNM, Liao P-C, Chew KW (2024) Deep eutectic solvents as sustainable extraction media for extraction of polysaccharides from natural sources: status, challenges and prospects. Carbohydr Polym 338:122199
125. Elencovan V, Joseph J, Yahaya N, Abdul Samad N, Raoov M, Lim V, Zain NNM (2022) Exploring a novel deep eutectic solvents combined with vortex assisted dispersive liquid–liquid microextraction and its toxicity for organophosphorus pesticides analysis from honey and fruit samples. Food Chem 368:130835
126. Barbayanov K, Timofeeva I, Bulatov A (2022) An effervescence-assisted dispersive liquid-liquid microextraction based on three-component deep eutectic solvent for the determination of fluoroquinolones in foods. Talanta 250:123709
127. Pochivalov A, Cherkashina K, Shishov A, Bulatov A (2021) Microextraction of sulfonamides from milk samples based on hydrophobic deep eutectic solvent formation by pH adjusting. J Mol Liq 339:116827
128. Sereshti H, Zarei-Hosseinabadi M, Soltani S, Taghizadeh M (2022) Green vortex-assisted emulsification microextraction using a ternary deep eutectic solvent for extraction of tetracyclines in infant formulas. Food Chem 396:133743
129. Li Q, Ji K, Tang N, Li Y, Gu X, Tang K (2023) Vortex-ultrasonic assisted dispersive liquid-liquid microextraction for seven sulfonamides of fish samples based on hydrophobic deep eutectic solvent and simultaneous detecting with HPLC-PDA. Microchem J 185:108269
130. Kurashov Y, Pochivalov A, Petrova A, Safonova E, Garmonov S, Bulatov A (2024) Supramolecular solvents based on hydrophobic natural deep eutectic solvents and primary amines for preconcentration and determination of enrofloxacin in milk. Talanta 279:126666
131. Mostafa A, Shaaban H, Alqarni AM, Alghamdi M, Alsultan S, Saleh Al-Saeed J, Alsaba S, AlMoslem A, Alshehry Y, Ahmad R (2022) Vortex-assisted dispersive liquid–liquid microextraction using thymol based natural deep eutectic solvent for trace analysis of sulfonamides in water samples: assessment of the greenness profile using AGREE metric, GAPI and analytical eco-scale. Microchem J 183:107976
132. Li T, Duan L, Du M, Shi L, Zhu W, Wang C, Fan J (2024) Hydrophobic deep eutectic solvent-based vortex assisted liquid–liquid microextraction for the determination of famciclovir in water samples coupled with high performance liquid chromatography. Microchem J 197:109710
133. Zhang J, Li S, Yao L, Han Y, Chen K, Qian M, Li Z, Lin H (2024) Cyclodextrin-based ternary supramolecular deep eutectic solvents for efficient extraction and analysis of trace quinolones and sulfonamides in wastewater by adjusting pH. Anal Chim Acta 1311:342714
134. Pochivalov A, Cherkashina K, Sudarkin A, Osmolowsky M, Osmolovskaya O, Krekhova F, Nugbienyo L, Bulatov A (2023) Liquid-liquid microextraction with hydrophobic deep eutectic solvent followed by magnetic phase separation for preconcentration of antibiotics. Talanta 252:123868
135. Díaz-Álvarez M, Martín-Esteban A (2022) Preparation and further evaluation of l-menthol-based natural deep eutectic solvents as supported liquid membrane for the hollow fiber liquid-phase microextraction of sulfonamides from environmental waters. Adv Sample Prep 4:100047

136. Meshcheva D, Krekhova F, Shishov A, Bulatov A (2024) Natural deep eutectic solvent for the simultaneous derivatization and microextraction of isoniazid from human plasma. Anal Chim Acta 1320:343007
137. Treder N, Kowal A, Roszkowska A, Bączek T (2024) A sustainable approach for the determination of pacritinib in biological samples with dispersive liquid-liquid microextraction and using hydrophobic natural deep eutectic solvents. Sustain Chem Pharm 42:101841
138. Han W-C, Zhang H-J, Chen J-B, Chen Y-Y, Wang W-J, Liu Y-W, Yang P, Yuan D-D, Chen D (2024) A green and rapid deep eutectic solvent dispersed liquid–liquid microextraction with magnetic particles-assisted retrieval method: proof-of-concept for the determination of antidepressants in biofluids. J Mol Liq 395:123875
139. Binjawhar DN, Mohammedsaeed W (2024) pH-switchable hydrophobic deep eutectic solvent-based liquid phase microextraction for detecting morphine and codeine in whole blood samples followed by HPLC-UV. J King Saud Univ Sci 36(8):103303
140. Magdy Saad A, Nasr JJM, El-Deen AK (2024) Utilizing an air-assisted liquid–liquid microextraction based on a natural hydrophobic deep eutectic solvent for the efficient determination of the anti-conjunctivitis olopatadine. J Mol Liq 410:125658
141. Pourvali Abarbakouh M, Faraji H, Shahbaazi H, Miralinaghi M (2024) (deep eutectic solvent-ionic liquid)-based ferrofluid a new class of magnetic colloids for determination of tamoxifen and its metabolites in human plasma samples. J Mol Liq 399:124421
142. Guo P, Xu M, Zhong F, Liu C, Cui X, Zhang J, Zhao M, Yang Z, Jia L, Yang C, Xue W, Fan D (2024) Molecularly imprinted solid-phase extraction combined with non-ionic hydrophobic deep eutectic solvents dispersed liquid-liquid microextraction for efficient enrichment and determination of the estrogens in serum samples. Talanta 269:125480
143. Nia NN, Hadjmohammadi MR (2023) Preparation of Ferrofluid from Adipic acid coated magnetic nanoparticles and hydrophobic deep eutectic solvent used in dispersive micro-solid-phase extraction: application for the pre-concentration and determination of clozapine in biological samples. J Chromatogr Sci 61(6):569–578
144. Al-Hashimi NN, Abed Alfattah H, El-Barghouthi MI, El-Sheikh AH, Ale-nezi HM, Sunjuk MS, Fahelelbom KM (2024) Solvent Bar microextraction method based on a natural deep eutectic solvent and multivariate optimization for determination of steroid hormones in urine and water. Appl Sci 14(11):4438
145. Khodayari P, Ebrahimzadeh H (2023) A green QuEChERS syringe filter based micro-solid phase extraction using hydrophobic natural deep eutectic solvent as immobilized sorbent for simultaneous analysis of five anti-diabetic drugs by HPLC-UV. Anal Chim Acta 1279:341765
146. Fathi Majd S, Heidari H, Razmi H (2023) A natural hydrophobic deep eutectic solvent-based microextraction method integrated with a spectrofluorimetric technique for the determination of daclatasvir as a direct-acting hepatitis C antiviral drug in plasma samples. Microchem J 192:108949
147. Dalmaz A, Sivrikaya Özak S (2024) Rapid and eco-friendly microextraction procedure based on green hydrophobic deep eutectic solvent for simultaneous determination of five sex hormones from cosmetic samples. Microchem J 207:111979
148. Al-Hashimi NN, Abed Alfattah H, El-Barghouthi MI, El-Sheikh AH, Ale-nezi HM, Sunjuk MS, Fahelelbom KM (2024) Solvent Bar microextraction method based on a natural deep eutectic solvent and multivariate optimization for determination of steroid hormones in urine and water. Appl Sci 14:4438

Chapter 12
Advances in Material Engineering Using Deep Eutectic Solvents

Suchandra Bhattacharjee, Tarnveer Kaur, and Vishwajit Chavda

12.1 Introduction

In the modern-day era, environmental pollution is a primary concern. To avoid toxic reagents and reactants in chemical reactions, researchers have progressively adopted green protocols, which mainly began to emerge in the mid-twentieth century [1]. Since solvents contribute significantly in chemical reactions, researchers have focused on developing environmentally friendly alternatives. The preparation of deep eutectic solvents (DESs) is a groundbreaking development in this emerging field. Formed by the union of hydrogen bond donor (HBD) and hydrogen bond acceptor (HBA) groups, these eutectic mixtures exhibit a melting temperature notably lower than that of their respective components. The most supportive factors of the solvents are enhanced sustainability and high efficiency. The solvents received outstanding attention after the trailblazing work reported by Abbott et al. in 2003 [2, 3].

DESs are classically formulated as an accumulation of a minimum of two components, in which one serves as a HBD while the other serves as a HBA. They can unite through hydrogen bonding to generate a novel eutectic phase. The new eutectic phase fabricated is characterized by a melting point that is always below 100 °C, noticeably less than the melting points of each constituent of the mixture. An easy and suitable

S. Bhattacharjee (✉)
University Center for Research and Development, Chandigarh University, Mohali, India
e-mail: suchandra.e18692@cumail.in

T. Kaur
Department of Basic & Applied Sciences, Sri Guru Granth Sahib World University, Fatehgarh Sahib, Punjab, India

V. Chavda
Department of Chemistry, Institute of Science, Humanities and Liberal Studies, Indus University, Shilaj, Ahmedabad, Gujarat, India

K. Kuperkar et al. (eds.), *Deep Eutectic Solvents*, Environmental Footprints and Eco-design of Products and Processes,
https://doi.org/10.1007/978-3-032-03801-2_12

protocol, eco-friendliness, cost-effectiveness, low toxicity, biodegradability, and renewability are the key advantages of eutectic solvents [4]. Furthermore, they exhibit several unusual properties, including a high potential for dissolving metal oxides, high tolerance to hygroscopicity, non-flammability, non-volatility, and high thermal stability [5]. The reaction carried out in DES does not require any expensive solvents for extraction, which in turn minimizes the number of reaction steps. Due to the aforementioned properties and advantages, DESs not only serve as green electrolytes but also as promising solvents for the synthesis of nanomaterials and polymers [6, 7].

Based on the nature of the components present, DESs are broadly sorted into four main types. DESs derived from the combination of metal chlorides (MCl_x) and quaternary ammonium salts, homologous to metal halide/imidazolium systems, are classified as Type I. Examples of Type I include EMIC (1-methyl-3-ethyl imidazolium chloride) combined with AgCl, CuCl, $ZnCl_2$, $SnCl_4$, $LaCl_3$, and many other metal chlorides. Type II DESs flourished in the range by utilizing hydrated metal halides and choline chloride (ChCl), offering air and moisture stability, as well as cost-effectiveness, for industrial use. The most extensively studied by researchers are Type III DESs, which involve choline chloride and HBDs such as amides, acids, and alcohols. They are water-compatible, biodegradable, affordable, and possess the capacity to dissolve a broad variety of metal species. Their tenable physical properties enabled them to be used in metal deposition, biodiesel purification, and biomass processing. Type IV DESs, composed of metal salts with organic HBDs such as urea and ethylene glycol (EG), facilitate the preparation of ambient-temperature eutectics. Choline-based DESs predominate due to the low toxicity of choline, as well as low cost and wide availability; they are produced on a large scale as a feed additive via an efficient, low-waste process. These DESs offer a sustainable alternative to conventional solvents with broad application potential in green chemistry and industrial processes [8, 9]. Figure 12.1 illustrates some examples of HBDs and HBAs commonly utilized for the preparation of DESs.

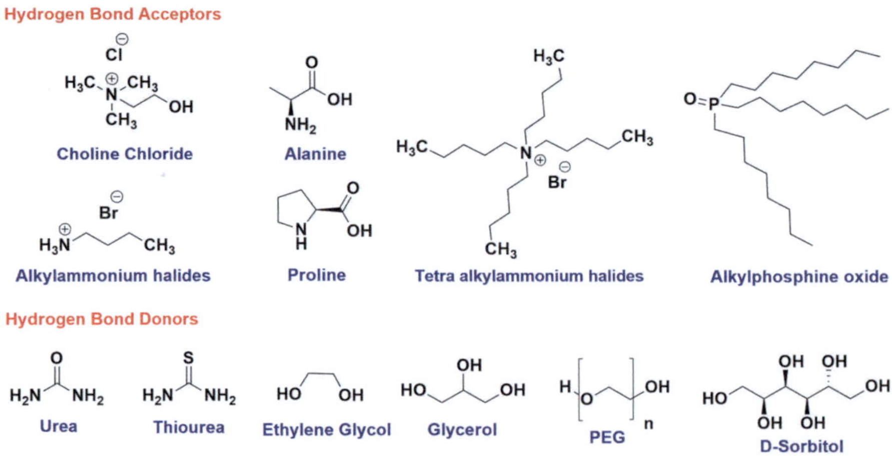

Fig. 12.1 Illustration of hydrogen bond donors (HBDs) and hydrogen bond acceptors (HBAs) utilized in DESs

12.2 Applications of Deep Eutectic Solvents in Advanced Materials

Due to their low toxicity, biodegradability, cost-effectiveness, thermal stability, and nonvolatility, DESs are a suitable alternative to traditional solvents and ionic liquids in numerous applications. They have achieved significant attention for their multi-fold applications in advanced materials, including electrodeposition, nanomaterials, and polymer synthesis (as shown in Fig. 12.2). In electrodeposition, DES facilitates the deposition of uniform metal and alloy films. They are also effective in metal recycling by selective leaching and in the synthesis of photocatalysts, thus improving efficiency by allowing precise control over nanostructure. For nanomaterials, DES provides excellent control over particle size, morphology, and crystallinity, useful for the construction of metal nanoparticles, oxides, and quantum dots. DESs act as both solvents and catalysts in polymer synthesis, promoting polymerization under mild and sustainable conditions. In this chapter, we will explore these applications in detail [10].

12.2.1 DES Serves as a Green Electrolyte

For versatile advanced material procedures, DES serves as an adaptable platform. Due to their low toxicity, tunability, and ability to function under mild conditions, they are highly suitable for green electrodeposition, efficient metal recovery, and the synthesis of high-performance photocatalysts. The quality of DESs to serve as both solvents and functionalizers for the synthesis of photocatalysts, as well as their enhanced activity for applications, including water splitting and pollutant degradation, is explored extensively in the following sections.

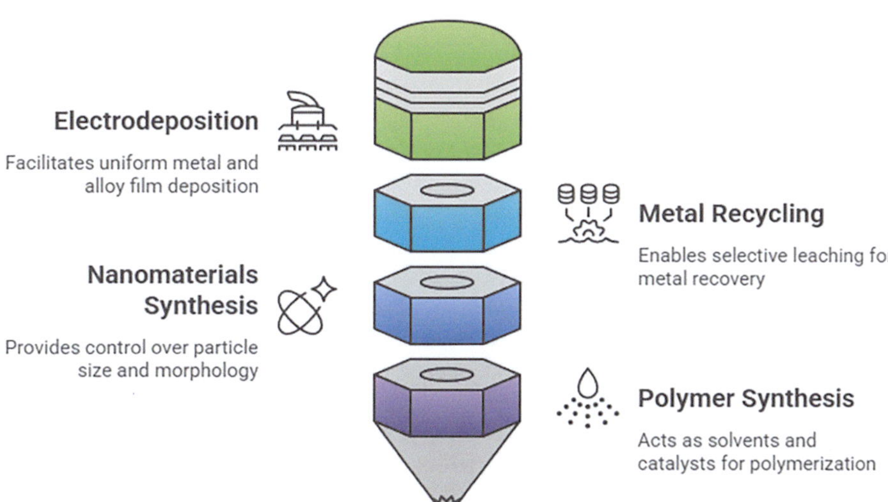

Fig. 12.2 Applications of DESs

12.2.1.1 DES-Mediated Electrodeposition

Electrodeposition of metals is a crucial technique with numerous applications, primarily in the industrial sector. Ryder and his coworkers, for the first time in 2009, reported the electrodeposition of copper using DES. The researchers prepared DES utilizing a mixture of ChCl and HBDs, including urea or EG. Both solvents showed that the deposition of copper was near about 100% [11]. The same group next conducted comparative studies of electrodeposition of nickel between three parallel nickel solutions, namely.

I. Watts nickel bath: $NiSO_4$, NaCl, and $B(OH)_3$ in water.
II. Aqueous solution: $NiCl_2.6H_2O$ in water.
III. DES: $NiCl_2.6H_2O$ in 1 ChCl: 2 EG (Ethaline).

The researcher observed that, under identical thermal and concentration parameters, the rates of electrodeposition of nickel from both the aqueous solutions and DES are parallel, indicating that conductivity and viscosity are not the rate-limiting factors for metal deposition. The study revealed a variation in the morphology of the metal deposit. The nickel, which is deposited from DES, exhibits a notably smooth and shiny surface, whereas the nickel deposited from two aqueous solutions is tedious and uneven in quality [12]. Later on, these two particular DESs, commonly known as reline and ethaline, were successfully used for the electrodeposition of various metals, including tin [13–15], zinc [16, 17], chromium [18, 19] and alloy coatings like zinc/ nickel [20], nickel/iron [21], and zinc/tin [22] etc.

Next, the researchers started investigating the additive effects of DES for the electrodeposition of metals. In 2017, Ryder and his group reported the effects of standard additives employed in nickel deposition from a simple DES composed of EG and ChCl, generally recognized as ethaline. The common additives considered for this study include nicotinic acid, boric acid, 5,5-dimethylhydantoin, and methyl-nicotinoate, respectively. It was observed that at low concentrations, these additives enabled the formation of shiny, firm, and sticky nickel coatings with high current efficiency. Traditionally, to obtain such coatings, complex water-based solutions of strong acids are required to control and maintain the rigorous process [23]. After 2 years, the same group reported the additive effects on the electrodeposition of zinc from DES-ethaline as an electrolyte. The effect of benzoquinone (BQ), nicotinic acid (NA), and boric acid (BA) as additives on the electrochemical deposition of Zn was investigated. It was observed that a mirror zinc coating on a copper substrate has been achieved in the presence of a small amount of NA. Both BA and BQ worked as brighteners for Zn deposition. Electrochemical experiments revealed that the morphologies of the Zn deposits were enhanced, resulting in a reinforced smoothness of the Zn coatings when additives were used in DES [24].

Frackowiak and coworkers proposed a novel application of DES, mainly reline, serving as the electrolyte for carbon-supported electrochemical capacitors (ECs). The study revealed that novel reline electrolytes are safe, eco-friendly, cost-effective, and biodegradable. Moreover, due to their anti-corrosion properties and high electrochemical stability, the neat reline and minimally aqueous reline offered fabulous cycling durability at 2.2 V, accompanied by excellent capacitance retention. It was observed that when 1% water was added to DES mixtures, the system became less viscous, which increased the conductivity of the electrolyte. It displayed the highest specific capacitance of approximately 157 F g^{-1}, along with improved energy storage capacity and cyclic durability. However, 5% aqueous mixtures with the DES represent insecure electrochemical activities due to the breakdown of the electrolyte [25].

In 2025, Chou et al. performed an outstanding study on the activity of zinc chloride-based DESs as eco-friendly electrolytes in organic electrochemical transistors (OECTs), utilizing a p-type conjugated polymer (CP), namely P3HT (poly(3-hexylthiophene-2,5-diyl)), as the active material. The researchers determined to examine two primary DES systems, namely DES1 and DES2. DES1 was composed of zinc chloride and acetamide, while DES2 was composed of zinc chloride and EG, with variants in DES2 incorporating potassium chloride. Among all the designed systems, DES2 displayed fantastic performance, showing comparatively lower threshold voltage and increased stability. The addition of KCl to DES2 boosts the response times and conductivity. This transformation is attributed to the strong hydrogen-bonding network present in DESs, which prevents ion hydration, thereby facilitating their more efficient doping into the polymer. As a consequence, OECTs with P3HT using DES2 as the electrolyte showed outstanding mobility-capacitance products, along with significant stability. Spectroscopic and morphological examinations supported the improved ion doping and reversibility in DES2, emphasizing DESs as promising, tunable, and sustainable electrolytes with strong potential in nonaqueous organic electronics [26].

12.2.1.2 Recycling of Metals

The exploration of lithium-ion batteries (LIB) in electronics is rapidly increasing in our everyday lives. As a result, scientists have been actively trying to develop efficient and eco-friendly methods to recycle the metal [27]. Ajayan and Ganguli came up with a great idea of recycling LIB with the help of DES, where they have revealed the fact that DES is not only a green solvent but also can efficiently extract metals from diverse complex chemical oxides, including lithium cobalt oxide (LCO; $LiCoO_2$) and lithium nickel manganese cobalt oxide (NMC; $LiNi_{1/3}Mn_{1/3}Co_{1/3}O_2$). This research demonstrated that DES, composed of ChCl and EG, can extract metals from LIB, achieving an excellent leaching efficiency for both cobalt and lithium of approximately greater than 90% [28]. The very next year, Peeters, Binnemans, and Riano proposed a novel DES with a composition of ChCl and citric acid in a 2:1 molar proportion, containing 35 wt% water, which is an efficient lixiviant for extracting cobalt and lithium from lithium-ion batteries. The copper and aluminum are present in LIBs as current collectors, which, under the influence of this DES, served as a reducing agent that facilitates the conversion of the oxidation state of cobalt from +3 to +2. Under the optimized leaching conditions, the authors reported that the leaching efficiency of DES is more than 98% for cobalt and 93% for lithium, respectively. The roles of copper and aluminum were further investigated, and it was observed that copper is a more efficacious reducing agent than aluminum, as it leads to the quantitative leaching of cobalt and lithium without the need for any additional chemical reducing agents. In contrast, aluminum reacted with protons to release hydrogen gas and, as a result, failed to serve as an effective reducing agent for cobalt. The author conducted additional controlled experiments to evaluate the efficiency of metal recovery in this DES system. When compared with conventional hydrochloric acid leaching, it was observed that DES-based leaching was slower; however, it notably did not release the harmful chlorine gas, which further strengthened the eco-friendliness of DES. Next, for the metal's reclamation in a nonaqueous solvent, the DES PLS (pregnant leach solution) was explored as the polar medium. Firstly, Copper(I/II) was recovered employing LIX 984, followed by cobalt(II) which was extracted using aliquat 336. Using 0.25 mol/L oxalic acid, each metal was effectively recovered from the organic phases as oxalates. Overall, the process afforded cobalt oxalate with 99.9% purity and 81% yield [29].

12.2.1.3 DES-Mediated Photocatalyst

A novel ternary heterojunction was successfully synthesized by utilizing DES, as reported by Jaihindh et al. in the year 2020. The scientist demonstrated the function of a ternary heterojunction as a highly efficient, metal-free oxygen evolution reaction (OER) photocatalyst for both acidic and alkaline media. Equimolar amounts of ChCl and citric acid were thermally treated at 80 °C for 30 min, and the colorless DESs were prepared. Initially, the flower-like structured p-n heterojunction BiOCl/$BiVO_4$ photocatalyst was developed through a simple wet chemical approach utilizing DESs. By ultrasonication and physical mixing of BiOCl/$BiVO_4$ (BOC/BVO) with g-C_3N_4 with different ratios, the researcher successfully prepared a ternary heterojunction (BOC/BVO/g-CN). The detachment capability of the photogenerated electron-hole pair and the prolonged lifespan of photogenerated charges are demonstrated by the synthesized ternary heterojunctions [30].

After the successful development of the ternary heterojunction (BOC/BVO/g-CN), Ren et al. (2022) published a research article in which they designed a ternary heterojunction comprising $BiVO_4$, BiOCl, and graphene quantum dots functionalized with sulfur and nitrogen (S,N-GQDS). The newly crafted ternary heterojunction was synthesized by the hydrothermal method using DES, which played a crucial role in determining the growth direction. SEM, TEM, XRD, FT-IR, XPS, PL spectroscopy, and UV-Vis DRS confirmed the composition and optical properties of synthesized heterojunction. The efficacy of the photocatalytic performance of the $BiVO_4$/BiOCl/S,N-GQDS hybrid materials was assessed via rhodamine B (RhB) degradation under the influence of visible light, a model organic pollutant. The highest degradation rate was observed in the sample doped with 3% S,N-GQDS, achieving 81% degradation within 5 h of exposure to visible light. The experimental results revealed the formation of a Z-scheme heterojunction, which effectively facilitates charge separation and controls the recombination of

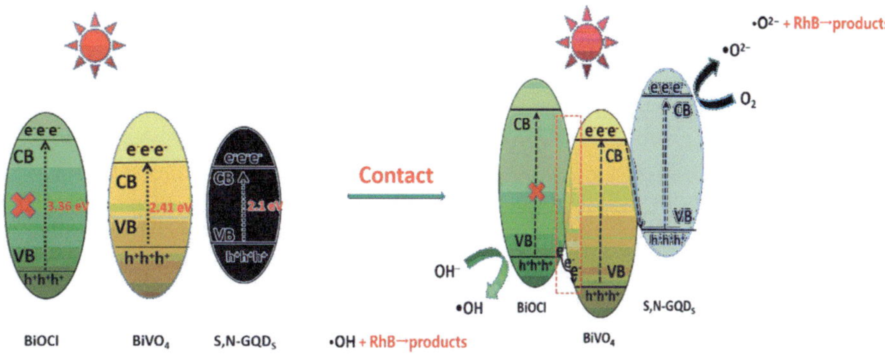

Fig. 12.3 Proposed mechanism for photocatalytic degradation of RhB by $BiVO_4$/BiOCl/S,N-GQDS. Reproduced from reference no. [31]

photogenerated electron-hole pairs. The enhanced charge mobility can be ascribed to the conductive S,N-GQDS and the formation of p-n heterojunctions between BiOCl (p-type) and $BiVO_4$ (n-type). Mechanically, the Z-scheme heterojunction permits the effective migration between electrons and holes. The holes in S,N-GQDS's valence band can efficiently combine with the electrons present in the conduction band of $BiVO_4$ (Fig. 12.3). As a result, strong oxidative holes are created in $BiVO_4$, and the reductive electrons present in S,N-GQDS effectively generate reactive oxygen species (•OH and $•O_2^-$). These reactive oxygen species were responsible for the degradation of rhodamine B, as confirmed by the researchers' reactive species trapping experiments. This research has significant implications for the resolution of water pollution [31].

12.2.2 Nanomaterial Synthesis in DES

For the synthesis of metallic nanoparticles, DES functions as an outstanding and efficacious solvent for the synthesis of nanoparticles using a very popular chemistry. This was first demonstrated by Sun and his coworkers, who illustrated the fabrication of gold nanoparticles with high-index facets exhibiting star, snowflake, and thorn-like morphologies in ChCl: urea-based DES without the use of surfactants or seeds [32]. After this groundbreaking research, many researchers have reported the synthesis of various nanomaterials using DES. Some of the remarkable syntheses are highlighted below.

Qiu and his group reported a novel protocol for synthesizing a series of hydroxide-based basic DES. They found it exceptionally efficient for producing transition metal oxide nanomaterials such as $NiCo_2O_4$, $MnCo_2O_4$, Co_3O_4, $NiMn_2O_4$, and $CoCu_2O_4$. Under moderate stipulation using these DES, the nanomaterials obtained with cubic ultrathin nanosheets have high pore volumes and large surface areas. The synthesized DESs were characterized through FT-IR, ^{1}H NMR (proton-NMR), and molecular dynamics (MD) simulation, confirming that the DESs are composed of a combination of strong hydroxides, such as LiOH, NaOH, and KOH, which interact as hydrogen-bonding acceptors. At the same time, a series of polyol compounds, including EG and Poly Ethylene Glycol (PEGs) variants (200 to 600), are HBDs [33].

Taghavi et al. reported the first DES-functionalized nano-photocatalyst n-TiO_2 P25@TDI@DES, which demonstrated its efficiency in converting benzyl alcohol to benzaldehyde in the presence of the oxidant $NaNO_3$ under blue LED irradiation. In the fundamental synthesis step, TiO_2-P25 nanoparticles were covalently functionalized using 2,4-toluene diisocyanate, an efficient and extremely reactive linker. The DES used in this process is prepared by adding $ZnCl_2$ and urea in a stoichiometric ratio of 1:4 at 100 °C; a consistent, distinct, and transparent liquid was obtained. Next, the synthesized DES was added to TiO_2-P25@TDI nanoparticles at 100 °C for 18 h. The resulting powder obtained was n-TiO_2P25@TDI@DES, which FTIR, XRD, FE-SEM, TGA, EDX, and DRS characterized. Scheme 12.1 represents the synthesis of the n-TiO_2P25@TDI@DES photocatalyst [34].

Wang and Tang, along with their group, synthesized a novel *Crassula perforata*-like TiO_2 with (001) and (101) planes under moderate reaction conditions, utilizing ChCl/oxalic acid DES as a solvent. To understand the role of H-bond donors and acceptors of DES on *Crassula perforata*-like TiO_2, the researcher investigated a variety of DES and successfully confirmed, using SEM, that TiO_2 appeared in various shapes as the DES changed. For example, when EG urea, glycerine, and oxalic acid are the H-bond donors, the appearances of TiO_2 nanomaterials are elongated, spherical, semi-spherical, and Crassula-like microstructures, respectively, which clearly illustrates that the kind of HBDs present in DES has an outstanding role in determining the shapes of the nanomaterials. Similarly, the H-bond acceptors also demonstrated their function. When betaine, ChCl, and acetylcholine chloride are H-bond acceptors, the appearance of TiO_2 samples is nanoneedle-assembly microspheres, quasi-peanuts, and quasi-*Crassula perforata*, respectively. The result reveals that the type of DES has a profound impact on the nanomaterial's appearance and morphology, which can be readily adjusted by modifying the hydrogen-bonding species present in DES. Next, it was observed that the rate of hydrogen generation through photocatalytic water decomposition by *Crassula perforata*-like TiO_2 is 18 times faster than that of commercially available P25 TiO_2, solely due to

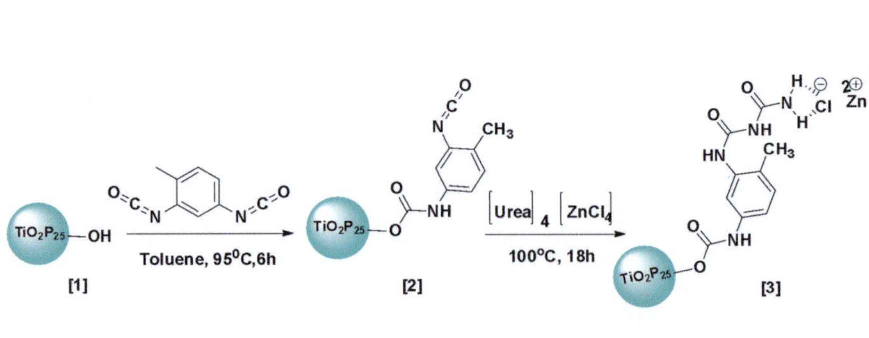

Scheme 12.1 Synthesis of n-TiO_2P25@TDI@DES photocatalyst

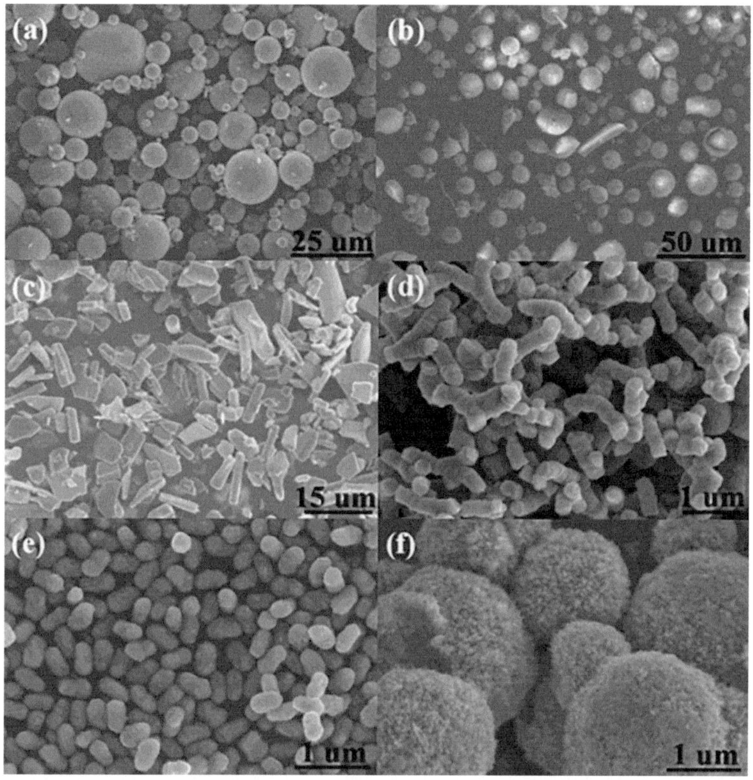

Fig. 12.4 (**a**–**d**) SEM images of TiO_2 samples synthesized using different HBDs: (**a**) urea, (**b**) glycerine, (**c**) EG, and (**d**) oxalic acid, with ChCl as the HBAs. (**e**–**g**) SEM images of TiO_2 samples prepared using oxalic acid as the HBDs and different HBAs: (**e**) ChCl, (**f**) betaine, and (**g**) chlorocholine chloride. (Reproduced from [35] with permission from the Elsevier)

its special shape. Figures 12.4 and Fig. 12.5 represent the SEM image of TiO_2 samples at various DESs [35].

In 2022, Haq et al. published a significant advancement for food technologists in detecting heavy metals, particularly lead (II), in food-grade oils by utilizing the silver nanoparticles impregnated with DES. Among the tested DES compositions, the mixture of a 1:2 molar ratio of ChCl and phenol showed the maximum extraction efficiency for lead (II). The protocol offers significant sensitivity, with a detection limit (LOD) of 0.28 μg/L and a quantification limit (LOQ) of 0.94 μg/L, effectively screening the full spectrum of lead content typically present in edible oils, from trace levels (ppb) to higher concentrations. Notably, the method is speedy, requiring only 6.5 min per sample. This enhanced efficiency can be ascribed to the superior sorption kinetics of silver nanoparticles, which significantly increase the sensitivity of the method [36].

In general, the section of the chapter emphasizes that DES functions as a sustainable alternative to customary solvents for the preparation of nanomaterials. DES contributes

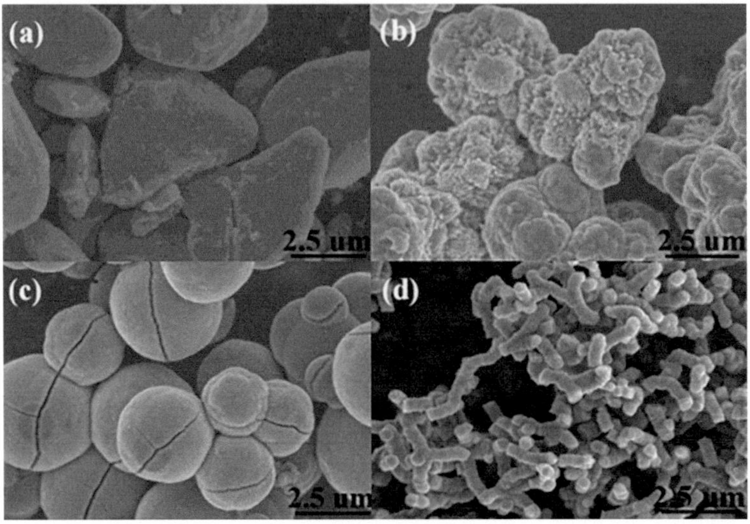

Fig. 12.5 SEM images of TiO_2 samples synthesized with: (**a**) Oxalic acid as the exclusive H-bond donor, (**b**) ChCl as the exclusive H-bond acceptor, (**c**) a physical mixture of oxalic acid and ChCl, and (**d**) DES formed from ChCl and oxalic acid. (Reproduced from [35] with permission from the Elsevier)

significantly to controlling nanoparticle growth and nucleation, thereby influencing the size, morphology, and crystallinity of the resulting nanomaterials. Additionally, they enable low-temperature synthetic pathways, thereby minimizing energy input. The discussion also includes their applications in analytical chemistry, such as the fast and sensitive determination of heavy metals in edible oils using DES-impregnated nanoparticles. The recent advancements highlighted in this section indicate that the distinct solvent characteristics, combined with their capacity for customization, make them highly suitable for next-generation green technologies.

12.2.3 DES Used for the Synthesis of Polymers

Already, we have seen DESs find applications in various fields. In this section, we will address DES-mediated polymer synthesis. Reversible addition-fragmentation chain transfer (RAFT) polymerization is a crucial method in the field of controlled polymerizations. Santha Kumar et al. explored the RAFT polymerization of 2-hydroxyethyl methacrylate (HEMA) and reported their kinetics in a DES composed of ChCl and urea. In the presence of DES, HEMA was polymerized by utilizing the RAFT agent 4-cyano-4-[(dodecyl sulfanyl thiocarbonyl)sulfanyl] pentanoic acid (CDCPA) and the thermal initiator 4,4′-Azobis(4-cyano valeric acid) (ACVA). The scientist observed that the kinetics of polymerization were significantly enhanced in DES compared to Dimethyl Formamide (DMF) and other ionic liquids, primarily due to the strong polarity and hydrogen-bonding capabilities of the

DES. The DES medium not only accelerated the polymerization rate but also facilitated higher monomer conversions, as confirmed by both Differential Scanning Calorimetry (DSC) and NMR analysis. Next, the synthesized polymer (p-HEMA) in the presence of DES, with methyl methacrylate (MMA) functions as a macro-RAFT agent for block copolymerization. The protocol led to a characteristic of polymerization-induced self-assembly (PISA). It fabricates self-assembled block copolymers (PHEMA-b-PMMA) with vesicular morphologies. This extraordinary work, in particular, highlights the versatility of DES not only for controlled radical polymerizations but also for synthesizing structured copolymers [37].

Inspired by this result, Pojman and his coworkers demonstrated the kinetics of photopolymerization of monomers within DESs using real-time Fourier transform-infrared spectroscopy (RT-FTIR). The experiments focused on DESs composed of ChCl combined with acrylic acid (AA), methacrylic acid (MAA), or isobutyric acid (IBA), with photopolymerization commenced by diphenyl (2,4,6-trimethylbenzoyl) phosphine oxide (TPO). The results revealed that polymerizable DESs, such as AA−ChCl and MAA − ChCl, exhibited significantly faster polymerization rates and higher conversions compared to pure monomers. Interestingly, enhanced polymerization rates were also observed in non-polymerizable DESs, such as MMA in IBA − ChCl, indicating that DESs can promote polymerization even when not chemically involved. The researchers revealed that the increased kinetics were influenced not only by increased viscosity but also by additional factors, including hydrogen bonding, monomer preorganization, and the highly polar environment of the DESs [38].

A system of a dozen monolithic copolymers was prepared by Ndizeye et al. across three distinct nonionic DESs, namely N-dimethylacetamide-acetamide (NMA-AA), N-dimethylacetamide-N-methylurea (NMA-NMU), and N-dimethylacetamide-N, N′-dimethylurea (NMA-NN′DMU). These monoliths were composed of three different types of crosslinked polymers: hydroxyethyl methacrylate-divinylbenzene (HEMA-DVB), hydroxyethyl methacrylate-ethylene glycol dimethyl methacrylate (HEMA-EGDMA), and methacrylic acid-1,4-bis(acryloyl)Piperazine (MAA-BAP) in three different nonionic DESs. For a comparative study, one polymer in the group was synthesized employing typical solvents, like toluene, acetone, and water. The results of the zeta potential revealed that the polymers afforded by nonionic DES have parallel superficial areas and hole sizes compared to those generated in conventional solvents, while also exhibiting a decreased surface charge. Moreover, the three nonionic DESs exhibited good recyclability, with possible recovery rates exceeding 80% [39].

In 2020, Cecone and his coworkers reported the novel protocol for synthesizing the water-soluble β-cyclodextrins (βCD)-based polymers having molecular weights varying from 15 k Da to 19 k Da, utilizing NADES (natural deep eutectic solvents) formulated from ChCl and citric acid in varying molar ratios such as 1:1, 1:1.5, and 1:2. The NADES not only served as a solvent but also contributed significantly to the fabrication of polymers. ChCl introduced positively charged quaternary ammonium functional groups in a one-step reaction, while the citric acid (CIT) present in NADES facilitated the linking of βCD units to form polymer chains. The resulting polymers were later converted into cross-linked water-insoluble βCD-based

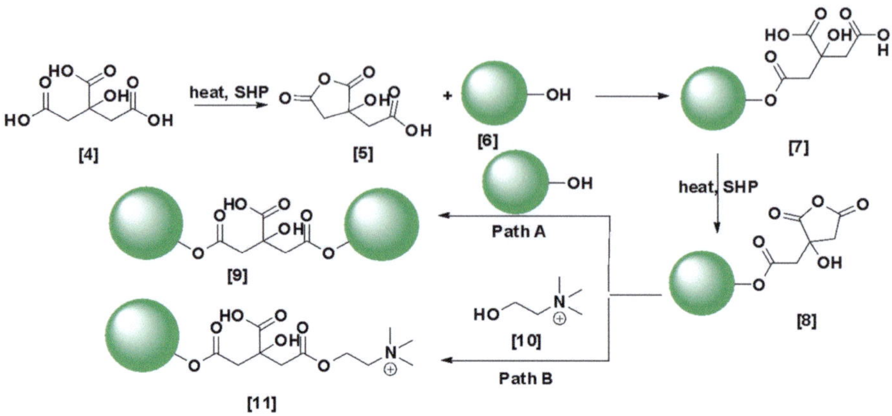

Scheme 12.2 The proposed reaction pathways during the polymer synthesis. (A) Development of anhydride intermediates [8]. (B) CIT acting as a linker between βCDs to extend the polymer chain (9). (C) CIT as a linker between βCD and CHO resulting in a quaternary ammonium functional group (11)

nanosponges through a thermal curing process. Observation showed that the structure and charge of polymers could be adjusted by adjusting the CHO:CIT ratio [40]. Scheme 12.2 represents the role of CIT in polymer synthesis.

Another groundbreaking research study conducted in 2020 reported the development of a novel DES that served as a medium for radical polymerization, leading to cross-linked polymers. Jaumaux and Liu et al. led this study. The novel eutectic mixtures were developed by mixing a solution of LiTFSI (Lithium bis(trifluoromethanesulfonyl)imide) and N-methylacetamide supplemented with fluoroethylene carbonate. These synthesized eutectics assist as a medium for the copolymerization of two monomers, namely pentaerythritol tetraacrylate (PETTA) and 2-(3-(6-methyl-4-oxo-1,4-dihydropyrimidin-2-yl)ureido)ethyl methacrylate (UPyMA). The outcome of the polymerization process is the formation of a cross-linked polymer network, which effectively encapsulates the DES-based electrolyte, forming a quasi-solid-state system that researchers have named DES-based self-healing polymer (DSP) electrolytes. This conceptualization exhibits several key features, including superior ionic conductivity, resistance to combustion, stable electrochemical behavior, and the ability to facilitate dendrite-free lithium plating. Most remarkably, leakage of electrolytes from the electrodes was decreased, which is one of the leading safety concerns for lithium metal battery technologies [41].

The second most prevalent natural polymer, after cellulose, is lignin. It is a complex organic biopolymer typically present in plant cell walls. In 2019, Hong et al. introduced a novel Type I DES composed of zinc chloride and acetamide in a 1:3 molar ratio, which efficiently dissolved lignin at concentrations of up to 16.7 wt% under moderate conditions at 100 °C. Although this newly generated DES allowed the lignin to dissolve, it altered the lignin's structure by splitting it into high- and low-molecular-weight polyphenols. By the addition of water, regenerated lignin (RL) with high molecular weight precipitates, which contain higher hydroxyl groups, whereas low-molecular-weight lignin persists in the DES-water phase. These structural changes were verified by FTIR, GPC, TGA, SEM, and NMR analysis, respectively. The properties of RL were evaluated by modifying them with phenol-formaldehyde (PF) resins. For this purpose, three types of resins were prepared: LPF (utilizing the original lignin), DLPF (using the complete lignin-DES solution), and RLPF (utilizing RL). It was observed that, among these, RLPF showed a higher bonding strength (1.28 MPa) and a shorter gel time (557 s) compared to the standard PF resin (1.32 MPa, 587 s), indicating that the efficiency of polymerization is enhanced. The shortest gel time (405 s) of DLPF was attributed to the catalytic effect of Zn(II) from the DES [42].

Overall, the examples demonstrated herein highlight the potential of DESs as a green, multifunctional platform for polymer synthesis and the development of functional, eco-friendly materials. These examples also illustrate how DESs serve as functional environments that influence polymerization mechanisms and kinetics. In addition, numerous studies have demonstrated that the polymers are designed to exhibit a controlled release profile for loaded molecules, signifying their suitability for applications such as drug delivery or environmental remediation. Type I DES, having a composition of $ZnCl_2$/acetamide (1:3 molar ratio), serves as an effective medium for lignin dissolution, utilizing the dissolved lignin into high-performance resins.

12.3 Conclusion and Future Prospects

In today's rapidly evolving research landscape, DESs have gained recognition as a promising class of sustainable solvents due to their remarkable advantages, including reduced toxicity, environmental degradability, economic viability, and ease of preparation. This book's chapters underscore their versatility across a range of advanced domains, including electrochemistry and materials science. This chapter is written to inform and inspire the young scientific community to explore, adapt, and contribute to existing knowledge and to promote the widespread adoption of DESs shortly. The broad range of chemical compounds dissolved by DES, while maintaining an eco-friendly approach, has prioritized DESs as an executable substitute to standard organic solvents and even ionic liquids in some cases. Furthermore, the industrial and environmental relevance is increased due to the customizable nature of DESs, which is achieved through simple variations in their components. Despite the significant advancements made in knowing and applying DESs, there are still some challenges that persist. Future research should be formulated based on the molecular-level interactions within DESs, enabling researchers to design systems rationally tailored to specific applications. A hybrid combination of DESs with other green solvents may also present new avenues for innovation. Additionally, ensuring the biodegradability of all synthesized DESs is crucial from the perspective of environmental safety. Bridging the gap between laboratory-based research and industrial implementation at scale remains a key objective. Continued efforts in these directions will be crucial in realizing the full potential of DESs across all scientific and technological domains.

Acknowledgments SB is thankful for the e-resource facilities provided by Chandigarh University. TK gratefully acknowledges the support of the library and e-resource facilities offered by Sri Guru Granth World University, Punjab, India. VC gratefully acknowledges the e-resource support made available by Indus University.

Conflict of Interest There is no conflict of interest.

Declaration Declaration of Generative AI and AI-Assisted Technologies in the Writing Process:

While preparing this chapter, the authors used the ChatGPT tool to improve language and readability for several statements in the manuscript. After using this service, the authors reviewed and edited the content as needed and take(s) full responsibility for the publication content.

References

1. Ahluwalia VK, Kidwai M (2004) New trends in green chemistry, 1st edn. Kluwer Academic Publishers Boston Dordrecht London and Anamaya Publishers, New Delhi
2. Abbott AP, Capper G, Davies DL, Rasheed RK, Tambyrajah V (2003) Novel solvent properties of choline chloride/urea mixtures. Chem Commun 1:70–71. https://doi.org/10.1039/B210714G
3. Abbott AP, Boothby D, Capper G, Davies DL, Rasheed RK (2004) Deep eutectic solvents formed between choline chloride and carboxylic acids: versatile alternatives to ionic liquids. J Am Chem Soc 126(29):9142–9147. https://doi.org/10.1021/ja048266j
4. Torregrosa-Crespo J, Marset X, Guillena G, Ramón DJ, Martínez-Espinosa RM (2019) New guidelines for testing "deep eutectic solvents" toxicity and their effects on the environment and living beings. Sci Total Environ 704:135382. https://doi.org/10.1016/j.scitotenv.2019.135382
5. Achkar TE, Greige-Gerges H, Fourmentin S (2021) Basics and properties of deep eutectic solvents: a review. Environ Chem Lett 19(4):3397–3408. https://doi.org/10.1007/s10311-021-01225-8
6. Hansen BB, Spittle S, Chen B, Poe D, Zhang Y, Klein JM, Horton A, Adhikari L, Zelovich T, Doherty BW, Gurkan B, Maginn EJ, Ragauskas A, Dadmun M, Zawodzinski TA, Baker GA, Tuckerman ME, Savinell RF, Sangoro JR (2020) Deep eutectic solvents: a review of fundamentals and applications. Chem Rev 121(3):1232–1285. https://doi.org/10.1021/acs.chemrev.0c00385
7. Zhang Y, Ni S, Wu R, Fu Y, Qin M, Willför S, Xu C (2022) Green fractionation approaches for isolation of biopolymers and the critical technical challenges. Ind Crop Prod 177:114451. https://doi.org/10.1016/j.indcrop.2021.114451
8. Smith EL, Abbott AP, Ryder KS (2014) Deep eutectic solvents (DESs) and their applications. Chem Rev 114(21):11060–11082. https://doi.org/10.1021/cr300162p
9. Ge X, Gu C, Wang X, Tu J (2017) Deep eutectic solvents (DESs)-derived advanced functional materials for energy and environmental applications: challenges, opportunities, and future vision. J Mater Chem A 5(18):8209–8229. https://doi.org/10.1039/c7ta01659j
10. Tomé LI, Baião V, Da Silva W, Brett CM (2017) Deep eutectic solvents for the production and application of new materials. Appl Mater Today 10:30–50. https://doi.org/10.1016/j.apmt.2017.11.005
11. Abbott AP, Ttaib KE, Frisch G, McKenzie KJ, Ryder KS (2009) Electrodeposition of copper composites from deep eutectic solvents based on choline chloride. Phys Chem Chem Phys 11(21):4269–4277. https://doi.org/10.1039/b817881j
12. Abbott AP, Ballantyne A, Harris RC, Juma JA, Ryder KS, Forrest G (2015) A comparative study of nickel electrodeposition using deep eutectic solvents and aqueous solutions. Electrochim Acta 176:718–726. https://doi.org/10.1016/j.electacta.2015.07.051

13. Hillman AR, Ryder KS, Zaleski CJ, Ferreira V, Beasley CA, Vieil E (2014) Application of the combined electrochemical quartz crystal microbalance and probe beam deflection technique in deep eutectic solvents. Electrochim Acta 135:42–51. https://doi.org/10.1016/j.electacta.2014.04.062
14. Ghosh S, Roy S (2014) Characterization of tin films synthesized from ethaline deep eutectic solvent. Mater Sci Eng B 190:104–110. https://doi.org/10.1016/j.mseb.2014.09.014
15. Salomé S, Pereira NM, Ferreira ES, Pereira CM, Silva A (2013) Tin electrodeposition from choline chloride based solvent: influence of the hydrogen bond donors. J Electroanal Chem 703:80–87. https://doi.org/10.1016/j.jelechem.2013.05.007
16. Abbott AP, Barron JC, Ryder KS (2009) Electrolytic deposition of Zn coatings from ionic liquids based on choline chloride. Trans Inst Met Finish 87(4):201–207. https://doi.org/10.1179/174591909x438857
17. Starykevich M, Salak A, Ivanou D, Lisenkov A, Zheludkevich M, Ferreira M (2015) Electrochemical deposition of zinc from deep eutectic solvent on barrier alumina layers. Electrochim Acta 170:284–291. https://doi.org/10.1016/j.electacta.2015.04.150
18. Zhang J, Gu C, Tong Y, Gou J, Wang X, Tu J (2015) Microstructure and corrosion behavior of Cr and Cr–P alloy coatings electrodeposited from a Cr(III) deep eutectic solvent. RSC Adv 5(87):71268–71277. https://doi.org/10.1039/c5ra13056e
19. Ferreira ES, Pereira C, Silva A (2013) Electrochemical studies of metallic chromium electrodeposition from a Cr(III) bath. J Electroanal Chem 707:52–58. https://doi.org/10.1016/j.jelechem.2013.08.005
20. Fashu S, Gu C, Zhang J, Huang M, Wang X, Tu J (2015) Effect of EDTA and NH_4Cl additives on electrodeposition of Zn–Ni films from choline chloride-based ionic liquid. Trans Nonferrous Met Soc China 25(6):2054–2064. https://doi.org/10.1016/s1003-6326(15)63815-8
21. Vo T, Hidalgo SDS, Chiang C (2019) Controllable electrodeposition of binary metal films from deep eutectic solvent as an efficient and durable catalyst for the oxygen evolution reaction. Dalton Trans 48(39):14748–14757. https://doi.org/10.1039/c9dt03028j
22. Abbott AP, Capper G, McKenzie KJ, Ryder KS (2006) Electrodeposition of zinc–tin alloys from deep eutectic solvents based on choline chloride. J Electroanal Chem 599(2):288–294. https://doi.org/10.1016/j.jelechem.2006.04.024
23. Abbott AP, Ballantyne A, Harris RC, Juma JA, Ryder KS (2017) Bright metal coatings from sustainable electrolytes: the effect of molecular additives on electrodeposition of nickel from a deep eutectic solvent. Phys Chem Chem Phys 19(4):3219–3231. https://doi.org/10.1039/c6cp08720e
24. Alesary HF, Cihangir S, Ballantyne AD, Harris RC, Weston DP, Abbott AP, Ryder KS (2019) Influence of additives on the electrodeposition of zinc from a deep eutectic solvent. Electrochim Acta 304:118–130. https://doi.org/10.1016/j.electacta.2019.02.090
25. Azmi S, Koudahi MF, Frackowiak E (2022) Reline deep eutectic solvent as a green electrolyte for electrochemical energy storage applications. Energy Environ Sci 15(3):1156–1171. https://doi.org/10.1039/d1ee02920g
26. Chou H, Li C, Huang Y, Chung C, Lin Y (2025) Revealing the potential functionality of deep eutectic solvents as a doping electrolyte in organic electrochemical transistors. Electrochim Acta 521:145912–145923. https://doi.org/10.1016/j.electacta.2025.145912
27. Suffia S, Dutta D (2023) Applications of deep eutectic solvents in metal recovery from E-wastes in a sustainable way. J Mol Liq 394:123738. https://doi.org/10.1016/j.molliq.2023.123738
28. Tran MK, Rodrigues MF, Kato K, Babu G, Ajayan PM (2019) Deep eutectic solvents for cathode recycling of Li-ion batteries. Nat Energy 4(4):339–345. https://doi.org/10.1038/s41560-019-0368-4
29. Peeters N, Binnemans K, Riaño S (2020) Solvometallurgical recovery of cobalt from lithium-ion battery cathode materials using deep-eutectic solvents. Green Chem 22(13):4210–4221. https://doi.org/10.1039/d0gc00940g
30. Jaihindh DP, Manikandan A, Chueh Y, Fu Y (2020) Deep eutectic solvent-assisted synthesis of ternary heterojunctions for the oxygen evolution reaction and photocatalysis. ChemSusChem 13(10):2726–2738. https://doi.org/10.1002/cssc.202000177

31. Ren H, Lv K, Liu W, Li P, Zhang Y, Lv Y (2022) Deep-eutectic-solvent-assisted synthesis of a Z-scheme BiVO4/BiOCl/S,N-GQDS heterojunction with enhanced photocatalytic degradation activity under visible-light irradiation. Micromachines 13(10):1604. https://doi.org/10.3390/mi13101604
32. Liao H, Jiang Y, Zhou Z, Chen S, Sun S (2008) Shape-controlled synthesis of gold nanoparticles in deep eutectic solvents for studies of structure–functionality relationships in electrocatalysis. Angew Chem Int Ed 47(47):9100–9103. https://doi.org/10.1002/anie.200803202
33. Chen J, Ali MC, Liu R, Munyemana JC, Li Z, Zhai H, Qiu H (2019) Basic deep eutectic solvents as reactant, template and solvents for ultra-fast preparation of transition metal oxide nanomaterials. Chin Chem Lett 31(6):1584–1587. https://doi.org/10.1016/j.cclet.2019.09.055
34. Taghavi S, Amoozadeh A, Nemati F (2020) The first report of deep eutectic solvent (DES) nano-photocatalyst (n-TiO2-P25@TDI@DES (urea: ZnCl2)) and its application on selective oxidation of benzyl alcohols to benzaldehydes. J Chem Technol Biotechnol 96(2):384–393. https://doi.org/10.1002/jctb.6550
35. Wang Q, Dong B, Zhao Y, Huang F, Xie J, Cui G, Tang B (2018) Controllable green synthesis of crassula peforata-like TiO_2 with high photocatalytic activity based on deep eutectic solvent (DES). Chem Eng J 348:811–819. https://doi.org/10.1016/j.cej.2018.05.020
36. Haq HU, Bibi R, Arain MB, Safi F, Ullah S, Castro-Muñoz R, Boczkaj G (2022) Deep eutectic solvent (DES) with silver nanoparticles (Ag-NPs) based assay for analysis of lead (II) in edible oils. Food Chem 379:132085. https://doi.org/10.1016/j.foodchem.2022.132085
37. Kumar ARSS, Singha NK (2019) RAFT polymerization of 2-hydroxyethyl methacrylate in a deep eutectic solvent. J Polym Sci Part A Polym Chem 57(23):2281–2286. https://doi.org/10.1002/pola.29527
38. Fazende KF, Gary DP, Mota-Morales JD, Pojman JA (2020) Kinetic studies of photopolymerization of monomer-containing deep eutectic solvents. Macromol. Chem Phys 221(6). https://doi.org/10.1002/macp.201900511
39. Ndizeye N, Suriyanarayanan S, Nicholls IA (2019) Polymer synthesis in non-ionic deep eutectic solvents. Polym Chem 10(39):5289–5295. https://doi.org/10.1039/c9py01039d
40. Cecone C, Hoti G, Krabicová I, Appleton SL, Caldera F, Bracco P, Zanetti M, Trotta F (2020) Sustainable synthesis of cyclodextrin-based polymers by exploiting natural deep eutectic solvents. Green Chem 22(17):5806–5814. https://doi.org/10.1039/d0gc02247k
41. Jaumaux P, Liu Q, Zhou D, Xu X, Wang T, Wang Y, Kang F, Li B, Wang G (2020) Deep-eutectic-solvent-based self-healing polymer electrolyte for safe and long-life lithium-metal batteries. Angew Chem Int Ed Engl 59(23):9134–9142. https://doi.org/10.1002/anie.202001793
42. Hong S, Sun X, Lian H, Pojman JA, Mota-Morales JD (2019) Zinc chloride/acetamide deep eutectic solvent-mediated fractionation of lignin produces high- and low-molecular-weight fillers for phenol-formaldehyde resins. J Appl Polym Sci 137:7. https://doi.org/10.1002/app.48385

MIX
Papier aus verantwortungsvollen Quellen
Paper from responsible sources
FSC® C105338

If you have any concerns about our products,
you can contact us on
ProductSafety@springernature.com

In case Publisher is established outside the EU,
the EU authorized representative is:
Springer Nature Customer Service Center GmbH
Europaplatz 3, 69115 Heidelberg, Germany

Printed by Libri Plureos GmbH
in Hamburg, Germany